DIE GRUNDLEHREN DER

MATHEMATISCHEN WISSENSCHAFTEN

IN EINZELDARSTELLUNGEN MIT BESONDERER
BERÜCKSICHTIGUNG DER ANWENDUNGSGEBIETE

HERAUSGEGEBEN VON

R. GRAMMEL · F. HIRZEBRUCH · E. HOPF
H. HOPF · W. MAAK · W. MAGNUS · F. K. SCHMIDT
K. STEIN · B. L. VAN DER WAERDEN

BAND 101

THE DIFFERENTIAL GEOMETRY OF FINSLER SPACES

BY

HANNO RUND

SPRINGER-VERLAG

BERLIN · GÖTTINGEN · HEIDELBERG

1959

THE DIFFERENTIAL GEOMETRY OF FINSLER SPACES

BY

DR. HANNO RUND

PROFESSOR OF APPLIED MATHEMATICS
IN THE UNIVERSITY OF NATAL

SPRINGER-VERLAG

BERLIN · GÖTTINGEN · HEIDELBERG

1959

ISBN 978-3-642-51612-2   ISBN 978-3-642-51610-8 (eBook)
DOI 10.1007/978-3-642-51610-8

© BY SPRINGER-VERLAG OHG.,
BERLIN · GÖTTINGEN · HEIDELBERG 1959
Reprint of the original edition 1959

BRÜHLSCHE UNIVERSITÄTSDRUCKEREI GIESSEN

Preface

The present monograph is motivated by two distinct aims. Firstly, an endeavour has been made to furnish a reasonably comprehensive account of the theory of Finsler spaces based on the methods of classical differential geometry. Secondly, it is hoped that this monograph may serve also as an introduction to a branch of differential geometry which is closely related to various topics in theoretical physics, notably analytical dynamics and geometrical optics. With this second object in mind, an attempt has been made to describe the basic aspects of the theory in some detail — even at the expense of conciseness — while in the more specialised sections of the later chapters, which might be of interest chiefly to the specialist, a more succinct style has been adopted.

The fact that there exist several fundamentally different points of view with regard to Finsler geometry has rendered the task of writing a coherent account a rather difficult one. This remark is relevant not only to the development of the subject on the basis of the tensor calculus, but is applicable in an even wider sense. The extensive work of H. BUSEMANN has opened up new avenues of approach to Finsler geometry which are independent of the methods of classical tensor analysis. In the latter sense, therefore, a full description of this approach does not fall within the scope of this treatise, although its fundamental significance cannot be doubted. Fortunately, a recent volume[1] covers this ground comprehensively and with far greater competence than could possibly be achieved in the present text. Consequently the theory of BUSEMANN is not included, and will be referred to only when it has a direct bearing upon a particular problem under discussion.

We shall thus restrict our attention to the methods of classical differential geometry which have prevailed in the literature of the subject up to the present time. The application of tensor methods has been dominated by the initial impetus given to the theory of Finsler spaces by L. BERWALD and E. CARTAN. Although there are marked differences between some of the basic concepts introduced by these authors, the last papers of BERWALD seem to indicate that both points of view have their rightful place within the general framework and may moreover be profitably combined. The present writer has endeavoured to give — at least in so far as it seemed feasible to do so — a unified treatment of these as well as of more recent theories, attempting, however, to

---

[1] BUSEMANN [10]. Numbers in square brackets refer to the bibliography.

preserve the spirit in which these theories were originally put forward. The inevitable result is the essentially classical character of this book: whether or not this requires an apology in our present age the writer has been unable to decide.

A glance at the table of contents will indicate the scope of this text. Unfortunately, it was not possible to include an account of the more recent generalisations, such as the geometries of CARTAN and KAWAGUCHI, without seriously curtailing the description of the theory of Finsler spaces, which, it was felt, should be avoided at all costs. Also, Professor E. T. DAVIES has indicated that a volume on general metric spaces which would deal with these topics is being planned by him, so that this deficiency may not be felt too seriously.

The same restriction applies to the bibliography (with the exception of a few necessary references with regard to the general mathematical apparatus and background). An attempt has been made to furnish a complete list of publications dealing with Finsler geometry (up to July, 1957). Any omissions are entirely unintentional, and practically all references are mentioned in the text at relevant stages.

Originally this book had been planned as a contribution to the „Ergebnisse“-series: however, after receiving the manuscript, the publishers suggested that it should be included in the „Grundlehren“-series. This required some alterations and additions in order to bring this work into line with the general tradition and character of earlier volumes of this series. In particular, the treatment of the more elementary parts was expanded. Nevertheless, this background accounts for the numerous footnotes referring to the original sources on which the text is based, as well as for the sections in small print in which brief summaries are given of results which could not conveniently be included in the main text. These footnotes and summaries are intended to serve chiefly as guides to the literature on Finsler spaces, and it is hoped that they may be of use to research workers in this field, but, on the other hand, they may be ignored entirely by any reader who does not intend to specialise in this direction.

A knowledge of the techniques of the tensor calculus and linear algebra is presupposed; and while an acquaintance with Riemannian geometry is highly desirable in view of frequent references to this subject, this requirement is not absolutely essential. The same applies to the classical differential geometry of curves and surfaces: a reader who is not familiar with the elements of this subject might be mystified by the motivation of the ideas and concepts introduced in the chapter dealing with subspaces.

In the revision of the proofs I have had, and wish to acknowledge most gratefully, the valuable assistance of my colleagues R. VAN DER

Borght, J. R. Vanstone and C. F. Templin. I am furthermore indebted to J. Abramovich (formerly of Toronto) for his translations of Russian texts. Above all, I should like to express my deep gratitude to Professor C. Y. Pauc, who not only suggested valuable improvements with regard to certain topics treated in this book, but whose advice and numerous discussions were also of inestimable benefit at a time when I first approached the subject matter of this monograph.

Finally, it gives me great pleasure to thank the publishers for their patience, their unfailing courtesy, and the cooperation which at all times they extended so willingly.

Cape Town, South Africa   H. Rund

December, 1958

Contents

Introduction

Chapter I: **Calculus of Variations. Minkowskian Spaces**

Chapter II: **Geodesics: Covariant Differentiation**

Chapter III: **The "Euclidean Connection" of E. CARTAN**

Chapter IV: **The Theory of Curvature**

Chapter V: **The Theory of Subspaces**

Chapter VI: **Miscellaneous Topics**

Hints to the reader

A reader who desires no more than a cursory acquaintance with the theory of Finsler spaces (in particular with a view to applications to theoretical physics) is advised to read in the following order, omitting all sections in small print: Chapter I, §§ 1—6; Chapter II, §§ 1—4; Chapter III, §§ 1—3; Chapter IV, §§ 1—4; Chapter V, § 1.

References to equations are of the form (N. M. P), where N and M indicate the corresponding chapter and section respectively. If N coincides with the chapter at hand it will be omitted.

A complete bibliography encompassing the literature concerning the generalisations of Finsler spaces, compiled by H. SCHUBERT, is to be found in the reprint of Finsler's thesis [1].

Introduction

The fundamental idea of a Finsler space may be traced back to the famous lecture of Riemann: „Über die Hypothesen, welche der Geometrie zugrunde liegen." In this memoir of 1854 Riemann discusses various possibilities by means of which an n-dimensional manifold may be endowed with a metric, and pays particular attention to a metric defined by the positive square root of a positive definite quadratic differential form. Thus the foundations of Riemannian geometry are laid; nevertheless, it is also suggested that the positive fourth root of a fourth order differential form might serve as a metric function. These functions have three properties in common: they are positive, homogeneous of the first degree in the differentials, and are also convex in the latter. It would seem natural, therefore, to introduce a further generalisation to the effect that the distance ds between two neighbouring points represented by the coordinates x^i and $x^i + dx^i$ be defined by some function $F(x^i, dx^i)$:

$$ds = F(x^i, dx^i), \qquad\qquad (i = 1, \ldots, n)$$

where this function satisfies these three properties.

It is remarkable that the first systematic study of manifolds endowed with such a metric was delayed by more than 60 years. It was an investigation of this kind which formed the subject matter of the thesis of Finsler in 1918, after whom such spaces were eventually named. It would appear that this new impulse was derived almost directly from the calculus of variations, with particular reference to the new geometrical background which was introduced by Carathéodory in connection with problems in parametric form. The kernel of these methods is the so-called indicatrix, while the property of convexity is of fundamental importance with regard to the necessary conditions for a minimum in the calculus of variations. In fact, the remarkable affinity between some aspects of differential geometry and the calculus of variations had been noticed some years prior to the publication of Finsler's thesis, in particular by Bliss, Landsberg and Blaschke. Both Bliss and Landsberg introduced (distinct) definitions of angle in terms of invariants of a parametric problem in the calculus of variations, while an analytic study of such invariants had been made by E. Noether and A. Underhill. Yet the geometrical theories of Bliss and Landsberg were developed against an euclidean background and cannot, therefore, be regarded as fulfilling the true objectives of the generalisation of

Riemann's proposal. Clearly, Finsler's thesis must be regarded as the first step in this direction.

A few years later, however, the general development took a curious turn away from the basic aspects and methods of the theory as developed by FINSLER. The latter did not make use of the tensor calculus, being guided in principle by the notions of the calculus of variations; and in 1925 the methods of the tensor calculus were applied to the theory independently but almost simultaneously by SYNGE, TAYLOR and BERWALD. It was found that the second derivatives of $\frac{1}{2} F^2 (x^i, d x^i)$ with respect to the differentials served admirably as components of a metric tensor in analogy with Riemannian geometry, and from the differential equations of the geodesics connection coefficients could be derived by means of which a generalisation of LEVI-CIVITA's parallel displacement could be defined. While the corresponding covariant derivatives as introduced by SYNGE and TAYLOR coincide, the theory of BERWALD shows a marked difference, in the sense that in his geometry the lemma of RICCI (which in Riemannian geometry implies the vanishing of the covariant derivative of the metric tensor) is no longer valid. Nevertheless, BERWALD continued to develop his theory with particular reference to the theory of curvature as well as to two-dimensional spaces. The significance of this work was enhanced by the advent of the general geometry of paths (a generalisation of the so-called Non-Riemannian geometry) due to DOUGLAS and KNEBELMAN, for the initial approach of BERWALD was such as to establish a close affinity between these branches of metric and non-metric differential geometry.

Again, the theory took a new and unexpected turn in 1934 when E. CARTAN published his tract on Finsler spaces. He showed that it was indeed possible to define connection coefficients and hence a covariant derivative such that the preservation of Ricci's lemma was ensured. On this basis CARTAN developed a theory of curvature, and practically all subsequent investigations concerning the geometry of Finsler spaces were dominated by this approach. Several mathematicians expressed the opinion that the theory had thus attained its final form. To a certain extent this was correct, but not altogether so, as we shall now indicate.

The above-mentioned theories make use of a certain device which basically involves the consideration of a space whose elements are not the points of the underlying manifold, but the line-elements of the latter, which form a $(2n - 1)$-dimensional variety. This facilitates the introduction of what CARTAN calls the "euclidean connection", which, by means of certain postulates, may be derived uniquely from the fundamental metric function $F(x^i, d x^i)$. The method also depends on the introduction of a so-called "element of support", namely, that at each point a previously assigned direction must be given, which then serves as directional argument in all functions depending on direction as well as position. Thus, for instance, the length of a vector and the vector obtained from it by an infinitesimal parallel displacement depend on the arbitrary choice of the element of support. It is this device which led to the development of Finsler geometry in terms of direct generalisations of the methods of Riemannian geometry.

It was felt, however, that the introduction of the element of support was undesirable from a geometrical point of view, while the natural link with the calculus of variations was seriously weakened. This view was expressed independently by several authors, in particular by VAGNER, BUSEMANN and the present writer. It was emphasised that the natural local metric of a Finsler space is a Minkowskian one, and that the arbitrary imposition of a euclidean metric would to some extent obscure some of the most interesting characteristics of the Finsler space. Thus at the beginning of the present decade further theories were put forward. The rejection of the use of the element of support, however desirable from a geometrical point of view, led to new difficulties: for instance, the natural orthogonality between two vectors is not in general symmetric, while the analytical difficulties are certainly enhanced, particularly since Ricci's lemma cannot be generalised as before. Fortunately, from the point of view of differential invariants, there exist marked similarities between all these theories, which is a perfectly natural phenomenon and could have been expected. It is in the application and in the interpretation of these invariants that the two points of view appear to be irreconcilable.

After this brief historical sketch it should be clear why the first chapter is devoted to some elementary discussion of the calculus of variations and to certain fundamental aspects of Minkowskian geometry, although no claim as regards completeness with respect to the latter is made. In the second chapter a set of connection coefficients will be derived in a purely analytical but natural manner, by means of which a parallel displacement independent of the element of support may be defined. This gives us an immediate access to Cartan's theory, practically without further calculation, but in view of the inestimable significance of the latter's approach, we have permitted ourselves to reproduce his fundamental postulates and their immediate consequences in some detail. This monograph could not have reflected the prevailing spirit of a large number of recent and contemporary publications had we failed to do so. This discussion forms the nucleus of the third chapter, in which the connection of BERWALD is also derived. A detailed comparison of the various covariant derivatives is given, chiefly because thus a latent but valuable unity amongst the different points of view is exposed. In the theory of curvature and its applications (Chapter IV) this unity is still partly maintained, but it will be seen that in the description of the theory of subspaces (Chapter V) sharp distinctions emerge with a certain inevitability. This is apparent also in the last chapter which is devoted to a number of more or less isolated topics.

As has been remarked in the preface, it was not found feasible to include an account of the various generalisations of Finsler geometry such as the geometry of CARTAN based on the notion of area or the spaces of KAWAGUCHI in which the metric function depends on higher derivatives of the directional argument. Naturally, it would have been possible to develop the complete theory from a more general point of view: for instance, we could well have begun with a fundamental metric function depending on covariant or contravariant vector densities, from which the geometry of Finsler spaces would have emerged as a special case. Nevertheless, it was felt that the consequent lack of geometrical clarity as well as the inevitable analytical complications would have repelled many a reader, while after merely a cursory perusal of the present text the reader should experience no difficulty whatsoever should he wish to study the original literature concerned with such generalisations.

THE DIFFERENTIAL GEOMETRY

OF FINSLER SPACES

Chapter I

Calculus of Variations. Minkowskian Spaces

The origin of the theory of Finsler spaces is to be found in the calculus of variations. In the present chapter, therefore, we shall formulate the simplest problems in the calculus of variations, and while no previous knowledge of this topic is presupposed, no attempt is made to cover the ground usually dealt with in standard treatises devoted to this subject. Only those developments which play a central role in the theory of Finsler spaces, such as the condition of LEGENDRE, are discussed in detail, and as far as possible from a geometrical point of view. It is indicated how a problem in the calculus of variations imposes a metric on the underlying manifolds; moreover, the local properties of such a metric are best described by the introduction of the so-called tangent spaces. Strictly speaking, the notion of a tangent space is independent of the imposition of a metric and should therefore have been introduced prior to the metric, but its significance is probably more easily understood in the light of the latter. Consequently, a large section of this chapter is devoted to the study of the metric properties of the tangent spaces, which will lead naturally to the definition of Minkowskian spaces. Finsler manifolds are, in fact, locally Minkowskian spaces. It is therefore necessary to formulate the definitions of the fundamental metric concepts such as length, angle, trigonometric ratios and areas in terms of a local Minkowskian geometry. Unfortunately, it frequently happens that distinct invariants of such a geometry may be regarded as generalisations of one and the same invariant of euclidean geometry: hence it is possible to define some metric entities (such as angle) in several distinct but equally meaningful ways. In the beginning the reader may find this state of affairs somewhat confusing: we are therefore compelled to analyse and to compare these different definitions of fundamental metric quantities. Some of these will be selected for use in subsequent chapters, while the rest will be excluded in order to avoid confusion.

§ 1. Problems in the Calculus of Variations in Parametric Form

Let R be a region of an n-dimensional space X_n which is covered completely by a coordinate system, such that any point of R is represented by a set of n independent variables $x^i (i = 1, 2, \ldots, n)$, which will be called the coordinates. It will be understood throughout that the coordinates are real. The extent to which the coordinate system covers the whole space X_n obviously depends on the topological character of X_n. We shall not touch upon such questions in the present monograph, nor shall we enter into a discussion of the axiomatic foundations of the notion of a space X_n; in connection with these questions the reader is referred to the treatise by VEBLEN and WHITEHEAD[1]. We shall restrict our attention to the region R, and when we refer to X_n in the sequel, it is to be understood that this restriction is implied.

A transformation of coordinates is represented by a set of n equations

$$x^{i'} = x^{i'}(x^1, \ \ldots, \ x^n), \qquad (i', j', \ldots = 1, \ldots, n), \qquad (1.1)$$

which indicate that the coordinates x^i of a point P of X_n are represented in the new coordinate system by new variables $x^{i'}$[2]. We shall assume[3] that the functions $x^{i'}$ of (1.1) are at least of class C^2 and that the Jacobian of the transformation (1.1) does not vanish identically; that is

$$\det\left(\frac{\partial x^{i'}}{\partial x^i}\right) \neq 0 . \qquad (1.2)$$

As a result of this assumption the x^i are expressible in terms of the $x^{i'}$:

$$x^i = x^i(x^{i'}) . \qquad (1.1\,\mathrm{a})$$

A set of points of R whose coordinates may be expressed as functions of a single parameter t is regarded as a curve of X_n. Thus the equations

$$x^i = x^i(t) \qquad (1.3)$$

define a curve C of X_n. If the functions (1.3) are of class C^1 we shall regard the entity whose components are given by

$$\dot{x}^i = \frac{d x^i}{dt} , \qquad (1.4)$$

as the tangent vector to C. We shall call the combination $(x^i, \dot{x}^i)$ a line-element of C. Two distinct points P_1 and P_2 on C are given by two distinct values t_1 and t_2 of the parameter t. Together with C we consider a second curve $\bar{C}$ of class C^1 represented by the equations

$$\bar{x}^i = \bar{x}^i(t) , \qquad (1.5)$$

---

[1] VEBLEN and WHITEHEAD [1].

[2] This is the kernel-index notation of SCHOUTEN [1]; thus a change of coordinate system is indicated by attaching a primed index as in (1.1). This notation enables us to distinguish coordinate transformations from mappings of the type: $x^i = x^i(\xi^1, \ldots, \xi^n)$, which establish a correspondence between two distinct points $P(x^i)$, $Q(\xi^i)$ with respect to the same coordinate system.

[3] We shall say that a function $f(x^i)$ is of class C^r if f is continuous in all its variables x^i and possesses continuous derivatives up to the rth order with respect to each x^i throughout the domain R.

such that $\overline{C}$ also joins the points P_1, P_2 which correspond to the same parameter values t_1 and t_2 on $\overline{C}$. This implies $\overline{x}^i(t_1) = x^i(t_1)$; $\overline{x}^i(t_2) = x^i(t_2)$. We shall say that $\overline{C}$ belongs to the close neighbourhood (ε, η) of C [1] if the $2n$ inequalities

$$|\overline{x}^i(t) - x^i(t)| < \varepsilon , \quad \left| \frac{d\overline{x}^i}{dt} - \frac{dx^i}{dt} \right| < \eta , \tag{1.6}$$

hold for all values of t in the interval $t_1 \leq t \leq t_2$.

Suppose now that we are given a function $F(x^i, \dot{x}^i)$ of the line-elements $(x^i, \dot{x}^i)$ of curves defined in R. We shall assume F to be of class C^5 in all its $2n$ arguments. Along the curves C and $\overline{C}$ we may then define the integrals

$$I = \int_{t_1}^{t_2} F(x^i, \dot{x}^i)\, dt , \quad \overline{I} = \int_{t_1}^{t_2} F\left(\overline{x}^i, \frac{d\overline{x}^i}{dt}\right) dt . \tag{1.7}$$

The curve C will be called an extremal of the fundamental function F if $I \leq \overline{I}$ (or $I \geq \overline{I}$) for all curves $\overline{C}$ belonging to the close neighbourhood of C within the region R.

The simplest problem in the calculus of variations is then the determination of all extremals belonging to a given fundamental function F [2].

The function F itself must satisfy certain conditions in order to ensure the existence of extremals. We shall deal with each of these conditions in turn.

Condition A: The function $F(x^i, \dot{x}^i)$ is positively homogeneous of degree 1 in the $\dot{x}^i$:

$$F(x^i, k\dot{x}^i) = kF(x^i, \dot{x}^i) , \quad \text{with} \quad k > 0 . \tag{1.8}$$

This condition is necessary and sufficient in order that the integral I given by (1.7) be independent of the choice of the parameter t [3]. For a large class of problems in the calculus of variations the value of I is independent of the direction in which we integrate along C. For this we would require that in addition to (1.8) we have

$$F(x^i, \dot{x}^i) = F(x^i, -\dot{x}^i) . \tag{A_1}$$

In the sequel, however, we shall not assume A_1 unless the contrary is stated explicitly [4].

Derivatives with respect to x^i or $\dot{x}^i$ will be denoted by corresponding suffixes. In virtue of our assumption (1.8) the functions $F_{\dot{x}^i}$, $F_{\dot{x}^i\dot{x}^j}$ will

---

[1] This is the notion of "engere Nachbarschaft"; CARATHÉODORY [1], p. 192. In connection with these definitions the reader may also consult BOLZA [1], Ch. V.

[2] Actually slightly less stringent conditions may be imposed on the curves of comparison. Cf. BOLZA [1], CARATHÉODORY [1].

[3] This is proved by CARATHÉODORY [1], p. 212—213.

[4] In connection with condition A_1, see BUSEMANN [3, 9].

be positively homogeneous of degree zero and -1 in the $\dot{x}^i$ respectively, as may be verified directly by differentiation of (1.8)[1]:

$$F_{\dot{x}^i}(x, k\dot{x}) = F_{\dot{x}^i}(x, \dot{x}) , \qquad (k > 0) , \tag{1.9}$$

$$F_{\dot{x}^i \dot{x}^j}(x, k\dot{x}) = k^{-1} F_{\dot{x}^i \dot{x}^j}(x, \dot{x}) , \qquad (k > 0) . \tag{1.10}$$

From Euler's theorem on homogeneous functions we then have[2]

$$F_{\dot{x}^i}(x, \dot{x}) \, \dot{x}^i = F(x, \dot{x}) , \tag{1.11}$$

$$F_{\dot{x}^i \dot{x}^j}(x, \dot{x}) \, \dot{x}^i = 0 . \tag{1.12}$$

From (1.12) we deduce the identity

$$\det |F_{\dot{x}^i \dot{x}^j}| = 0 . \tag{1.13}$$

We remark that condition A is not as stringent as it appears to be: in general a non-homogeneous function of the line-element can be made homogeneous by introducing an additional dimension.

If a fundamental function $F(x^i, \dot{x}^i)$ is defined for all line-elements in the region R, it would be natural to regard F as defining a distance in X_n: for instance the "length" of the curve $\overline{C}$ between the points P_1 and P_2 could well be defined by the second of the integrals (1.7). More precisely, if $A(x^i)$ and $B(x^i + dx^i)$ are two neighbouring points of R, the distance ds between them is defined by

$$ds = F(x^j, dx^j) . \tag{1.14}$$

Since F is homogeneous of first degree in the dx^i, this would lead to the required integral. In this manner a metric is imposed on our X_n[3].

If, in particular, the function F is of the form

$$F(x^i, dx^i) = [g_{ij}(x^k) \, dx^i \, dx^j]^{\frac{1}{2}} , \tag{1.15}$$

where the $g_{ij}(x^k)$ are coefficients independent of the dx^i, the metric defined by F is the metric of a *Riemannian space*.

The space X_n is called a *Finsler space* if the fundamental function $F(x^i, \dot{x}^i)$ defining the metric (1.14) satisfies further conditions. It is natural to assume that all distances are positive; hence we stipulate

---

[1] In future we shall sometimes denote arguments without attaching indices, e. g. $F(x, \dot{x})$ is supposed to represent $F(x^k, \dot{x}^k)$.

[2] We adopt the usual summation convention, which implies that a repeated index is summed over its specified range. Thus the left-hand of (1.11) stands for $\sum\limits_{i=1}^{n} F_{\dot{x}^i}(x, \dot{x}) \, \dot{x}^i$.

[3] A more precise formulation is reserved for § 2.

Condition B: the function $F(x^i, \dot{x}^i)$ is positive if not all $\dot{x}^i$ vanish simultaneously[1]:

$$F(x^i, \dot{x}^i) > 0\,, \quad \text{with} \quad \sum_i (\dot{x}^i)^2 \neq 0\,. \tag{1.16}$$

Finally, in the calculus of variations a further condition has to be imposed: the so-called condition of LEGENDRE. For our present purpose it is advantageous to formulate this condition in a manner which differs slightly from the formulation prevalent in text-books on the calculus of variations; however, we shall show that the two formulations are equivalent.

Condition C: The quadratic form

$$F^2_{\dot{x}^i \dot{x}^j}(x, \dot{x})\, \xi^i\, \xi^j \equiv \frac{\partial F^2(x, \dot{x})}{\partial \dot{x}^i\, \partial \dot{x}^j}\, \xi^i\, \xi^j \tag{1.17}$$

is assumed to be positive definite for all variables ξ^i [i. e. only positive values are assumed by (1.17) unless the ξ^i vanish simultaneously] and for any line element $(x^i, \dot{x}^i)$ which appears as the argument in the coefficients of the form (1.17). We note that in view of (1.13) we are forced to choose F^2 rather than the function F in the formulation of Condition C.

An immediate consequence of this condition is the inequality

$$(F^2_{\dot{x}^i \dot{x}^j}(x, \dot{x})\, \xi^i\, \eta^j)^2 \leq (F^2_{\dot{x}^i \dot{x}^j}(x, \dot{x})\, \xi^i\, \xi^j)\,(F^2_{\dot{x}^h \dot{x}^k}(x, \dot{x})\, \eta^h\, \eta^k)\,, \tag{1.18}$$

which holds for any pair of variables ξ^i, η^i and for any line element $(x^k, \dot{x}^k)$, while the equality sign occurs if and only if there exists a

---

[1] Again, this condition is not too restrictive, for under certain circumstances it is possible to replace an integrand $F(x^i, \dot{x}^i)$ which assumes both positive and negative values by an "equivalent" integrand which is positive throughout a domain. If $S(x^k)$ is any function of class C^1 depending only on the x^k, the expression $S_{x^i}(x^k)\, d x^i$ is an exact differential: and if this expression is integrated along a curve C joining two points P_1, P_2 of the region R, the result $S(P_2) - S(P_1)$ is independent of the path. Hence if we add the terms $S_{x^i}(x^k)\, \dot{x}^i$ to the integrand of the integral (1.7), the resulting extremals are identical to those of the original integral, i. e. from the point of view of the calculus of variations the two problems defined by the integrands $F(x^k, \dot{x}^k)$ and $F^*(x^k, \dot{x}^k) = F(x^k, \dot{x}^k) + S_{x^i}(x^k)\, \dot{x}^i$ are *equivalent*. But it may be possible to choose the function $S(x^k)$ such that F^* is positive even though F may change sign. In fact, a sufficient condition ensuring that such a construction may be carried out is that there exists at some point $P(x^i_{(0)})$ of R a line-element $(x^i_{(0)}, \dot{x}^i_{(0)})$ such that

(a) $F(x^k_{(0)}, \dot{x}^k) - F_{\dot{x}^i}(x^k_{(0)}, \dot{x}^k_{(0)})\, \dot{x}^i > 0$

for all values $\dot{x}^i \neq k\, \dot{x}^i_{(0)}$ ($k > 0$). If this condition is satisfied, a function $F^*(x^k, \dot{x}^k)$ may be found such that $F^*(x^k, \dot{x}^k) > 0$ for all line-elements of a neighbourhood of $P(x^i_{(0)})$ (CARATHÉODORY [1], p. 243). The converse of this result is also valid, while if the function $F(x^k, \dot{x}^k)$ satisfies condition A_1, the inequality (a) is sufficient to ensure that F itself be positive, so that for this class of functions the construction is superfluous (RUND [16]). An alternative form of the condition (a) is given by DAMKÖHLER and HOPF [1, 2].

relation $\xi^i = \lambda \eta^i$ with some factor of proportionality λ. The proof of (1.18) follows directly from the fact that the following quadratic expression in μ, namely

$$F^2_{\dot{x}^i \dot{x}^j}(x, \dot{x}) \, (\xi^i + \mu \, \eta^i) \, (\xi^j + \mu \eta^j) = F^2_{\dot{x}^i \dot{x}^j} \xi^i \, \xi^j + 2 \, \mu \, F^2_{\dot{x}^i \dot{x}^j} \xi^i \, \eta^j +$$
$$+ \mu^2 F^2_{\dot{x}^i \dot{x}^j} \eta^i \, \eta^j$$

cannot, in view of condition C, possess real zeros unless $\xi^i + \mu \eta^i = 0$. The inequality (1.18) simply expresses the fact that the corresponding discriminant is negative.

Clearly we have the identity

$$\tfrac{1}{2} F^2_{\dot{x}^i \dot{x}^j}(x, \dot{x}) = F_{\dot{x}^i}(x, \dot{x}) \, F_{\dot{x}^j}(x, \dot{x}) + F(x, \dot{x}) \, F_{\dot{x}^i \dot{x}^j}(x, \dot{x}) , \quad (1.19)$$

and hence, from (1.11) and (1.12)

$$\tfrac{1}{2} F^2_{\dot{x}^i \dot{x}^j}(x, \dot{x}) \, \dot{x}^j = F(x, \dot{x}) \, F_{\dot{x}^i}(x, \dot{x}) , \qquad (1.19\,\mathrm{a})$$

and

$$\tfrac{1}{2} F^2_{\dot{x}^i \dot{x}^j}(x, \dot{x}) \, \dot{x}^i \, \dot{x}^j = F^2(x, \dot{x}) , \qquad (1.19\,\mathrm{b})$$

from which it follows that (1.19) may be expressed in the form

$$\tfrac{1}{2} F^2_{\dot{x}^i \dot{x}^j}(x, \dot{x}) = \tfrac{1}{4} F^{-2}(x, \dot{x}) \, F^2_{\dot{x}^i \dot{x}^h}(x, \dot{x}) \, \dot{x}^h \, F^2_{\dot{x}^j \dot{x}^k}(x, \dot{x}) \, \dot{x}^k +$$
$$+ F(x, \dot{x}) \, F_{\dot{x}^i \dot{x}^j}(x, \dot{x}) . \qquad (1.19\,\mathrm{c})$$

Thus for any set of variables ξ^i we have

$$F_{\dot{x}^i \dot{x}^j}(x, \dot{x}) \, \xi^i \, \xi^j = [2F(x, \dot{x})]^{-1} \, \{F^2_{\dot{x}^i \dot{x}^j}(x, \dot{x}) \, \xi^i \, \xi^j -$$
$$- \tfrac{1}{2} F^{-2}(x, \dot{x}) \, (F^2_{\dot{x}^i \dot{x}^j}(x, \dot{x}) \, \dot{x}^i \, \xi^j)^2\} .$$

To the last term on the right-hand side of this equation we apply the inequality (1.18) with $\eta^i = \dot{x}^i$, and on observing (1.19 b) we find that

$$F_{\dot{x}^i \dot{x}^j}(x, \dot{x}) \, \xi^i \, \xi^j \geqq 0 , \qquad (1.20)$$

the equality sign occurring if and only if the ξ^i are proportional to the $\dot{x}^i$, which is in accordance with equation (1.12).

The quadratic form (1.20) is thus positive indefinite, the rank of the matrix $\|F_{\dot{x}^i \dot{x}^j}\|$ being $(n-1)$. From a well-known theorem on convex functions it follows *that the function $F(x^k, \dot{x}^k)$ is convex in the $\dot{x}^k$* [1].

---

[1] Consider a function $f(u^i)$ of n variables $u^1, \ldots, u^n$. A domain D in the space of the u^i is said to be convex if it contains the whole of the segment of a straight line which connects any two of its points. The function $f(\xi^i)$ is said to be *convex in D* if it is defined everywhere in D and if the inequality

(a) $$f \left(\frac{u^i_{(1)} + u^i_{(2)}}{2} \right) \leqq \tfrac{1}{2} \left(f(u^i_{(1)}) + f(u^i_{(2)}) \right)$$

is satisfied for all pairs of points $u^i_{(1)}, u^i_{(2)}$ of D (HARDY, LITTLEWOOD and PÓLYA [1], p. 79). It is *strictly convex* if the inequality holds for all pairs of distinct points $u^i_{(1)}, u^i_{(2)}$. If $f(u^i)$ is convex and continuous it may be shown that it satisfies the more general inequality

(b) $$f((1 - \theta) \, u^i_{(1)} + \theta \, u^i_{(2)}) \leqq (1 - \theta) \, f(u^i_{(1)}) + \theta \, f(u^i_{(2)})$$

We shall now verify this result by proving the following more precise assertion:

If $(x^i, \dot{x}^i)$, $(x^i, \ddot{\bar{x}}^i)$ are any two line-elements attached to the same "centre" x^i, then the following inequality is satisfied:

$$F(x^i, \dot{x}^i + \ddot{\bar{x}}^i) \leq F(x^i, \dot{x}^i) + F(x^i, \ddot{\bar{x}}^i) , \tag{1.21}$$

where the equality sign occurs if and only if there exists a relation of the form $\dot{x}^i = v\,\ddot{\bar{x}}^i$ with $v > 0$.

Proof: Let $(x^i, \dot{x}^i)$, (x^i, ξ^i) be any two line elements (with a common centre x^i) for which neither all the $\dot{x}^i$ nor all the ξ^i vanish simultaneously. In terms of the second mean-value theorem we have

$$F(x, \xi) = F(x, \dot{x}) + F_{\dot{x}^i}(x, \dot{x}) (\xi^i - \dot{x}^i) + \Phi_{ij}(\xi^i - \dot{x}^i)(\xi^j - \dot{x}^j) ,$$

where we have written

$$\Phi_{ij} = \tfrac{1}{2} F_{\dot{x}^i \dot{x}^j}(x^k, \theta\,\dot{x}^k + (1 - \theta)\,\xi^k) , \qquad 0 < \theta < 1 .$$

As a result of (1.11) the above expansion reduces to the form

$$F(x, \xi) = F_{\dot{x}^i}(x, \dot{x})\,\xi^i + \Phi_{ij}(\xi^i - \dot{x}^i)(\xi^j - \dot{x}^j) . \tag{1.22}$$

Strictly speaking, we must exclude the possibility that $\xi^i = \lambda\,\dot{x}^i$ for a negative number λ, for it is easily verified that in this case the expression $\theta\,\dot{x}^k + (1 - \theta)\,\xi^k$ could vanish, rendering (1.22) meaningless. With this restriction in mind let us consider the quadratic form defined by

$$\Phi = \Phi_{ij}(\xi^i - \dot{x}^i)(\xi^j - \dot{x}^j) .$$

In view of (1.20) this form is non-negative and vanishes if and only if there exists a number ϱ such that

$$\xi^i - \dot{x}^i = \varrho\,(\theta\dot{x}^i + (1 - \theta)\,\xi^i) ,$$

or

$$\xi^i(1 + \varrho\theta - \varrho) = \dot{x}^i(1 + \varrho\theta) .$$

Now the two numbers $(1 + \varrho\theta) - \varrho$ and $(1 + \varrho\theta)$ cannot vanish simultaneously irrespective of whether $\varrho = 0$ or $\varrho \neq 0$: thus neither may vanish, for otherwise all the $\dot{x}^i$ or all the ξ^i would vanish contrary to our hypothesis. Thus, subject to the above-mentioned restriction, Φ is non-negative and vanishes if and only if

$$\xi^i = k\,\dot{x}^i \quad \text{with} \quad (k > 0) . \tag{1.23}$$

---

for any value of θ satisfying $0 \leq \theta \leq 1$. The inequality (b) is often taken as definition of convexity of f (BONNESEN and FENCHEL [1], p. 18).

Our subsequent discussion does not depend on the theorem on convex functions referred to in the text above: instead we shall derive a slightly different theorem which is more closely adapted to our needs, especially since our proof will involve several formulae which will be useful at a later stage.

Note that if the function f is homogeneous of the first degree in the u^i, the inequality (a) is equivalent to

$$f(u^i_{(1)} + u^i_{(2)}) \leq f(u^i_{(1)}) + f(u^i_{(2)}) .$$

Returning to the expansion (1.22), we may now deduce the following fundamental inequality:

$$F(x, \xi) \geqq F_{\dot{x}^i}(x, \dot{x})\, \xi^i , \qquad (1.24)$$

the equality sign occurring if and only if $\xi^i = k\,\dot{x}^i$ with $k > 0$. That this statement is true also for the excluded case $\xi^i = \lambda\,\dot{x}^i$ with $\lambda < 0$ is easily verified by direct substitution. For, putting $\xi^i = \lambda\,\dot{x}^i$ on the left-hand side of (1.24) we obtain $F(x, \lambda\dot{x})$, which is positive in view of assumption B, while the right-hand side becomes $\lambda F_{\dot{x}^i}(x, \dot{x})\,\dot{x}^i = \lambda F(x,\dot{x})$, which is negative for $\lambda < 0$.

We may now apply (1.24) to two distinct pairs of line-elements, viz. $(x^i, \dot{x}^i + \overset{\rightharpoonup}{x}{}^i)$, $(x^i, \dot{x}^i)$ and $(x^i, \dot{x}^i + \overset{\rightharpoonup}{x}{}^i)$, $(x^i, \overset{\rightharpoonup}{x}{}^i)$ in turn, obtaining

$$F_{\dot{x}^i}(x^k, \dot{x}^k + \overset{\rightharpoonup}{x}{}^k)\, \dot{x}^i \leqq F(x^i, \dot{x}^i)$$

(with equality if and only if $k_1\,\dot{x}^i = \dot{x}^i + \overset{\rightharpoonup}{x}{}^i$, $k_1 > 0$), together with

$$F_{\dot{x}^i}(x^k, \dot{x}^k + \overset{\rightharpoonup}{x}{}^k)\, \overset{\rightharpoonup}{x}{}^i \leqq F(x^i, \overset{\rightharpoonup}{x}{}^i)$$

(with equality if and only if $k_2\overset{\rightharpoonup}{x}{}^i = \dot{x}^i + \overset{\rightharpoonup}{x}{}^i$, $k_2 > 0$). On adding these two inequalities, taking into account (1.11), we obtain the desired relation (1.21), thus proving our assertion.

It should be noted, however, that the convexity of F does not conversely imply condition C; in § 3 we shall study an example which satisfies equation (1.21) but for which the quadratic form (1.17) is positive indefinite.

Finally, let us express condition C in the form in which it is more commonly known in the calculus of variations. From the theory of quadratic forms we know that condition C implies[1] that the determinant

$$g(x, \dot{x}) \equiv \det \left|\tfrac{1}{2} F^2_{\dot{x}^i\dot{x}^j}(x, \dot{x})\right| > 0 \qquad (1.25)$$

for all line-elements $(x^k, \dot{x}^k)$. This determinant may be expressed in terms of the determinant (1.13); an elementary calculation[2] based on equations (1.19) and (1.12) gives

$$\begin{aligned}
g &= \left| F F_{\dot{x}^i \dot{x}^j} + F_{\dot{x}^i} F_{\dot{x}^j} \right| \\[4pt]
&= \begin{vmatrix} F F_{\dot{x}^i \dot{x}^j} + F_{\dot{x}^i} F_{\dot{x}^j} , & F_{\dot{x}^i} \\ 0 & , 1 \end{vmatrix} \\[4pt]
&= \begin{vmatrix} F F_{\dot{x}^i \dot{x}^j} , & F_{\dot{x}^i} \\ - F_{\dot{x}^j} , & 1 \end{vmatrix} \\[4pt]
&= - F^{-1} \begin{vmatrix} F F_{\dot{x}^i \dot{x}^j} , & F_{\dot{x}^i} \\ F F_{\dot{x}^j} , & -F \end{vmatrix} ,
\end{aligned}$$

---

[1] Cf. Carathéodory [1], p. 178.
[2] Taylor [2], p. 261.

or

$$g(x, \dot{x}) = -F^{n-1} D(x, \dot{x}), \tag{1.26}$$

where

$$D(x, \dot{x}) = \begin{vmatrix} F_{\dot{x}^i \dot{x}^j}, & F_{\dot{x}^i} \\ F_{\dot{x}^j}, & 0 \end{vmatrix}. \tag{1.26a}$$

Thus, in view of equation (1.25) condition C leads to the inequality

$$D(x, \dot{x}) < 0, \tag{1.27}$$

which is the form in which the condition of LEGENDRE is commonly expressed[1].

Conversely, one may show that if (1.27) holds, the rank of the matrix $\|F_{\dot{x}^i \dot{x}^j}\|$ is $n - 1$, and hence it is found that the quadratic form (1.20) is positive indefinite[2]. On applying equation (1.19) one immediately deduces condition C. The latter is thus equivalent to the condition (1.27) of LEGENDRE. If (1.27) is satisfied for all line-elements of a region, the problem in the calculus of variations corresponding to $F(x, \dot{x})$ is said to be *regular* in this region.

§ 2. The Tangent Space. The Indicatrix

In the previous section we have used the notion of a line-element $(x^i, \dot{x}^i)$ in a somewhat loose manner, without giving a clear indication of the nature of the association envisaged between the two sets of n variables x^i and n variables $\dot{x}^i$. This association is best described by the introduction of the so-called "tangent spaces".

Let us consider a change of local coordinates as represented by equations (1.1). Along the curve (1.3), referred to an invariant parameter t, the new components of the tangent vector $\dot{x}^{i'} = dx^{i'}/dt$ are obtained by differentiating the relations

$$x^{i'} = x^{i'}(x^i(t))$$

with respect to t, which gives

$$\dot{x}^{i'} = \frac{\partial x^{i'}}{\partial x^i} \dot{x}^i, \tag{2.1}$$

or, in terms of differentials,

$$dx^{i'} = \frac{\partial x^{i'}}{\partial x^i} dx^i. \tag{2.1a}$$

---

[1] CARATHÉODORY [1], p. 216; BLISS [3], p. 107.

[2] The intermediate steps are fully set out in the treatises of CARATHÉODORY and BLISS. See also § 3, where it will be seen that the determinant (1.27) is proportional to the determinant

$$\begin{vmatrix} F_{\dot{x}^i \dot{x}^j}, & \dot{x}^i \\ \dot{x}^j, & 0 \end{vmatrix}.$$

We remark that condition (1.27) implies that the quadratic form $F_{\dot{x}^i \dot{x}^j}(x, \dot{x})\, \xi^i \xi^j$ is positive definite subject to the condition $F_{\dot{x}^i}(x, \dot{x})\, \xi^i = 0$.

Interpreting the dx^i as the "components" of a displacement in X_n from a point $P(x^i)$ to a point $Q(x^i + dx^i)$ (irrespective of any curve), we see that equation (2.1a) represents the law of transformation of the components of such displacements. If the point $P(x^i)$ is fixed, which implies that the coefficients $\partial x^{i'}/\partial x^i$ of the transformation (2.1a) are fixed, while the dx^i are variable, the relation (2.1a) represents a *linear transformation* of the dx^i onto the $dx^{i'}$. Naturally the same is true for the variables $\dot{x}^i$ and $\dot{x}^{i'}$ in the transformation (2.1). This suggests that entities of this kind may be interpreted as the elements of an n-dimensional linear vector space.

In the language of the tensor calculus, a system of n quantities X^i whose transformation law under (1.1) is equivalent to that of the $\dot{x}^i$ [equation (2.1)], is called a contravariant vector, "attached" to the point $P(x^i)$ of X_n, the individual X^i representing its "components". Such contravariant vectors constitute the elements of our new vector space. Alternatively, the latter may be defined by the totality of all contravariant vectors attached to $P(x^i)$ (or by the totality of all differentials at P)[1]. Clearly, with each point $P(x^i)$ of X_n, which we shall call the *underlying manifold* X_n, such a vector space is to be associated, the coordinates x^i of P entering the coefficients of the linear transformation (2.1). This vector space is to be denoted by $T_n(P)$ [or $T_n(x^i)$] and will be referred to as the "tangent space" attached to $P(x^i)$.

The elements of $T_n(P)$ will generally be denoted by $\dot{x}^i$ in the sequel (but not always: we shall sometimes use ξ^i, η^i, ..., X^i, ...), so that the dot on $\dot{x}^i$ need not necessarily represent the result of a differentiation with respect to some parameter t (this should always be clear from the context). Thus the reader may interpret the elements $\dot{x}^i$ or dx^i of $T_n(P)$ as "directions at P".

Further, since the transformation (2.1) is homogeneous, one may regard the tangent spaces as "centred" affine spaces, the centre or origin corresponding to the values $\dot{x}^1 = 0$, $\dot{x}^2 = 0$, ..., $\dot{x}^n = 0$. In view of the analogy with the tangent planes of ordinary differential geometry it is sometimes of assistance to imagine that the origin of $T_n(P)$ coincides with the point P of X_n, i. e. that the origin is the "point of contact".

From our construction it is clear that *a transformation of the type* (1.1) *in the underlying manifold* X_n *induces a linear transformation* (2.1) *in each of the tangent spaces.*

Let us denote by O the origin of a given tangent space $T_n(P)$ attached to a point P of X_n. Any vector η^i of $T_n(P)$ may be interpreted by means of the directed segment or displacement OQ in $T_n(P)$, the point Q

---

[1] Cf. for instance, WHITEHEAD and VEBLEN [1], p. 61, who call this vector space *the tangent space of differentials.* Compare also SCHOUTEN [1], p. 69; LICHNEROWICZ [8], p. 4.

having coordinates η^i with respect to the coordinate system of $T_n(P)$ (which, as we have seen, depends on the choice of coordinates in the underlying manifold X_n). Thus contravariant vectors may be represented not only by displacements but also by "points" in T_n. If ζ^i is another vector of $T_n(P)$, corresponding to a displacement OS, the displacement QS in $T_n(P)$ has components $\zeta^i - \eta^i$: the usual laws of vector addition, subtraction and multiplication by scalars are applicable to the elements of $T_n(P)$.

In particular, two vectors ξ^i, η^i of $T_n(P)$ are said to be *parallel* if their components are proportional: $\xi^i = \lambda \eta^i (\lambda \neq 0)$.

Summarising: when dealing with a line-element $(x^i, \dot{x}^i)$, the x^i refer to some point P in X_n, while the $\dot{x}^i$ refer to an element of the tangent space $T_n(P)$ attached to P. When we study a given fixed tangent space, the x^i are to be regarded as constants, while the $\dot{x}^i$ are variables. The relationships between distinct tangent spaces attached to neighbouring points of X_n will be considered in subsequent chapters.

Naturally, the notion of tangent space is independent of the metric (if any) imposed on the underlying manifold X_n. However, we shall now show that the introduction, by means of (1.14), of a metric in X_n gives rise to a metric or measurement of length in each tangent space in a unique manner.

Consider, then, the function $F(x^i, \dot{x}^i)$, defined for all line-elements $(x^i, \dot{x}^i)$ over the region R of X_n and satisfying conditions A to C of § 1. We shall, in the present section, restrict our attention to a fixed point $P(x^i)$ of X_n and the corresponding $T_n(P)$. The equation

$$F(x^i, \dot{x}^i) = 1 \qquad (x^i \text{ fixed}, \dot{x}^i \text{ variable}) \qquad (2.2)$$

will represent an $(n-1)$-dimensional locus in $T_n(P)$, i. e. a hypersurface. We shall now show that *the elements $\dot{x}^i$ (or points) of $T_n(P)$ satisfying the inequality*

$$F(x^i, \dot{x}^i) \leq 1 \qquad (x^i \text{ fixed}, \dot{x}^i \text{ variable}) \qquad (2.3)$$

are interior or boundary points of a convex body whose closed boundary is given by (2.2)[1]. This property is a direct consequence of assumptions A to C. Firstly, we note that any directed radius vector OR from the origin of $T_n(P)$ intersects the hypersurface (2.2) in one and only one point. For from assumption B and equation (1.8) it follows that each radius vector does in fact intersect the hypersurface, while if two vectors $\lambda \dot{x}^i, \mu \dot{x}^i (\lambda > 0, \mu > 0, \lambda \neq \mu)$ issuing from O and coinciding in direction with OR were to satisfy (2.2), we would conclude from (1.8) that $(\lambda - \mu) F(x, \dot{x}) = 0$, which would contradict assumption B. Secondly, let $\dot{x}^i, \ddot{x}^i$ be any two points satisfying (2.3) [we may regard $\dot{x}^i$ and $\ddot{x}^i$ as

---

[1] BONNESEN and FENCHEL [1], p. 22.

vectors of $T_n(P)$ issuing from O, whose end-points have coordinates $\dot{x}^i$, $\ddot{x}^i$]. The set (2.3) is convex if any point ξ^i lying on the join of $\dot{x}^i$ and $\ddot{x}^i$ between $\dot{x}^i$ and $\ddot{x}^i$ also satisfies (2.3). But such a point ξ^i may be represented in the form $\xi^i = (1 - \theta)\,\dot{x}^i + \theta\ddot{x}^i$, with $0 \leq \theta \leq 1$. We then have in view of (1.21)

$$F(x^i, \xi^i) \leq F(x^i, (1 - \theta)\,\dot{x}^i) + F(x^i, \theta\ddot{x}^i)\,,$$

or, using (1.8),

$$F(x^i, \xi^i) \leq (1 - \theta)\,F(x^i, \dot{x}^i) + \theta\,F(x^i, \ddot{x}^i)\,.$$

Since, by hypothesis, $F(x^i, \dot{x}^i) \leq 1, F(x^i, \ddot{x}^i) \leq 1$, it follows that $F(x^i, \xi^i) \leq 1$, i. e. the point ξ^i satisfies (2.3), which proves our assertion.

We may now define the length of an arbitrary vector OQ of $T_n(P)$. If the coordinates of Q are η^i (in which case we denote the vector OQ by η^i) the length (or norm) of η^i is given by

$$|\eta| = F(x^i, \eta^i)\,. \tag{2.4}$$

Geometrically, this is simply the ratio OQ/OR where R is the point in which OQ (produced if necessary) intersects the hypersurface (2.2). The length of a vector QS joining two arbitrary points $Q(\eta^i)$, $S(\zeta^i)$ is given by $F(x^i, \zeta^i - \eta^i)$; for if we construct the parallelogram $OQSS'$, QS and OS' are equal and opposite sides, while the coordinates of S' will be $\zeta^i - \eta^i$. We shall thus call $F(x^i, \dot{x}^i)$ the *metric function*.

It is clear, therefore, that the hypersurface (2.2) plays the role of the unit sphere in the geometry of the vector space $T_n(P)$. Following CARATHÉODORY[1], who first introduced this hypersurface in the calculus of variations, we shall call it the *indicatrix*.

Having thus defined a metric in each tangent space, we may conversely assert that the metric in a tangent space $T_n(P)$ determines the local metric of X_n in the immediate vicinity of the point $P(x^i)$. For, strictly speaking, the expression $F(x^i, d\,x^i)$ represents the length of a vector $d\,x^i$ of the tangent space attached to P, but when we use equation (1.14) we interpret this expression as an element ds of length (or displacement) in X_n, namely as a first-order approximation.

A vector space whose metric is defined by a function which satisfies the conditions A to C is called a *Minkowskian space*, since MINKOWSKI introduced a metric function of this type for number-theoretical purposes[2]. However, while MINKOWSKI always presupposed a euclidean

---

[1] CARATHÉODORY [2, 3]; MINKOWSKI [1].

[2] Strictly speaking, the correct definition of a Minkowskian space is based on the convexity condition (1.21) rather than on the more stringent condition C; so that the Minkowskian spaces considered in the sequel are slightly less general than those defined according to the orthodox definition. See also § 8 below.

background, we shall not do so here. As a consequence the theory developed in the following pages will differ considerably from Minkowski's point of view. For an axiomatic treatment of Minkowskian spaces as distinct from other metric spaces we refer the reader to BLUMENTHAL and BUSEMANN[1].

The indicatrix is *symmetric* about the origin if the additional condition A_1 is imposed. Clearly this property corresponds to the axiom of strong monodromy[2]: if $F(x^i, \dot{x}^i - \ddot{x}^i)$ denotes the Minkowskian distance between two distinct points $\dot{x}^i$ and $\ddot{x}^i$ of T_n, then $F(x^i, \dot{x}^i - \ddot{x}^i) = F(x^i, \ddot{x}^i - \dot{x}^i)$. We shall indicate distinctly where such symmetry properties are required.

Equation (1.21) represents the triangle inequality. It follows that the shortest distance joining two points is given by the line-segment joining them.

Conversely, suppose we are given in $T_n(P)$ an arbitrary closed hypersurface $f(x^i, \dot{x}^i) = 1$, centred at the origin O, where we assume only that f is non-negative and positively homogeneous of degree 1 in the $\dot{x}^i$. This again defines a metric in the sense that the distance $\varrho(a, b)$ between two points a, b of $T_n(P)$ is given by $f(x^i, \eta^i)$, where the η^i are the components of the vector $a\,b$. Now let the points a, b, c form an arbitrary triangle in $T_n(P)$. At O we construct the vectors $OA = a\,b$, $OB = b\,c, OC = a\,c$. The intersections of these vectors, produced if necessary, with the given hypersurface are denoted by A', B', C' respectively, while $B'A'$ and OC' intersect in a point S. (Note that triangle $a\,b\,c$ defines a plane, so that our construction is basically a two-dimensional one.) We may state the following

Theorem[3]*:* In the triangle $a\,b\,c$ the sum of the lengths of two sides: $\varrho(a, b) + \varrho(b, c)$ is greater, equal to or less than the length $\varrho(a, c)$ of the third side according as S is an interior point, a boundary point or an exterior point of the region J enclosed by the hypersurface $f(x^i, \dot{x}^i) = 1$.

Proof: By definition, $OA = \varrho(a, b)\,OA', OB = \varrho(b, c)\,OB', OC = \varrho(a, c)\,OC'$; thus, since $a\,b + b\,c = a\,c$, we have

$$\varrho(a, b)\,OA' + \varrho(b, c)\,OB' = \varrho(a, c)\,OC'.$$

But $OA' = OS + SA'$, and $OB' = OS - B'S$; hence

$$(\varrho(a, b) + \varrho(b, c))\,OS + \varrho(a, b)\,SA' - \varrho(b, c)\,B'S = \varrho(a, c)\,OC'. \qquad (2.5)$$

From our construction if follows that triangle OAC is equivalent to the figure obtained by a direct translation of triangle $a\,b\,c$. Thus if we construct $A'D$ parallel to AC to meet OC in D, $A'D$ will be parallel to OB. Triangles $OB'S, DA'S$ are similar, so that $\varrho(B'S)/\varrho(SA') = \varrho(OB')/\varrho(A'D)$. But $\varrho(A'D)/\varrho(AC) = \varrho(OA')/\varrho(OA)$, while $AC = OB$. Hence

$$\varrho(B'S)/\varrho(SA') = \varrho(OB')/\varrho(OB) \cdot \varrho(OA)/\varrho(OA') = [\varrho(b\,c)]^{-1}\,\varrho(a\,b)$$

---

[1] BLUMENTHAL [1]; BUSEMANN [1, 2, 3, 4, 10].

[2] HAMEL [1].

[3] This theorem is given by GOLAB and HÄRLEN [1], p. 389. However, these authors base their proof on a superimposed euclidean metric. Since we intend to systematically avoid any such procedure we have given an alternative proof. The reader is also referred to this paper for further details concerning so-called pseudo-Minkowskian spaces, i. e. spaces whose metric function need not necessarily be convex everywhere.

by construction. Since B', S, A' are collinear, we therefore have $\varrho(b, c)\, B'S = \varrho(a, b)\, SA'$. Also, since O, S, C' are collinear, we may write $OC' = \varepsilon \cdot OS$. Equation (2.5) now becomes

$$(\varrho(a, b) + \varrho(b, c))\, OS = \varepsilon\, \varrho(a, c)\, OS\, . \tag{2.6}$$

Since C' is a boundary point of the given hypersurface it follows that $\varepsilon \gtreqless 1$ according as S is an interior point, a boundary point or an external point of J. Equation (2.6) therefore implies the theorem.

The region J is said to be nowhere concave if there exists a hyperplane K at each boundary point Q of J such that Q is contained in K, while J is situated entirely on one side of K. If, in addition, K contains no other points of J except Q, J is said to be everywhere convex. Thus a point S on the join of two boundary points A', B' is either a boundary or an interior point if J is nowhere concave. Thus the preceding theorem implies the theorem of Minkowski[1]: A metric function f satisfies the triangle inequality if and only if the hypersurface $f(x^i, \dot{x}^i) = 1$ is nowhere concave; while the equality sign implies collinearity only if the hypersurface is everywhere convex.

A generalisation of the notion of a Minkowskian space may be obtained if condition B is dropped (apart from the relaxation of differentiability assumptions). The corresponding indicatrices may then be hypersurfaces extending to infinity. It is shown by Alt [1] that the validity of the triangle inequality is a necessary and sufficient condition for the indicatrix to be projectively convex, which means that the indicatrix may be mapped by means of a collineation onto a (not necessarily bounded) convex hypersurface. These results are amplified by Pauc [1, 2] and applied to problems in the calculus of variations in parametric form. In this manner the condition of Legendre for the regularity (or quasi-regularity[2]) of a problem is obtained, even in the indefinite case. Similar results are generalised by Aronszajn [1] to infinitely-dimensional vector spaces, which may — in this context — be regarded as Banach spaces with an indefinite norm. Further applications of the notion of quasi-regularity to the calculus of variations are discussed by Busemann and Mayer [1], §§ 3—4. It would appear that the results of these authors would form the basis of a systematic theory of Finsler spaces for which conditions B and C are suitably relaxed.

Vagner [13] also treats the indicatrix from a more general point of view. Initially an arbitrary hypersurface is considered, and any vector issuing from the origin whose oriented direction is such that the latter intersects the hypersurface is called a "measurable" vector[3].

§ 3. The Metric Tensor and the Osculating Indicatrix

Let $\dot{x}^i$ represent an arbitrary vector issuing from the origin O in the tangent space $T_n(P)$. We define a set of quantities g_{ij} by the equations

$$g_{ij}(x, \dot{x}) = \frac{1}{2}\frac{\partial^2 F^2(x, \dot{x})}{\partial \dot{x}^i\, \partial \dot{x}^j}\, , \tag{3.1}$$

---

[1] Minkowski [1], §§ 6, 16, 17.

[2] For the definition of quasi-regularity see Pauc [2], p. 30. See also Choquet [1] and [2].

[3] Vagner [13], § 2. For the definition of the "singularity" class of an indicatrix see Ch. V, § 8.

and it is easily verified by means of (2.1) that the g_{ij} form the components of a covariant tensor of rank 2[1]. In view of the homogeneity condition (1.8) the function F^2 is positively homogeneous of second degree in the $\dot{x}^i$, and hence we have from equation (1.19b)

$$F^2(x, \dot{x}) = g_{ij}(x, \dot{x})\, \dot{x}^i\, \dot{x}^j \,. \tag{3.2}$$

Thus, according to our definition (2.4) all lengths in $T_n(P)$ may be expressed in terms of the g_{ij}, which we shall regard as the components of the *metric tensor* of $T_n(P)$. The equation (2.2) of the indicatrix of $T_n(P)$ may now be written in the form:

$$F^2(x, \dot{x}) = g_{ij}(x, \dot{x})\, \dot{x}^i\, \dot{x}^j = 1 \,. \tag{3.3}$$

Also, from the definition (3.1) we deduce that the $g_{ij}(x, \dot{x})$ are positively homogeneous of degree zero in the $\dot{x}^i$ and symmetric in their indices. We may therefore construct the following useful tensor[2]

$$C_{ijk}(x, \dot{x}) = \frac{1}{2}\frac{\partial g_{ij}(x, \dot{x})}{\partial \dot{x}^k} = \frac{1}{4}\frac{\partial^3 F^2(x, \dot{x})}{\partial \dot{x}^i\, \partial \dot{x}^j\, \partial \dot{x}^k} \,, \tag{3.4}$$

which is positively homogeneous of degree -1 and symmetric in all three of its indices. The following identities (resulting from homogeneity) will be repeatedly applied in the sequel:

$$C_{ijk}(x, \dot{x})\, \dot{x}^i = C_{ijk}(x, \dot{x})\, \dot{x}^j = C_{ijk}(x, \dot{x})\, \dot{x}^k = 0 \,, \tag{3.5}$$

together with

$$\frac{\partial C_{ijk}(x, \dot{x})}{\partial x^h}\, \dot{x}^i = \frac{\partial C_{ijk}(x, \dot{x})}{\partial x^h}\, \dot{x}^j = \frac{\partial C_{ijk}(x, \dot{x})}{\partial x^h}\, \dot{x}^k = 0 \,. \tag{3.6}$$

For future reference we might, at this stage, express the equation of the tangent hyperplane to the indicatrix (2.2) at a given point $\dot{x}^i_{(0)}$ of $T_n(P)$ in terms of the metric tensor (3.1). As in ordinary analytic geometry the equation of this hyperplane is[3]

$$F_{\dot{x}^i}(x^k, \dot{x}^k_{(0)})\, (\dot{x}^i - \dot{x}^i_{(0)}) = 0 \,. \tag{3.7}$$

---

[1] We note that under the transformation (1.1) we always have $\partial \dot{x}^{i\prime}/\partial \dot{x}^i = \partial x^{i\prime}/\partial x^i$ in view of (2.1), so that differentiation of a tensor with respect to a direction $\dot{x}^i$ leads to a new tensor whose covariant valency is increased by one. The same does not, of course, apply to differentiation with respect to a positional coordinate x^i except when a scalar quantity $\Phi(x^i)$ is being differentiated.

[2] CARTAN [1], p. 11. The application of the tensor calculus to the calculus of variations is discussed in great detail by DE DONDER [1], and DUSCHEK and MAYER [1]. In this connection we should also mention TUCKER [1], BOSQUET [1, 2], and VEBLEN [2]. An attempt to apply the tensor calculus to non-homogeneous problems in the calculus of variations is made by JOHNSON [1].

[3] This is easily verified directly: for when (2.2) is written in a parametric representation $\dot{x}^i = \dot{x}^i(x^j, u^\alpha)$, $(\alpha = 1, \ldots, n-1)$, the $(n-1)$ quantities $\partial \dot{x}^i/\partial u^\alpha$ defined at each point of (2.2) span the desired hyperplane. On differentiating the identity $F(x^k, \dot{x}^k(x^j, u^\alpha)) = 1$, which results from (2.2), with respect to u^α, the result follows immediately.

In virtue of (1.19a) and (3.1) it will be seen that this equation becomes

$$g_{ij}(x^k, \dot{x}^k_{(0)})\, \dot{x}^i_{(0)}\, \dot{x}^j = 1 \,. \tag{3.8}$$

If we express the distance ds between two points x^i and $x^i + dx^i$ of X_n as given by (1.14) in terms of the metric tensor we have

$$ds^2 = g_{ij}(x, dx)\, dx^i\, dx^j \,. \tag{3.9}$$

A Riemannian space with positive definite metric is therefore a particular case of our metric space and corresponds to a metric function F whose metric tensor g_{ij} is independent of direction. The indicatrix (3.3) in any tangent space $T_n(P)$ then assumes the form $g_{ij}(x)\, \dot{x}^i\, \dot{x}^j = 1$, the components of the metric tensor assuming fixed values (for a given coordinate system) in $T_n(P)$. The latter equation, however, simply represents a quadric hypersurface, and since the quadratic form on the left-hand side is positive definite according to our assumption C, this hypersurface will be an $(n-1)$-dimensional ellipsoid. By means of an affine transformation in $T_n(P)$ this can be transformed into the form $\sum_{i=1}^{n}(\dot{x}^i)^2 = 1$; in other words, the quadric indicatrix is — apart from an affine transformation — equivalent to the $(n-1)$-dimensional unit sphere of euclidean geometry. Since in every case the metric of X_n is determined locally by the metric of its tangent spaces, we see that the metric of a Riemannian space is locally euclidean.

In the same manner we may regard the metric of the Finsler space as being locally Minkowskian.

In the latter case the indicatrix is not in general a quadric hypersurface. Nevertheless, it is possible to construct a quadric hypersurface for each fixed direction $\dot{x}^i_{(0)}$ by means of the equation

$$Q^2 = g_{ij}(x^k, \dot{x}^k_{(0)})\, \dot{x}^i\, \dot{x}^j = 1 \,. \tag{3.10}$$

In this equation the x^k, $\dot{x}^k_{(0)}$ are regarded as being fixed, while the $\dot{x}^i$ are the running variables. This hypersurface is called the osculating indicatrix[1], since it has contact of the second order with the indicatrix (3.3) at $\dot{x}^i = \dot{x}^i_{(0)}$. This property follows immediately from the fact that the second derivatives of F and Q in (3.3) and (3.10) respectively are equal when $\dot{x}^i = \dot{x}^i_{(0)}$.

This construction has influenced the development of Minkowskian geometry to a very great extent. *Instead of letting the indicatrix (3.3) play the role of the unit sphere, one may construct an osculating quadric (3.10) for each direction $\dot{x}^i_{(0)}$ in $T_n(P)$ and thus obtain a euclidean metric for each line-element $(x^i, \dot{x}^i_{(0)})$.* In the following chapters we shall indicate how the acceptance or non-acceptance of this construction leads to

---

[1] FINSLER [1], p. 42.

entirely different approaches to Finsler geometry. A systematic development of Minkowskian geometry by means of a euclidean metric attached to each direction is given by VARGA [1].

The construction of the osculating indicatrix depends directly on condition C and would not always be possible if we were to replace the latter condition by the more general convexity condition (1.21). The following 2-dimensional example will help to clarify the position. Consider the function

$$F(\dot{x}_1, \dot{x}_2) = (\dot{x}_1^m + \dot{x}_2^m)^{\frac{1}{m}}, \tag{3.11}$$

where m is a positive even integer ≥ 2 [1]. It is easily verified that the function (3.11) satisfies the conditions A, B and the convexity condition (1.21). Using (3.1) we find

$$g_{11} = \dot{x}_1^{m-2}(\dot{x}_1^m + \dot{x}_2^m)^{\frac{2}{m}-2}\,[\dot{x}_1^m + (m-1)\,\dot{x}_2^m]\,,$$

$$g_{12} = (2-m)\,(\dot{x}_1\dot{x}_2)^{m-1}(\dot{x}_1^m + \dot{x}_2^m)^{\frac{2}{m}-2}\,, \tag{3.12}$$

$$g_{22} = \dot{x}_2^{m-2}(\dot{x}_1^m + \dot{x}_2^m)^{\frac{2}{m}-2}[(m-1)\,\dot{x}_1^m + \dot{x}_2^m]\,.$$

At the point $\dot{x}_1 = 1$, $\dot{x}_2 = 0$ on the indicatrix g_{12} and g_{22} will vanish, while $g_{11} = 1$, unless $m = 2$. The equation (3.10) of the osculating indicatrix at this point is therefore $\dot{x}_1^2 = 1$, i. e. we do not obtain an ellipse, but a pair of straight lines, namely the tangent lines to the indicatrix at the points $(-1, 0)$ and $(+1, 0)$ as may be seen from (3.8). A similar phenomenon evidently occurs at the points $(0, 1)$ and $(0, -1)$. The reason for this peculiarity is of course due to the fact that the determinant of the g_{ij}, namely

$$g = \frac{(m-1)\,(\dot{x}_1\dot{x}_2)^{m-2}}{(\dot{x}_1^m + \dot{x}_2^m)^{2-\frac{4}{m}}} \tag{3.13}$$

vanishes for $\dot{x}_1 = 1$, $\dot{x}_2 = 0$ unless $m = 2$ (in which case we have a euclidean metric).

It will be seen, however, that many of the results which we shall deduce on the basis of condition C will also hold under the more general convexity assumption (1.21). *This fact should be stressed:* we shall return to the same example in due course.

The condition C may be expressed in a slightly different form as follows. If we return to equation (1.12) we see that if F^{ij} represents the cofactor of $F_{\dot{x}^i\dot{x}^j}$ in the determinant $|F_{\dot{x}^i\dot{x}^j}|$ we have

$$\frac{\dot{x}^1}{F^{i1}} = \frac{\dot{x}^2}{F^{i2}} = \cdots = \frac{\dot{x}^n}{F^{in}}\,. \tag{3.14}$$

Therefore there must exist a function $F_1(x, \dot{x})$ such that

$$\dot{x}^i\,\dot{x}^j\,F_1 = F^{ij}\,. \tag{3.15}$$

If we multiply this equation by $F_{\dot{x}^i}F_{\dot{x}^j}$ (summing over i and j) we obtain in view of (1.11)

$$F_1 = F^{-2}\,F^{ij}F_{\dot{x}^i}F_{\dot{x}^j}\,,$$

---

[1] In fact, RIEMANN himself suggested this example with $m = 4$ ([1], p. 278). We shall write the indices indicating components as subscripts in this particular example in order to avoid confusion with the exponential symbols.

or[1]

$$F_1 = -F^{-2} \begin{vmatrix} F_{\dot{x}^i \dot{x}^j}, & F_{\dot{x}^i} \\ F_{\dot{x}^j}, & 0 \end{vmatrix}. \tag{3.16}$$

On multiplying (3.15) by $\dot{x}^i \dot{x}^j$ we find similarly

$$F_1 = -\left\{ \sum_{i=1}^{n} (\dot{x}^i)^2 \right\}^{-2} \begin{vmatrix} F_{\dot{x}^i \dot{x}^j}, & \dot{x}^i \\ \dot{x}^j, & 0 \end{vmatrix}. \tag{3.17}$$

This is the form of the function F_1 as introduced by WEIERSTRASS in his theory of parametric problems in the calculus of variations. Such a problem is regular in a region if and only if the function F_1 is everywhere positive in that region[2]. This follows directly from (1.26a) and (1.27).

§ 4. The Dual Tangent Space. The Figuratrix

We shall continue with our consideration of the tangent space $T_n(P)$ attached to the point $P(x^i)$ of X_n, so that the positional coordinates x^i are still to be regarded as fixed.

Using the metric tensor g_{ij} of (3.1) we may associate with each arbitrary contravariant vector $\dot{x}^i$ of $T_n(P)$ a covariant vector y_i defined by the relation

$$y_i = g_{ij}(x, \dot{x})\, \dot{x}^j, \tag{4.1}$$

where it is to be noted that the directional argument in the g_{ij} must coincide with the vector $\dot{x}^i$ under consideration. Thus (4.1) assigns a set of values y_i to each point $\dot{x}^i$ of $T_n(P)$. This correspondence is the analytical representation of a singularly striking geometrical state of affairs. Firstly, we observe that the y_i may be regarded as the positional coordinates of points in a second space $T_n'(P)$[3], which we shall call the *dual* tangent space of X_n at P. Clearly $T_n'(P)$ is the totality of all covariant vectors attached to X_n at P; and again any coordinate system in X_n determines a unique coordinate system in $T_n'(P)$. The transformation (1.1) induces a linear transformation

$$y_{i'} = \frac{\partial x^i}{\partial x^{i'}}\, y_i, \tag{4.2}$$

in $T_n'(P)$, as may be easily verified by means of (4.1).

---

[1] Compare, for instance, PERRON [1], p. 99.

[2] Cf. CARATHÉODORY [1], p. 216 and 245.

[3] This is evident from the fact that the relation (4.1) possesses an inverse: for from the implicit function theorem it follows that we may solve equations (4.1) for the $\dot{x}^i$ as functions of the y_i provided

$$\det \left| \frac{\partial}{\partial \dot{x}^j}\, (g_{ih}(x, \dot{x})\, \dot{x}^h) \right| \neq 0.$$

But from (3.4) and (3.5) we may deduce that the elements of this determinant are the $g_{ij}(x, \dot{x})$, so that the required condition is satisfied in virtue of (1.25).

On the other hand it is well known[1] that any covariant vector u_i may be represented by means of a hyperplane (or a pair of hyperplanes)

$$u_i \, \dot{x}^i = \text{const.} , \tag{4.3}$$

in $T_n(P)$ itself. By varying the constant in this equation we obtain a family of parallel hyperplanes. Let us assign a value to this constant such that the hyperplane (4.3) is tangent to the hypersurface

$$F^2 \left(x^k, \dot{x}^k \right) = g_{ij} \left(x^k, \dot{x}^k \right) \dot{x}^i \, \dot{x}^j = r^2 , \tag{4.4}$$

where $r > 0$ is fixed, so that (4.4) represents a hypersurface homothetic to the indicatrix (3.3). Naturally the value of the constant in (4.3) depends on the values of the u_i and r; the equation to the tangent hyperplane will therefore be of the form

$$u_i \, \dot{x}^i = r H \left(x^k, u_k \right) , \tag{4.5}$$

where H is some function of the x^k, u_k [since the g_{ij} are functions of the x^k even though the latter are fixed in $T_n(P)$ and $T'_n(P)$][2]. Naturally, if we replace the u_k by proportional components $u_k^* = \lambda \, u_k$ with $\lambda > 0$, we obtain for the same hyperplane the equation (4.5) with the u_k^* replacing the u_k. It follows therefore that

$$H \left(x^k, \lambda \, u_k \right) = \lambda H \left(x^k, u_k \right) , \quad (\lambda > 0) , \tag{4.6}$$

i. e. the function H is positively homogeneous of degree 1 in the u_k.

The hyperplane (4.5) divides the space $T_n(P)$ into two regions S_1 and S_2. It has one point in common with the hypersurface (4.4), namely the point of contact. In view of our convexity condition (1.21) all other points of this hypersurface lie entirely on one side of the hyperplane (4.5), i. e. they are contained entirely in either of the regions S_1 and S_2. The hyperplane (4.5) is said to be a *supporting plane* of the hypersurface (4.4), while the function H is called the *function of support* of the hypersurface[3].

For all points interior to and on the hypersurface (4.4) we have

$$u_i \, \dot{x}^i \leqq r H \left(x^k, u_k \right) , \tag{4.7}$$

so that $r H \left(x^k, u_k \right)$ may be defined alternatively as the least upper bound of $(u_i \, \dot{x}^i)$ subject to the condition $F(x^k, \dot{x}^k) \leqq r$. Conversely, the set of all points satisfying the inequality (4.7) (for any choice of the u_i) is

---

[1] SCHOUTEN [1], p. 7.

[2] The fact that the constant on the right-hand side of (4.3) is directly proportional to r [thus justifying the notation of (4.5)] is easily established by considering distinct values r_1 and r_2 of r in the construction (4.4).

[3] Compare BONNESEN and FENCHEL [1], p. 4 and 23, where the case $r = 1$ is treated only. For our purpose, however, it is often useful to consider hypersurfaces homothetic to the indicatrix.

the set for which $F(x^i, \dot{x}^i) \leqq r$. Since we have assumed $r > 0$ it follows directly that

$$H(x^k, u_k) > 0 \qquad (4.8)$$

unless all the u_k vanish simultaneously.

Let $\dot{x}^i_{(0)}$ represent the coordinates of the point of contact of (4.5) with (4.4). Analogously to (3.8) we may write the equation of the tangent hyperplane in the form

$$g_{ij}(x^k, \dot{x}^k_{(0)})\, \dot{x}^i_{(0)}\, \dot{x}^j = r^2\,,$$

or, using (4.1), if $y_i^{(0)}$ denotes the components of the covariant vector of $T'_n(P)$ corresponding to $\dot{x}^i_{(0)}$ of $T_n(P)$,

$$y_i^{(0)}\, \dot{x}^i = r^2\,. \qquad (4.9)$$

Since the equations (4.5) and (4.9) represent the same hyperplanes we have

$$y_i^{(0)} = \frac{r\, u_i}{H(x^k, u_k)}\,, \qquad (4.10)$$

so that the u_i are determined by the $y_i^{(0)}$ to within a positive factor λ in view of (4.6). Let us choose $u_i^{(0)} = \lambda u_i$ such that $H(x^k, u_k^{(0)}) = r$. Then we have $u_i^{(0)} = y_i^{(0)}$ in consequence of (4.10), and the equation (4.9) to the tangent hyperplane becomes

$$y_i^{(0)}\, \dot{x}^i = H^2(x^k, y_k^{(0)})\,. \qquad (4.11)$$

In particular, the point $\dot{x}^i_{(0)}$ lies on the hyperplane, so that we have from (3.2), (4.1) and (4.11)

$$H^2(x^k, y_k^{(0)}) = y_i^{(0)}\, \dot{x}^i_{(0)} = g_{ij}(x^k, \dot{x}^k_{(0)})\, \dot{x}^i_{(0)}\, \dot{x}^j_{(0)} = F^2(x^k, \dot{x}^k_{(0)})\,, \qquad (4.12)$$

this being a perfectly general identity involving corresponding vectors (or points) $\dot{x}^i_{(0)}$ and $y_i^{(0)}$ of $T_n(P)$ and $T'_n(P)$ respectively.

Since, by definition, the right hand side of (4.12) represents the square of the length of the vector $\dot{x}^k_{(0)}$, it is natural to regard the left-hand side $H^2(x^k, y_k^{(0)})$ (which does not involve the $\dot{x}^k_{(0)}$ explicitly) as the square of the length of the vector $y_i^{(0)}$ of $T'_n(P)$. *Thus we shall regard the function of support of the indicatrix (or of homothetic hypersurfaces) in $T_n(P)$ as the metric function of $T'_n(P)$*[1]. In $T'_n(P)$ we may construct the hypersurface defining unit covariant vectors:

$$H(x^k, y_k) = 1\,. \qquad (4.13)$$

Again this is a closed, convex hypersurface. This may be proved as for the case of the indicatrix, for we have seen that apart from convexity $H(x^k, y_k)$ satisfies the same conditions as $F(x^k, \dot{x}^k)$. The latter condition is easily proved as follows[2]: If u_i, v_i are arbitrary vectors of $T'_n(P)$ it

---

[1] RUND [1], p. 60—62.

[2] BONNESEN and FENCHEL [1], p. 24.

follows from (4.7) that we have for all points satisfying $F(x^i, \dot{x}^i) \leqq r$,

$$u_i \dot{x}^i \leqq r H(x^k, u_k) , \qquad v_i \dot{x}^i \leqq r H(x^k, v_k) ,$$

or

$$(u_i + v_i) \dot{x}^i \leqq r \{H(x^k, u_k) + H(x^k, v_k)\} .$$

But since $r H(x^k, u_k + v_k)$ is the least upper bound of $(u_i + v_i) \dot{x}^i$, attained by a definite point of the set $F(x^i, \dot{x}^i) \leqq r$, we must have

$$H(x^k, u_k + v_k) \leqq H(x^k, u_k) + H(x^k, v_k) . \tag{4.14}$$

The hypersurface (4.13) of $T'_n(P)$ thus plays the role of the unit sphere in $T'_n(P)$ and thus induces a Minkowskian metric in this space. It is called the *Figuratrix*.

We thus have a *completely symmetrical reciprocity* between $T_n(P)$ and $T'_n(P)$, for it may be shown conversely that $F(x, \dot{x})$ is the function of support of the figuratrix.

The figuratrix was first discovered by MINKOWSKI in connection with his theory of convex bodies (presupposing a euclidean background). The nomenclature is due to HADAMARD, who re-discovered it in virtue of its importance in the calculus of variations[1].

If a euclidean background is introduced, one may define the euclidean unit sphere $\sum\limits_{i=1}^{n} (\dot{x}^i)^2 = 1$ in $T_n(P)$, and it is easy to show that the figuratrix results from the indicatrix by means of transformation by reciprocal polars with respect to the euclidean unit sphere[2]. In contradistinction to our method CARATHÉODORY writes the equation to the figuratrix in the form $\mathfrak{H}(x, y) = 0$. This difference is due to a basically different approach to the treatment of Hamiltonian functions of parametric problems in the calculus of variations. We shall return to this point in the next section.

BLASCHKE [1] gives a detailed treatment of the figuratrix in two and three dimensions, again superimposing a euclidean metric. For the two-dimensional case it is shown that the function F_1 (§ 3) represents the radius of curvature of the figuratrix, while the Weierstrass $\mathscr{E}$-function[3] admits a simple interpretation, which we shall discuss in § 6 for the n-dimensional case. It is pointed out also that the figuratrix may be more useful than the indicatrix from the point of view of the calculus of variations when condition B is relaxed. In the three-dimensional case the function F_1 is the reciprocal of the Gaussian curvature of the figuratrix. This leads to a lucid geometrical interpretation of some of the necessary conditions in the calculus of variations in terms of the principal radii of curvature. In the same paper BLASCHKE discusses the generalisation of the notion of the figuratrix to the case of multiple integrals. Almost the same ground is covered independently in a paper by RIDER [1]. With the help of the analytical apparatus described by BONNESEN and FENCHEL [1] the results of BLASCHKE may be generalised directly to apply to the n-dimensional case. The figuratrix is useful also in applications to geometrical optics[4].

---

[1] MINKOWSKI [2], § 8; HADAMARD [1], p. 92.

[2] CARATHÉODORY [1], pp. 246—248.

[3] CARATHÉODORY [1], p. 224.

[4] CARATHÉODORY [5]; SYNGE [4].

§ 5. The Hamiltonian Function

Since the function $H(x^k, y_k)$ is positively homogeneous of degree 1 in the y_k, we have the identities

$$H(x, y) = H_{y_i}(x, y)\, y_i \,, \tag{5.1}$$

and

$$H_{y_i y_j}(x, y)\, y_i = 0 \,, \tag{5.2}$$

so that the determinant

$$\det \left| \frac{\partial^2 H}{\partial y_i \, \partial y_j} \right| = 0 \tag{5.3}$$

in analogy to (1.13).

We may now define a contravariant metric tensor corresponding to a covariant vector y_i by putting

$$g^{ij}(x, y) = \frac{1}{2} \frac{\partial^2 H^2(x, y)}{\partial y_i \, \partial y_j} \,. \tag{5.4}$$

Again the tensor character of the g^{ij} is proved by direct differentiation of the invariant H and application of (4.2). Since H^2 is positively homogeneous of degree 2 in the y_i, it follows from Euler's theorem on homogeneous functions that

$$H^2(x, y) = g^{ij}(x, y)\, y_i\, y_j \,. \tag{5.5}$$

Now let $\dot{x}^i$ and y_i be two vectors of $T_n(P)$ and $T'_n(P)$ respectively corresponding to each other according to (4.1). Using (3.1) and (1.19a) we deduce that the correspondence (4.1) may be written in the form

$$y_i = F(x, \dot{x}) \frac{\partial F(x, \dot{x})}{\partial \dot{x}^i} \,. \tag{5.6}$$

Since, as we have seen above, the relations between $T_n(P)$ and $T'_n(P)$ are completely symmetrical, we have similarly

$$\dot{x}^i = H(x, y) \frac{\partial H(x, y)}{\partial y_i} \,. \tag{5.7}$$

But from (5.4) we see that

$$g^{ij}(x, y) = H_{y_i}(x, y) \cdot H_{y_j}(x, y) + H(x, y)\, H_{y_i y_j}(x, y) \,.$$

Multiplying this equation by y_j and using (5.1) and (5.2) we deduce that equation (5.7) is equivalent to

$$\dot{x}^i = g^{ij}(x, y)\, y_j \,. \tag{5.8}$$

This, then, is the inverse of relation (4.1).

Differentiating (5.7) with respect to $\dot{x}^k$ we find[1]

$$\delta^i_k = \left(H_{y_i} H_{y_j} + H H_{y_i y_j} \right) \frac{\partial y_j}{\partial \dot{x}^k} \,,$$

---

[1] Here δ^i_k stands for the KRONECKER symbol: $\delta^i_k = 0$ if $i \neq k$, $\delta^i_k = 1$ if $i = k$.

and in view of (5.7), (4.1) and (3.5) this reduces to

$$g_{ij}(x, \dot{x})\, g^{ik}(x, y) = \delta_j^k \; . \tag{5.9}$$

Again, the $g^{ij}(x^k, y_k)$ are homogeneous of degree zero in the y_k. Hence we have the identities:

$$\frac{\partial g^{ik}(x, y)}{\partial y^j}\, y_j = \frac{\partial g^{ik}(x, y)}{\partial y_j}\, y_k = 0 \; . \tag{5.10}$$

Thus, when we differentiate (5.9) with respect to x^h and multiply by $\dot{x}^j\, y_k$, we have

$$\frac{\partial g_{ij}(x, \dot{x})}{\partial x^h}\, g^{ik}(x, y)\, \dot{x}^j\, y_k = -g_{ij}(x, \dot{x}) \cdot \frac{\partial g^{ik}(x, y)}{\partial x^h}\, \dot{x}^j\, y_k \; ,$$

or, using (4.1) and (5.8),

$$\frac{\partial g_{ik}(x, \dot{x})}{\partial x^h}\, \dot{x}^i\, \dot{x}^k = - \frac{\partial g^{ik}(x, y)}{\partial x^h}\, y_i\, y_k \; .$$

From (3.2) and (5.5) we therefore deduce

$$\frac{\partial F(x, \dot{x})}{\partial x^h} = - \frac{\partial H(x, y)}{\partial x^h} \; , \tag{5.11}$$

where — in analogy to the canonical equations of mechanics — we regard the $(x^i, \dot{x}^i)$ and the (x^i, y_i) as the independent variables of the left-hand and right-hand sides respectively[1]. Furthermore, if y_i is a unit vector, i. e. if

$$H(x^k, y_k) = 1 \; , \tag{5.12}$$

equations (5.6) and (5.7) become

$$y_i = \frac{\partial F}{\partial \dot{x}^i} \; , \qquad \dot{x}^i = \frac{\partial H}{\partial y_i} \; . \tag{5.13}$$

The treatment of the function $H(x, y)$ as outlined in this and in the preceding section was purposely given in order to exhibit clearly the geometrical nature of this function, and in particular the fact that it serves as function of support to the indicatrix. The following, purely analytical treatment is considerably shorter: for the sake of completeness we shall sketch it briefly. First of all we observe that under the conditions which we have imposed on the function $F(x, \dot{x})$, the differentiability properties of $H(x, y)$ with respect to the y_i present no difficulties: in fact, this problem is treated from a far more general point of view by Bonnesen and Fenchel[2]. Now, in view of (3.1), the expressions (4.1) defining the y_i are positively homogeneous of degree 1 in the $\dot{x}^i$. Observing condition (1.25) we see that we may solve the equations (4.1) for the $\dot{x}^i$ in terms of the y_i, obtaining

$$\dot{x}^i = \psi^i(x^k, y_k) \; , \tag{5.7a}$$

these functions being positively homogeneous of degree one in the y_k. Substituting back in (4.1) we have an identity in the (x^k, y_k), and on differentiating the latter with respect to y_k we find

$$\delta_i^k = 2\, C_{ijh}(x, \psi)\, \psi^{hk}\, \psi^j + g_{ij}\, \psi^{jk} \; ,$$

---

[1] Rund [2], p. 210.

[2] Bonnesen and Fenchel [1], p. 26.

where we have put $\psi^{jk} = \partial \psi^j / \partial y^k$. As a result of (3.5) the first term on the right-hand side of this equation vanishes identically; hence the ψ^{jk} are the elements of the matrix inverse to the symmetric matrix $\|g_{ij}\|$:

$$g_{ij}(x, \psi) \, \psi^{jk}(x, y) = \delta_i^k \, . \tag{5.9a}$$

The alternative definition of $H(x, y)$ results from the direct substitution of (5.7a) in $F(x, \dot{x})$, i. e. we write

$$H(x, y) = F(x, \psi(x, y)) \, .$$

On differentiating with respect to y_i and using (1.19a) and (3.1) we find

$$\frac{\partial H}{\partial y_i} = \frac{\partial F}{\partial \dot{x}^k} \, \psi^{ki} = F^{-1} g_{kh} \, \psi^h \, \psi^{ki} \, ,$$

which, together with (5.7a) and (5.9a) yields

$$\frac{\partial H}{\partial y_i} = F^{-1} \dot{x}^i \, , \tag{5.7b}$$

which is equivalent to (5.7). In conjunction with (5.9) it follows from (5.9a) that $\psi^{ik} = g^{ik}$ as defined by (5.2). This method also enables one to establish the differentiability of $H(x, y)$ in all its arguments without any difficulty.

The reader will immediately recognise the close similarity between the canonical relations of mechanics and our equations (5.11) and (5.13), provided that the function H of support is interpreted as a Hamiltonian function. We may remark that condition (5.12) can always be enforced by a suitable choice of parameter s of arc: in fact if the function F is regarded as the Lagrangian function of a dynamical system, it can be shown that the "normalisation" (5.12) is equivalent to the energy equation for the case of a conservative system[1]. In the sequel we shall therefore regard H as the Hamiltonian associated with the problem in the calculus of variations arising out of the function F. This is in direct contrast to the method devised by CARATHÉODORY[2] for the study of parametric problems, where a Hamiltonian equation is postulated instead of a unique Hamiltonian function. The latter method is also used by DIRAC [1]: as a consequence one has to contend with an undetermined Hamiltonian function and hence with undetermined multipliers in both these theories. It is clear that the present method, which provides a unique Hamiltonian in a natural geometrical manner, does not suffer from such disadvantages.

Furthermore, we now have a lucid geometrical interpretation of the relations (4.1) and (5.8). These equations provide us with a correspondence between the two tangent spaces $T_n(P)$ and $T_n'(P)$, or between the indicatrix and the figuratrix, while from the point of view of mechanics they represent the correspondence between the generalised components of velocity and the canonical momenta of the dynamical system. It

---

[1] RUND [2].

[2] CARATHÉODORY [1], p. 218.

should also be pointed out that the functions F and H may be obtained from each other by means of LEGENDRE transformations[1].

The question arises as to whether the above relations between $T_n(P)$ and $T'_n(P)$ hold also under the less restrictive convexity condition (1.21), or whether condition C is essential. In this connection, it is instructive to glance back at our two-dimensional example (3.11). Using (3.12) we have from (4.1)

$$y_\alpha = \dot{x}_\alpha^{m-1}(\dot{x}_1^m + \dot{x}_2^m)^{\frac{2}{m}-1} , \qquad (\alpha,\ \beta = 1,\ 2) . \tag{5.14}$$

The points $(0, 1)$ and $(1, 0)$ of $T_2(P)$ at which the determinant of equation (3.13) vanishes correspond to points $(0, 1)$, $(1, 0)$ of $T'_2(P)$. The Hamiltonian function turns out to be

$$H(y_1,\ y_2) = \left(y_1^{\frac{m}{m-1}} + y_2^{\frac{m}{m-1}}\right)^{\frac{m-1}{m}} , \tag{5.15}$$

and one may verify that the figuratrix (4.13) is a closed convex curve in $T'_2(P)$ as a result of our assumption that m is a positive even integer. Equation (5.7) is easily seen to be an identity by direct differentiation of (5.15). The metric tensor (5.4) assumes the following form

$$\begin{aligned}
g^{11} &= H^{-\frac{2}{m-1}}\, y_1^{\frac{2-m}{m-1}}\left[y_1^{\frac{m}{m-1}} + \frac{1}{m-1}\, y_2^{\frac{m}{m-1}}\right],\\[2mm]
g^{12} &= H^{-\frac{2}{m-1}}\left(\frac{m-2}{m-1}\right)(y_1 y_2)^{\frac{1}{m-1}},\\[2mm]
g^{22} &= H^{-\frac{2}{m-1}}\, y_2^{\frac{2-m}{m-1}}\left[\frac{1}{m-1}\, y_1^{\frac{m}{m-1}} + y_2^{\frac{m}{m-1}}\right].
\end{aligned} \tag{5.16}$$

With the help of these equations, together with (5.12), the verification of (5.8) and (5.9) is a matter of straight-forward calculation. These are seen to hold for all values of y_1 and y_2. For although $g^{11} \to \infty$ when $y_1 \to 0$ $(m > 2)$, we see that

$$g^{11} y_1 = H^{-\frac{2}{m-1}}\, y_1^{\frac{1}{m-1}}\left[y_1^{\frac{m}{m-1}} + \frac{1}{m-1}\, y_2^{\frac{m}{m-1}}\right] \to 0$$

as $y_1 \to 0$, and similarly

$$g^{11} g_{12} = -\,(m-2)\, H^{-\frac{2m}{m-1}}\left[y_1^{\frac{m+1}{m-1}}\, y_2 + \frac{1}{m-1}\, y_1^{\frac{1}{m-1}}\, y_2^{\frac{2m-1}{m-1}}\right] \to 0$$

as $y_1 \to 0$, while the expression

$$g^{11} g_{11} = H^{-\frac{2m}{m-1}}\left[y_1^{\frac{2m}{m-1}} + (m-1)\,(y_1 y_2)^{\frac{m}{m-1}} + \frac{1}{m-1}\,(y_1 y_2)^{\frac{m}{m-1}} + y_2^{\frac{2m}{m-1}}\right] \to 1 .$$

By symmetry, the point $(1, 0)$ may be treated in the same manner.

It is clear, therefore, that in principle the theory in its present form is valid despite the fact that the determinant g vanishes at certain points. On the other hand, if we wish to use the osculating indicatrix (or an osculating figuratrix) these points will have to be excluded, for we would then have to consider quantities of the type $g^{11}(x,\ y)\, \eta_1$ (with $\eta_1 \neq y_1$) which would tend to infinity when $y_1 \to 0$. The root of the difficulty lies, of course, in the fact that $H(x,\ y)$ need not necessarily be twice differentiable in the y_i at all points.

---

[1] VELTE [1].

§ 6. The Trigonometric Functions and Orthogonality

In our study of Finsler spaces we shall have to use trigonometric functions and angles defined at an arbitrary point $P(x^i)$. Clearly such definitions will have to be given in terms of the local metric, i. e. in terms of the metric of the Minkowskian tangent space $T_n(P)$ at the point under consideration. It is, therefore, a necessary preliminary step for us to discuss these notions for the case of a Minkowskian space. Unfortunately, there exist a number of distinct invariants in a Minkowskian space all of which reduce to the same classical euclidean invariant if the Minkowskian space degenerates into a euclidean space. Consequently distinct definitions of the trigonometric functions and of angles have appeared in the literature concerning Minkowskian and Finsler spaces. In the present section we shall discuss the various points of view as regards these questions, treating the indicatrix as our primary tool.

We shall deal first with the notion of orthogonality. In euclidean geometry any direction parallel to a plane tangent to a sphere is orthogonal to the radius vector corresponding to the point of contact of the plane with the sphere. Since we regard the indicatrix as a generalised sphere in Minkowskian space we are naturally led to the following definition: Let ξ^i be an arbitrary vector of length $|\xi| \equiv F(x, \xi)$, issuing from the origin O of $T_n(P)$. We construct the hypersurface $I_\xi \colon F(x^i, \dot{x}^i) = |\xi|$, which is homothetic to the indicatrix (i. e. the generalised "sphere of radius $|\xi|$"). In analogy to (3.8) the equation of the tangent hyperplane to I_ξ at ξ^i is

$$g_{ij}(x^k, \xi^k)\, \xi^i\, \dot{x}^j = |\xi|^2 . \tag{6.1}$$

Any vector η^i contained in or parallel to this hyperplane is said to be *orthogonal or normal with respect to* ξ^i. Since such a vector η^i may be represented by the difference of two vectors $\dot{x}^i_{(2)}$ and $\dot{x}^i_{(1)}$ both of which satisfy (6.1), we see that the condition that η^i be normal with respect to ξ^i in $T_n(P)$ may be expressed by the relation

$$g_{ij}(x^k, \xi^k)\, \xi^i\, \eta^j = 0 . \tag{6.2}$$

We remark that *orthogonality is not a symmetrical relationship* between the vectors ξ^i and η^i. In general, this would be the case if the coefficients $g_{ij}(x^k, \xi^k)$ on the left-hand side of (6.2) were independent of the arguments ξ^k, which would imply that the metric of $T_n(P)$ is simply a euclidean metric[1]. It will be evident in the sequel that this lack of

---

[1] More precisely, one is led to formulate the question as to which conditions must be satisfied by the fundamental function $F(x, \dot{x})$ of a regular problem in the calculus of variations in order that the transversality condition be symmetric. This question has been answered by BLASCHKE [3] for the case $n = 3$ as follows: A necessary and sufficient condition that the required symmetry property holds is

symmetry gives rise to many of the essential distinctive features between a locally euclidean and a locally Minkowskian geometry.

In view of (4.1) and (5.6) — applied to ξ^i — it follows that the orthogonality relation (6.2) may be written in the form

$$\frac{\partial F(x^k, \xi^k)}{\partial \dot{x}^i} \, \eta^i = 0 \, . \tag{6.3}$$

But this is the *transversality condition* of the calculus of variations in parametric form[1]: hence our generalised notion of orthogonality is identical with that of transversality.

We may now define the cosine corresponding to two arbitrary directions ξ^i and η^i issuing from the origin of $T_n(P)$. Let η^i (produced if necessary) cut the hyperplane (6.1) at Q, and let us denote the point of contact of (6.1) with the hypersurface I_ξ by R. Thus, in accordance with our definition, the vector RQ is normal with respect to OR, and we may therefore define

$$\cos(\xi, \eta) = \pm \, |OR|/|OQ| \, , \tag{6.4a}$$

where the right-hand side refers to the ratio of the corresponding Minkowskian lengths and where the $+$ or $-$ sign is chosen according as the points η^i and R lie on the same or on opposite sides of the hyperplane through O parallel to the hyperplane (6.1). The coordinates of Q may be determined by means of equation (6.1); we then find the following

---

that $F(x, \dot{x})$ be of the form

$$F(x, \dot{x}) = [a_{ij}(x) \, \dot{x}^i \, \dot{x}^j]^{1/2} \, .$$

This result is proved by BLASCHKE on the basis of the following theorem: A closed convex surface which has the property that it touches every circumscribed cylinder along a plane curve must necessarily be an ellipsoid. See also BOHNENBLUST [1].

It should be noted, however, that the theorem is *not* true for $n = 2$; a certain class of closed plane curves discussed by RADON [1] furnishes a counter-example. These curves share with the ellipses the property of possessing conjugate diameters (defined in terms of pairs of parallel tangents). A characterisation of such curves is given by LAUGWITZ [1], p. 239.

[1] BOLZA [1], p. 303; CARATHÉODORY [1], p. 248. It is not difficult to show that transversality is identical with the (euclidean) notion of orthogonality if and only if $F(x, \dot{x})$ is of the form

$$F(x, \dot{x}) = G(x) \left[\sum_i (\dot{x}^i)^2 \right]^{1/2} ,$$

which implies a locally euclidean metric. See MANCILL [1]; BEKE [1]. For the non-parametric case the problem is discussed by LA PAZ [1, 2], p. 461. Conversely, the latter author investigates the problem of finding all suitable metric functions $F(x, \dot{x})$ if the transversality conditions have been previously prescribed.

analytical expression[1]

$$\cos(\xi, \eta) = \frac{g_{ij}(x^k, \xi^k)\, \xi^i\, \eta^j}{F(x^k, \xi^k)\, F(x^k, \eta^k)} \,.\tag{6.4}$$

Again, this function is not symmetrical in its arguments ξ^i and η^i. From (6.2) it follows that $\cos(\xi, \eta) = 0$ if the vector η^i is normal with respect to ξ^i. Since the g_{ij} are homogeneous of degree zero in their directional arguments we deduce from (6.4) that $\cos(\xi, \eta)$ is independent of the lengths of ξ^i and η^i. It is easily verified that $-1 \leqq \cos(\xi, \eta) \leqq 1$ as a result of equations (1.18) and (3.1). It should be pointed out, however, that the cosine (6.4) is to be regarded as a function of two directions, and not as a function of a single variable corresponding to an angle as in euclidean geometry.

In the dual tangent space $T_n'(P)$ there are two covariant vectors corresponding to ξ^i and η^i of $T_n(P)$. It is natural to carry out an analogous construction in $T_n'(P)$, using a hypersurface homothetic to the figuratrix. If we denote the cosine obtained in this manner by $\mathrm{Cos}(\xi, \eta)$ it may be shown that[2]

$$\mathrm{Cos}(\xi, \eta) = \cos(\eta, \xi)\,.$$

The difference between the indicatrix and the figuratrix thus accounts for the lack of symmetry of the cosine.

Let the equation

$$\alpha_i\, \dot{x}^i = 1 \tag{6.5}$$

represent an arbitrary hyperplane of $T_n(P)$. Such a hyperplane possesses a unique normal direction as a consequence of our definition; for we can always construct a hypersurface homothetic to the indicatrix which will touch the hyperplane in some point S. The direction OS will represent the normal direction. Again, as a result of convexity, OS will be the shortest distance joining O to the hyperplane. If $T(\dot{x}^i)$ is an arbitrary point not in the hyperplane, we may construct a straight line TM parallel to OS to meet the hyperplane in the point M. An elementary calculation shows that the normal distance $|TM|$ of T from the hyperplane (6.5) is given by

$$|TM| = \pm\left|\frac{\alpha_i\, \dot{x}^i - 1}{H(x, \alpha)}\right|, \tag{6.6}$$

the — or + sign being taken according as T lies on the opposite or on the same side of the hyperplane as the origin. In particular, the normal distance of the hyperplane from the origin is $H^{-1}(x, \alpha)$.

As an immediate application of this formula let us consider two distinct points $A(\dot{x}^i_{(1)})$ and $B(\dot{x}^i_{(0)})$ on the indicatrix $F(x, \dot{x}) = 1$. If $y_i^{(0)}$

---

[1] Strictly speaking, when the — sign is applicable in (6.4a), the equivalence of (6.4) and (6.4a) depends on condition A_1. We shall, however, regard (6.4) as the general definition of cosine.

[2] RUND [3], p. 17.

is the element of $T'_n(P)$ corresponding to $\dot{x}^i_{(0)}$, we have $\dot{x}^i_{(0)}\, y^{(0)}_i = 1$ as a result of (4.12), so that the equation (4.11) of the tangent hyperplane to the indicatrix at B is $y^{(0)}_i\, \dot{x}^i = 1$. We construct AC parallel to OB to intersect this hyperplane at C. Then, by (6.6), the normal distance $|AC|$ is

$$1 - y^{(0)}_i\, \dot{x}^i_{(1)} = 1 - g_{ij}(x, \dot{x}_{(0)})\, \dot{x}^i_{(0)}\, \dot{x}^j_{(1)} = 1 - \cos(\dot{x}_{(0)}, \dot{x}_{(1)}) ,$$

the second two terms resulting from (4.1) and (6.4), where we have taken into account that $\dot{x}^i_{(0)}$ and $\dot{x}^i_{(1)}$ are unit vectors. But the ratio $|AC|/|OB|$ is precisely the "Weierstrass excess-function" $\mathscr{E}(x^k, x^k_{(0)}, \dot{x}^k_{(1)})$ of the calculus of variations[1]. Since, by construction, $|OB| = 1$, we have in view of (5.6):

$$\mathscr{E}(x^k, \dot{x}^k_{(0)}, \dot{x}^k_{(1)}) = F(x^k, \dot{x}^k_{(1)}) - F_{\dot{x}^i}(x^k, \dot{x}^k_{(0)})\, \dot{x}^i_{(1)} = 1 - \cos(\dot{x}_{(0)}, \dot{x}_{(1)}). \quad (6.7)$$

This equation may also be used to show that the definition (6.4) of the cosine is identical with that given by Finsler[2]. The latter, however, considers the cosine corresponding to two directions as a function of an angle φ between these directions. In essence Finsler's definition of cosine is formulated in a similar manner, except that a limiting process is used. The angle φ itself is then defined by postulating the usual series expansion of $\cos\varphi$ in terms of φ. We shall return to this point later. The limiting process used by Finsler in turn leads to equation (6.7).

On the other hand, given an arbitrary fixed direction $\dot{x}^i$, we may construct the osculating indicatrix corresponding to $\dot{x}^i$ and define the cosine corresponding to two directions ξ^i and η^i in an analogous geometrical manner. Instead of the right-hand side of (6.4) one would obtain the expression

$$\frac{g_{ij}(x^k, \dot{x}^k)\, \xi^i\, \eta^j}{[g_{ij}(x^k, \dot{x}^k)\, \xi^i\, \xi^j]^{1/2}\, [g_{ij}(x^k, \dot{x}^k)\, \eta^i\, \eta^j]^{1/2}} . \quad (6.8)$$

This is the alternative definition of cosine as proposed independently by Berwald and Synge[3]. Clearly this expression is symmetrical in ξ^i and η^i, but it suffers from the disadvantage that it depends on the original choice of the direction $\dot{x}^i$, the so-called element of support. Hence (6.8) is used in those theories of Finsler spaces which are based on the latter notion.

An entirely different approach to the problem of defining trigonometric functions in symmetric spaces (cf. condition A_1) is due to Buse-

---

[1] Carathéodory [1], p. 223, equation (262.1), and p. 244. This function was originally introduced from a purely analytical point of view in connection with necessary conditions for extrema in the calculus of variations. Its equivalence with the geometrical definition given here was pointed out by Carathéodory [2, 3], using a euclidean background. It is clear that the convexity of the indicatrix implies that $\mathscr{E} > 0$ for all sets of unequal values of $\dot{x}^i_{(1)}$ and $\dot{x}^i_{(0)}$ and conversely. This follows directly from equation (1.24). See also Whitehead [3].

[2] Finsler [1], p. 39.

[3] Berwald [1], p. 217; [2], p. 56; Synge [1], p. 65.

MANN, this approach being based on the notion of an associated euclidean metrisation[1]. Since these definitions are closely concerned with Busemann's approach to the problem of determining a suitable measure of area in Minkowskian spaces, we shall defer the discussion of these notions to the section (§ 8) on area. Also, at this stage it would be natural to introduce a Minkowskian sine-function. Such a function is defined by BUSEMANN, where, in contrast to the cosine function defined above, less stringent differentiability assumptions concerning the indicatrix may be adopted.

An alternative definition is suggested directly by the definition (6.4a) of the cosine. Let OR and OQ be two vectors of $T_n(P)$ corresponding to ξ^i and η^i, and let Q' be the point of intersection of OQ (produced if necessary) with the tangent plane to I_ξ at R. We define[2]:

$$\sin(\xi^i, \eta^i) = \pm \frac{|RQ'|}{|OQ'|}, \tag{6.9a}$$

where the $+$ or $-$ sign is chosen according as the rotation RQ' about O is counter-clockwise or clockwise. This function does not admit an analytical representation as simple as (6.4) for the cosine. Nor do we have $\sin^2(\xi, \eta) + \cos^2(\xi, \eta) = 1$: as BUSEMANN[3] points out, such a relation implies the theorem of PYTHAGORAS which does not generally hold in Minkowskian spaces. If we denote the vector RQ' by ζ^i, it follows from (6.4a) and (6.4) that

$$\zeta^i = \frac{F(x, \xi)}{F(x, \eta)} \sec(\xi, \eta)\, \eta^i - \xi^i, \tag{6.9}$$

while in view of (6.9a) the relation between the Minkowskian sine and cosine is found to be

$$\tag{6.10}$$

$$\sin^2(\xi, \eta) = g_{ij}(x, \zeta) \left\{ \frac{\eta^i \eta^j}{F^2(x, \eta)} - \frac{2\eta^i \xi^j}{F(x, \eta)\,F(x, \xi)} \cos(\xi, \eta) + \frac{\xi^i \xi^j}{F^2(x, \xi)} \cos^2(\xi, \eta) \right\}.$$

Again, this function is independent of the lengths of ξ^i and η^i, and is not symmetric with respect to its directional arguments. If we assume that the indicatrix is symmetric (condition A_1) it follows from (6.9) that if η^i is normal with respect to ξ^i, then $\sin(\xi, \eta) = \pm 1$. The relation (6.10) is not very useful, the real significance of the sine function being due to the fact that it allows us to develop addition formulae for a Minkowskian trigonometry: Let ξ^i, η^i, $\dot{x}^i$ represent three coplanar points on the indicatrix, such that the rotation about O from ξ^i to η^i to $\dot{x}^i$ is counter-clockwise. The following "addition" formulae may then be proved:

$$\cos(\xi, \dot{x}) = \cos(\xi, \eta) \cos(\eta, \dot{x}) - \frac{\sin(\eta, \dot{x})}{\sin(\eta, \xi)} \{1 - \cos(\xi, \eta) \cos(\eta, \xi)\}, \tag{6.11}$$

$$\sin(\xi, \dot{x}) = \frac{\sin(\xi, \eta)}{\sin(\eta, \xi)} \{\sin(\eta, \xi) \cos(\eta, \dot{x}) + \cos(\eta, \xi) \sin(\eta, \dot{x})\}. \tag{6.12}$$

With the aid of these equations one may easily find generalisations of the formulae of euclidean trigonometry: for instance, if $\dot{x}^i$ is normal with respect to η^i

---

[1] BUSEMANN [4], § 2; [5], § 2.

[2] RUND [3], § 10.

[3] BUSEMANN [4], p. 162.

we would have

$$\sin(\xi, \dot{x}) = \frac{\sin(\xi, \eta)}{\sin(\eta, \xi)} \cos(\eta, \xi) \, ,$$

corresponding to the identity $\sin\left(\alpha + \dfrac{\pi}{2}\right) = \cos\alpha$. For the cosine and sine corresponding to two neighbouring unit vectors ξ^i and $\xi^i + \delta\xi^i$ one may easily establish the following relations:

$$\cos(\xi^i, \xi^i + \delta\xi^i) = 1 - \tfrac{1}{2} g_{ij}(x, \xi)\, \delta\xi^i\, \delta\xi^j + O(\delta\xi^3) \, , \tag{6.13}$$

together with

$$\sin(\xi^i, \xi^i + \delta\xi^i) = [g_{ij}(x, \delta\xi)\, \delta\xi^i\, \delta\xi^j]^{1/2} + O(\delta\xi^2) \, . \tag{6.14}$$

These relations have some significance with regard to the definition of angle. In conclusion we may remark that the equations (6.11) to (6.14) may be used to determine the derivatives of the sine and cosine.

§ 7. Definitions of Angle

If we continue to regard the indicatrix as representing the unit sphere of the Minkowskian space $T_n(P)$, the following definition of angle seems to be the most natural generalisation of the euclidean notion of angular measure. Two arbitrary, non-parallel vectors ξ^i, η^i issuing from P define a two-dimensional linear subspace T_2, namely the set of points of $T_n(P)$ whose coordinates are expressible in the form $\dot{x}^i = \lambda\,\xi^i + \mu\,\eta^i$, where λ, μ are variable parameters. The intersection of T_2 with the indicatrix is a closed convex curve I_2 contained in T_2. This curve plays the role of the "unit circle" on T_2. Let ξ^i and η^i (produced if necessary) intersect I_2 in points a and b. Then the shorter of the two (Minkowskian) arc-lengths cut off from I_2 by a and b represents a possible definition of angle. In particular, let ξ^i and $\xi^i + d\xi^i$ be two neighbouring vectors of T_2, both of equal length. From the equality of their lengths we immediately deduce that

$$g_{ij}(x^k, \xi^k)\, \xi^i\, d\xi^j = 0 \tag{7.1}$$

in view of (3.5). Thus $d\xi^i$ is normal with respect to ξ^i (or transversal to ξ^i) in the sense of equation (6.2). The angle $d\theta$ between the vectors ξ^i and $\xi^i + d\xi^i$ is defined to be

$$d\theta = [g_{ij}(x^k, d\xi^k)\, d\xi^i\, d\xi^j]^{1/2} \cdot F^{-1}(x, \xi) \, , \tag{7.2}$$

for $d\xi^i/F(x, \xi)$ will represent the element of arc cut off from the indicatrix I_2 by the two vectors. A finite angle will be obtained by integrating the expression (7.2) over a finite arc of I_2.

This definition is actually equivalent to a definition of angle given by Bliss[1] for the case of a two-dimensional variational problem against a euclidean background. The integrand of the problem is given in the form $f(x_1, x_2, \tau) \sqrt{\dot{x}_1^2 + \dot{x}_2^2}$ where x_1, x_2 represent the points of a Cartesian plane and τ is the (euclidean) angle between the x_1-axis and the direction defined by the vector $\dot{x}_1$, $\dot{x}_2$. The corresponding

---

[1] Bliss [1], p. 190; the form (7.2) for the angle is given by Rund [1], p. 63.

integral defines "generalised lengths" in the (x_1, x_2)-plane, and curves which yield shortest generalised lengths are the extremals. Let the tangent at a point x_1, x_2 on an extremal make an angle τ with the x_1-axis. The (euclidean) angle $\bar{\tau}$, defined by the (euclidean) trigonometric functions

$$\cos\bar{\tau} = \frac{f\sin\tau + f_\tau\cos\tau}{\sqrt{f^2 + f_\tau^2}}\,, \qquad \sin\bar{\tau} = \frac{-f\cos\tau + f_\tau\sin\tau}{\sqrt{f^2 + f_\tau^2}}\,, \tag{7.3}$$

where the arguments of f and $f_\tau \equiv \partial f/\partial\tau$ are (x_1, x_2, τ), represents the direction transversal to the extremal at (x_1, x_2). A simple calculation shows that this notion of transversality is equivalent to the one defined by equation (6.3). Now let OA', OA be two extremals intersecting in O, such that the generalised arc lengths $OA = OA' = r$. Denote by l the generalised arc-length of a curve AA' transversal to both OA and OA'. The generalised angle θ between OA and OA' is then defined by

$$\theta = \lim_{r\to 0}\frac{l}{r}\,. \tag{7.4}$$

The transversality condition (7.1) shows directly that the definition (7.2) reduces to (7.4) for the special case under consideration. In terms of the euclidean geometry of the (x_1, x_2)-plane one may derive the following formula for the angle θ between two directions whose euclidean angles with the x_1-axis are τ_1 and τ_2:

$$\theta = \int_{\tau_1}^{\tau_2}\frac{f(x_1, x_2, \bar{\tau})}{[f(x_1, x_2, \tau)]^2}\sqrt{f^2 + f_\tau^2}\,d\tau\,, \tag{7.5}$$

where $\bar{\tau}$ is the direction transversal to τ as given by equations (7.3)[1].

The principal advantage of the definition of angle as given above is the fact that it is an additive function in the sense that if ξ^i, η^i, ζ^i represent three coplanar vectors issuing from O, then the sum of the angles defined by the directions ξ^i, η^i and η^i, ζ^i is equal to the angle defined by ξ^i, ζ^i. Certain disadvantages will be discussed when we have given the alternative definitions of angle according to FINSLER and LANDSBERG.

FINSLER begins with the consideration of equation (6.7)[2]. It is assumed that two unit vectors ξ^i and η^i define a scalar function φ such that

$$\cos(\xi, \eta) = 1 - \frac{\varphi^2}{2!} + \frac{\varphi^4}{4!} + \cdots\,, \tag{7.6}$$

the cosine being given by (6.7). If we expand the term $F(x^k, \eta^k)$ appearing in the expression (6.7) for the $\mathscr{E}$-function in a Taylor series about ξ^k, we find that

$$\mathscr{E}(x^k, \xi^k, \eta^k) = \tfrac{1}{2}\widetilde{F}_{\dot{x}^i\dot{x}^j}(\xi^i - \eta^i)(\xi^j - \eta^j)\,, \tag{7.7}$$

where

$$\widetilde{F}_{\dot{x}^i\dot{x}^j} = F_{\dot{x}^i\dot{x}^j}(x^k, \theta\,\xi^k + (1-\theta)\,\eta^k)\,, \qquad 0 < \theta < 1\,.$$

---

[1] BLISS [1], p. 192.

[2] FINSLER [1], p. 39.

In virtue of (6.7), (7.6), (7.7) one may therefore put

$$\varphi^2 = F_{\dot{x}^i \dot{x}^j}(x^k, \xi^k)(\xi^i - \eta^i)(\xi^j - \eta^j)(1 - \Lambda),\qquad(7.8)$$

where $\Lambda \to 0$ when $\eta^i \to \xi^i$. In particular, if $\eta^i = \xi^i + d\xi^i$ (unit vectors), the angle $d\varphi$ between these vectors is defined by

$$d\varphi^2 = F_{\dot{x}^i \dot{x}^j}(x, \xi)\, d\xi^i\, d\xi^j .\qquad(7.9)$$

Geometrically this definition may be interpreted as follows: From (3.1) and (1.19c) we have

$$g_{ij}(x, \xi) = g_{ih}(x, \xi)\, g_{jk}(x, \xi)\, \xi^h\, \xi^k + F_{\dot{x}^i \dot{x}^j}(x, \xi),$$

since $F(x^k, \xi^k) = 1$; on multiplying this equation by $d\xi^i\, d\xi^j$ and taking into account (7.1) we find as a result of definition (7.9)

$$d\varphi^2 = g_{ij}(x, \xi)\, d\xi^i\, d\xi^j ,\qquad(7.10)$$

neglecting quantities of smaller order of magnitude. This represents the square of the arc-length cut off from the osculating indicatrix I_2^* by the vectors ξ^i and $\xi^i + d\xi^i$, (measured by I_2^*), where I_2^* is the quadric obtained by the intersection of the two-dimensional linear space T_2 defined by ξ^i and $\xi^i + d\xi^i$ with the osculating indicatrix corresponding to the direction ξ^i.

Furthermore, using (7.7) we may write equation (7.9) in the form

$$\tfrac{1}{2} d\varphi^2 = \mathscr{E}(x^k, \xi^k, \xi^k + d\xi^k) .\qquad(7.11)$$

This relation had already been found by LANDSBERG in connection with two-dimensional problems in the calculus of variations, again subject to a euclidean background[1]. The equation (7.11) is not, however, the original definition of LANDSBERG; the definition proper is given by the relation

$$\varphi^* = \int_{\xi^i_{(0)}}^{\xi^i_{(1)}} (\xi^1\, d\xi^2 - \xi^2\, d\xi^1)\, [F_1(x, \xi)]^{1/2} ,\qquad(7.12)$$

this expression representing the angle between two unit vectors $\xi^i_{(0)}, \xi^i_{(1)}$ where the function F_1 is defined in a manner equivalent to equation (3.15). LANDSBERG was led to the definition (7.12) by considering exact invariant differentials arising from a two-dimensional problem in the calculus of variations. Equation (7.11), which results from (7.12), establishes the equivalence of the definitions of LANDSBERG and FINSLER for *infinitesimal* angles.

FINSLER himself pointed out in a later publication[2] that the definition (7.8) does not represent an additive measure of angle. However, an additive definition

---

[1] This relation allows us to interpret $d\varphi^2$ on the actual indicatrix: for by definition (§ 6) the right-hand side of (7.11) corresponds to the Minkowskian distance of the point $\xi^i + d\xi^i$ on the indicatrix from the tangent-hyperplane through ξ^i. Equations (7.11) and (7.12) are given in LANDSBERG [2], p. 338 and p. 331 respectively.

[2] FINSLER [2], p. 8.

may be derived from (7.8) by means of the following process. Let $\varphi\,(\xi_{(0)},\,\xi_{(1)})$ be the angle between two unit vectors according to (7.8). The additive function $\overline{\varphi}\,(\xi_{(0)},\,\xi_{(1)})$ is then obtained by putting

$$\overline{\varphi}\,(\xi_{(0)},\,\xi_{(1)}) = \int\limits_{\xi_{(0)}}^{\xi_{(1)}} \left(\frac{\partial\varphi(\eta,\,\xi)}{\partial\xi}\right)_{\eta\,=\,\xi} d\xi\,. \tag{7.13}$$

This function has the property that $\overline{\varphi} \to \varphi$ when $\xi_{(1)} \to \xi_{(0)}$.

When ξ^i is not a unit vector, the equation (7.9) should be replaced by

$$d\varphi^2 = F^{-2}(x,\,\xi)\,g_{ij}(x,\,\xi)\,d\xi^i\,d\xi^j = F^{-1}(x,\,\xi)\,F_{\dot{x}^i\dot{x}^j}(x,\,\xi)\,d\xi^i\,d\xi^j\,. \tag{7.14}$$

This is the form used by CARTAN in accordance with his method based on the osculating indicatrix[1].

According to (6.13) and (6.14) we have the following relations between two neighbouring unit vectors ξ^i and $\xi^i + d\xi^i$:

$$\cos(\xi^i,\,\xi^i + d\xi^i) = 1 - \tfrac{1}{2}\,d\varphi^2 + O(d\varphi^3)\,, \tag{7.15}$$

and

$$\sin(\xi^i,\,\xi^i + d\xi^i) = d\theta + O(d\theta^2) \tag{7.15a}$$

where $d\theta$ and $d\varphi$ are given by the alternative definitions (7.2) and (7.10).

Since in virtue of (3.2) the right-hand side of (7.2) may be written as $F(x,\,d\xi)\,F^{-1}(x,\,\xi)$, we see that this definition requires that $F(x,\,\dot{x})$ be of class C^1 only, whereas the definition (7.10) requires that $F(x,\,\dot{x})$ be of class C^2. A fundamental difference is that (7.2) involves the global nature of the indicatrix, for the length $F(x,\,d\xi)$ depends on the shape of the indicatrix in the direction $d\xi$ (which is transversal to ξ), while (7.10) involves the indicatrix only in the immediate neighbourhood of the direction ξ (i. e. the osculating indicatrix corresponding to this direction)[2].

In both cases an extremely difficult question arises when one wishes to discuss the angle corresponding to a half rotation in a plane (corresponding to the euclidean π) or a complete rotation. According to definition (7.2) this involves the total length of the indicatrix in the 2-dimensional case. In a space of arbitrary many dimensions this leads however to a predicament. The angle between two parallel opposite directions may be different for different two-dimensional planes T_2 containing these directions, for the length of the indicatrix I_2 depends on the choice of T_2, which is not uniquely defined. Similar remarks apply to the definition

---

[1] CARTAN [1], p. 14; see also DELENS [1, 2].

[2] These facts are pointed out by GOLAB [1], p. 79. It is also shown by GOLAB [9] that the angle of LANDSBERG is invariant under a transformation by reciprocal polars. This result has some bearing on the conformal geometry of Finsler spaces (see Ch. VI, § 2). Furthermore, it is indicated by BIELECKI and GOLAB [1] that for $n = 2$ the definition of angle as given by FINSLER possesses the properties of additivity and invariance under interchange of the sides of the angle (i. e. symmetry) only when the Finsler space is Riemannian. A further discussion concerning the definition of FINSLER and the convexity of the indicatrix is given by GOLAB [11].

(7.10). Thus whichever definition is taken, there will be different numbers corresponding to the number π of euclidean geometry, these numbers depending on the choice of the two-dimensional linear subspace. For this reason a normalised angle θ' is suggested[1]: This is θ as defined by (7.2) divided by the total length l_2 of the circumference of the indicatrix of T_2. Then the normalised angle corresponding to a complete rotation is always unity, irrespective of the choice of T_2, while if in addition the symmetry condition A_1 holds, the angle corresponding to a half rotation will always be $\frac{1}{2}$.

On the other hand, by definition,

$$l_2 = \oint \sqrt{g_{ij}(x, dx)\, dx^i\, dx^j}\,, \tag{7.16}$$

the integral being taken around the indicatrix, so that for the two-dimensional Minkowskian space T_2 the analogue of the number π is

$$\Pi = \tfrac{1}{2}\, l_2\,. \tag{7.17}$$

A comprehensive study has been made by GOLAB[2] with respect to the bounds within which Π must lie. We shall briefly describe the principal results concerning these bounds; and for the sake of the present discussion we shall momentarily relax our differentiability conditions. We shall prove the following theorem[3]: If the indicatrix I_2 possesses a centre of symmetry (condition A_1) then

$$3 \le \Pi \le 4\,. \tag{7.18}$$

In order to prove the left-hand inequality, consider a fixed unit vector $\bar{\xi}^i$ issuing from O. Let η^i be a variable unit vector, also attached to O, which may rotate from $\bar{\xi}^i$ to $-\bar{\xi}^i$. During this rotation the value of the function $F(x, \eta - \bar{\xi})$ takes on values ranging from 0 to 2. Denote that position η^i for which $F(x, \eta - \bar{\xi}) = 1$ by $\bar{\eta}^i$, so that the end-point of the vector with components $\bar{\zeta}^i = \bar{\eta}^i - \bar{\xi}^i$ issuing from O also lies on the indicatrix. The end-points of the vectors $\pm\,\bar{\xi}^i$, $\pm\,\bar{\eta}^i$, $\pm\,\bar{\zeta}^i$ thus define a hexagon which is inscribed in the indicatrix, the sides of this hexagon being of unit length. But from the triangle inequality it follows that the length l_2 of the circumference of the indicatrix is not less than the length of the circumference of an inscribed hexagon, i. e. $6 \le l_2$, from which the first part of (7.18) follows directly.

For the proof of the right-hand inequality of (7.18) we make use of the fact that every closed convex curve with a centre of symmetry possesses at least one pair of conjugate diameters $\bar{\xi}^i_{(1)}$, $\bar{\xi}^i_{(2)}$[4]. Applying this result to the indicatrix I_2 we see that these directions define an inscribed parallelogram P_i and a circumscribed parallelo-

---

* * *

[1] RUND [1], p. 63.

[2] GOLAB [2, 3].

[3] GOLAB [2], p. 70 et seq. The proof given here is essentially similar to a proof due to LAUGWITZ [1].

[4] This result is due to FUNK [1], whose proof is based on the assumption that the curve is of class C^2. However, LAUGWITZ [1], p. 238, constructs a proof which is independent of such differentiability assumptions, the method being similar to that used by RADON [1] in connection with the investigation of the supporting lines of closed convex curves.

3*

gram P_e, the latter being formed by the pairs of parallel tangents to the indicatrix at the extremities of the conjugate diameters. We remark that P_i may be regarded as the indicatrix J_2 of a second Minkowskian metric defined on T_2. By construction, the length of any vector of T_2 measured with respect to J_2 will not be less than its length as measured with respect to I_2. In particular, l_2 is not greater than the length of the circumference of I_2 measured with respect to J_2, and in turn this is not greater than the length of the circumference of P_e measured with respect to J_2. But clearly the latter length is simply 8 by construction, so that $l_2 \leqq 8$. This proves the second part of the theorem.

Furthermore, we note that the bounds of Π as given by (7.18) may in fact be attained. In fact, let ε be a constant such that $0 \leqq \varepsilon \leqq 1$, and construct the hexagon whose vertices with respect to a linear coordinate system in T_2 are given by

$$(1, 1), \quad (-1, -1), \quad (1, 0), \quad (-1, 0), \quad (\varepsilon, -1), \quad (-\varepsilon, +1).$$

If this hexagon is regarded as the indicatrix of T_2, it follows that the length of its circumference is $6 + 2\varepsilon$, so that in this case $\Pi = 3 + \varepsilon$. The bounds of (7.18) are obtained for the figures for which $\varepsilon = 0$ and $\varepsilon = 1$.

Proceeding to the case of non-symmetric indicatrices, we should mention the following construction, also due to GOLAB. Let K_2 be any closed convex curve in T_2, and let m be any point in the region enclosed by K_2. By regarding m as the centre of K_2, the latter may be used to define a Minkowskian metric on T_2, which, in particular, defines the length L_2 of K_2. But clearly this length will, in general, depend on the choice of the centre m: hence we write $L_2(m)$. It can then be shown that[1]

$$\Pi = \tfrac{1}{2} L_2(m) \geqq 2 + \frac{1}{\sqrt{2}}, \tag{7.19}$$

while Π has in general no upper bound.

However, let L_2^0 represent the minimum value of $L_2(m)$ with respect to all possible positions of m within the region bounded by K_2, the latter being fixed. Then the following inequality holds[2]:

$$\Pi^0 \equiv \tfrac{1}{2} L_2^0 \leqq 12. \tag{7.19a}$$

In conclusion let us consider the example of a triangular indicatrix with vertices a, b, c, the centre being an arbitrary point m within the triangle. Construct the segment $a'' m a'$ parallel to $a b$, intersecting $a c$ and $b c$ in a'' and a' respectively. Similar segments $b'' m b'$, $c'' m c'$ are drawn parallel to $b c$ and $c a$ respectively. The (Minkowskian) lengths of the sides of the triangle satisfy the similarity relations

$$\frac{ab}{ca} = \frac{m a'}{c'' m} \quad \text{or} \quad \frac{m a'}{ab} = \frac{c'' m}{ca} = \frac{cb'}{ca},$$

with similar identities for $m b'$ and $m c'$. Thus

$$\frac{m a'}{a b} + \frac{m b'}{b c} + \frac{m c'}{c a} = \frac{c b' + b' a'' + a'' a}{c a} = \frac{c a}{c a} = 1. \tag{7.20}$$

But, by definition $L_2(m)$ is given by

$$L_2(m) = \frac{a b}{m a'} + \frac{b c}{m b'} + \frac{c a}{m c'},$$

---

[1] GOLAB [2], p. 56 et seq.
[2] GOLAB [2], p. 61.

or, if we put $\dfrac{m\,a'}{a\,b} = \lambda,\ \dfrac{m\,b'}{b\,c} = \mu$, it follows from (7.20) that

$$L_2(m) = \lambda^{-1} + \mu^{-1} + (1 - \lambda - \mu)^{-1}.$$

On differentiating we find that for a minimum L_2^0 of $L_2(m)$ we must have

$$\lambda = \mu = (1 - \lambda - \mu) \quad \text{or} \quad \lambda = \mu = \tfrac{1}{3}.$$

Hence $L_2^0 = 9$ for a triangular indicatrix. Clearly $L_2(m) \to \infty$ when m tends to one of a', b', c', which illustrates the result that $L_2(m)$ has no upper bound.

In connection with investigations concerning generalised forms of the Gauss-Bonnet theorem and its consequences, BUSEMANN[1] remarks that for many problems it is not necessary to insist on a particular angular measure, and requires only that such a measure should satisfy the following conditions:

(a) it must be additive for angles with the same vertex;

(b) "straight" angles have a constant measure (Π).

We remark that the normalised angle introduced above satisfies BUSEMANN's requirements.

§ 8. Area and Volume

The quest for a suitable definition of area or volume in Minkowskian spaces (and hence also in Finsler spaces) raises a number of fairly complex issues, and many of the definitions which have been suggested suffer from serious drawbacks. Since we shall not have occasion to use the notion of volume very frequently in the sequel, the present discussion will consist of a very brief account of the most important definitions that have been suggested, this account being inserted chiefly for the sake of completeness.

For a three-dimensional Finsler space F_3 (independently of any differentiability assumptions) a two-dimensional measure ("area") was introduced by CHOQUET[2]. This measure was used by CHOQUET to extend to F_3 some of the concepts of euclidean vector analysis. A systematic study of p-dimensional measures in n-dimensional Minkowskian spaces, generalising Choquet's measure (for the Minkowskian space T_3) and based on measure-theoretical considerations was made by BUSEMANN[3].

In order to be able to clearly formulate the relevant definitions, it is necessary that we should briefly touch upon a certain fundamental aspect of BUSEMANN's approach to Minkowskian geometry, namely the notion of the "associated euclidean spaces". In this connection it is probably advisable to abandon — for the moment — the tensor notation and to re-define Minkowskian spaces without differentiability require-ments. Let $F(\dot{x})$ be a real function defined over an n-dimensional vector space T_n. It is assumed (1) that $F(\dot{x})$ is positive for $\dot{x} \neq 0$ (null-vector); (2) that $F(\dot{x})$ is symmetric: $F(\dot{x}) = F(-\dot{x})$; (3) that $F(\dot{x})$ is positively

---

[1] BUSEMANN [6], p. 280.

[2] BOULIGAND and CHOQUET [1].

[3] BUSEMANN [5].

homogeneous of the first degree, and (4) that $F(\dot{x})$ is convex. The space T_n is then endowed with a *Minkowskian norm* defined by $||\dot{x}||_F = F(\dot{x})$ [1].

If we superimpose a euclidean norm $||\dot{x}||_E$ on T_n (giving rise to a space E_n), we have, for $\dot{x} \neq 0$,

$$\frac{||\dot{x}||_F}{||\dot{x}||_E} = F(\xi) , \qquad (8.1)$$

where

$$\xi = \frac{\dot{x}}{||\dot{x}||_E} \qquad (8.2)$$

is to be interpreted as the unit vector in the direction of $\dot{x}$ with respect to the euclidean metric [2].

The Minkowskian and euclidean distances are thus related by a factor of proportionality which depends only on the direction of the vector under consideration. Clearly the Minkowskian space can be derived from different euclidean spaces, all of which are related to each other by non-degenerate affine transformations. These spaces are called the *associated euclidean spaces* of the Minkowskian space. Those concepts and theorems which are independent of the choice of the associated space have an intrinsic significance with respect to the Minkowskian space [3].

The Minkowskian norm is invariant under translations: $\dot{z} = \dot{x} + a$, and since the latter operates in a simply transitive way on the space it follows from the theory of Haar measure that up to a constant factor at most one measure exists which is invariant under translations [4]. The Lebesgue measure $\lambda_n(M)$ of a Lebesgue measurable set M in the euclidean space depends, however, on the choice of the associated space, and hence we are compelled to seek a factor σ which is such that the measure $|M|_n = \sigma \lambda_n(M)$ satisfies the independence requirement.

---

[1] The four conditions outlined above are, of course, more general than those imposed on the metric function in § 1, except for the symmetry condition (2), which is equivalent to condition A_1 of § 1. For the purpose of the present discussion we shall thus assume that A_1 is satisfied, although this is not generally supposed in subsequent chapters. The function $F(\dot{x})$ plays the same role as our metric function $F(x^j, \dot{x}^j)$, the x^j being constant for any Minkowskian space "tangent" to the Finsler space at the point x^j.

[2] $F(\xi)$ is MENGER's „Abstandskoeffizient" (cf. PAUC [2], p. 38).

[3] BUSEMANN [4], § 1; [5], § 2.

[4] For the theory of HAAR measure in locally compact topological groups, which is beyond the scope of the present volume, see WEIL [1] or HAAR's original paper [1]. We refer to HAAR measure in the n-dimensional vector space T_n if we regard the latter as an abelian additive group provided with its natural topology (defined by any euclidean or Minkowskian norm). In the present simple case of T_n the procedure of LEBESGUE [1] shows that the LEBESGUE measure is, up to a constant factor, the only measure invariant under translation. The euclidean metric is used only for normalisation.

The following general definition of σ will serve this purpose. Let V_r be an r-dimensional hyperplane of T_n, and denote by $\lambda_r(M)$ the r-dimensional Lebesgue measure (in an associated euclidean space) of a Lebesgue measurable set M in V_r. Denote by $U(V_r)$ the set in which the linear r-space parallel to V_r and passing through the origin (centre of the indicatrix) intersects the "solid" indicatrix $F(\dot{x}) \leqq 1$. The Lebesgue measure of $U(V_r)$ being $\lambda_r(U(V_r))$, we put

$$\sigma(V_r) = \frac{\omega^{(r)}}{\lambda_r(U(V_r))}\,, \qquad (8.3)$$

where $\omega^{(r)}$ is the volume of the euclidean r-dimensional unit sphere:

$$\omega^{(r)} = \frac{\pi^{\frac{r}{2}}}{\Gamma\left(\dfrac{r}{2} + 1\right)}\,. \qquad (8.4)$$

The factor (8.3) possesses the required properties. The Minkowskian measure of the set M of V_r is thus given by

$$|M|_r = \sigma(V_r)\, \lambda_r(M)\,. \qquad (8.5)$$

There are other reasons which lead to the same choice of σ[1]; the resulting notion of volume is amply justified by the successful formulation and solution of a large number of problems of Minkowskian geometry, of which the isoperimetric problems are probably the most outstanding[2].

The quantities defined above may be used to give rise to a sine-function as follows. Let two linear subspaces V_m and V_r, of dimension m and r respectively, intersect in a p-dimensional linear subspace V_p, where $p \geqq 0$ (which implies that V_p is not empty). The relative positions of V_m and V_r as sets in an associated euclidean space can be described by a single angle only if[3]

$$\text{Min}\,(m, r) = p + 1\,, \qquad (8.6)$$

so that the ordinary sine of this angle — to be denoted by $\sin(V_m, V_r)$ — is defined only if (8.6) is satisfied. Also, the dimension q of the linear subspace V_q of lowest dimension which contains V_m and V_r is

$$q = m + r - p\,.$$

---

[1] BUSEMANN [5], p. 243 et seq.

[2] See, for instance, BUSEMANN [13].

[3] Compare, for instance, SOMERVILLE [1].

The Minkowskian sine-function $\mathrm{sm}\,(V_m, V_r)$ may then be defined by[1]

$$\mathrm{sm}\,(V_m, V_r) = \sin\,(V_m, V_r)\,\frac{\sigma(V_p)\,\sigma(V_q)}{\sigma(V_m)\,\sigma(V_r)}\,, \tag{8.7}$$

where $\sigma(V_p) = 1$ if $p = 0$, and $\sigma(V_q) = 1$ if one of the linear subspaces V_m or V_r contains the other.

Again, this sine-function cannot be regarded as a function of a real variable, i. e. of an angle. Nevertheless, it possesses many properties of the sine function of euclidean geometry. For instance, a law of sines for triangles may be deduced very easily (assuming $m = 1$, $r = 1$, $p = 0$), while the Minkowskian sine- and cosine-functions (6.4) are related to each other[2].

It might be instructive to glance at the 2-dimensional case. We may choose a fixed direction through the origin to serve as "polar axis", so that if we introduce an associated euclidean metric, giving rise to an E_2, we may define the corresponding euclidean angle ψ between any radius vector and this axis. Noting (8.2), let us put $G(\psi) = F(\xi)$, the components of ξ being $\cos\psi$ and $\sin\psi$. The equation $F(\dot{x}) = 1$ to the indicatrix may be written in polar form as

$$r = \frac{1}{G(\psi)}\,.$$

Hence the factor σ, defined by (8.3) and depending on the choice of E_2, is given for $r = 2$ by

$$\sigma = \pi\left[\int_0^\pi G^{-2}(\psi)\,d\psi\right]^{-1}. \tag{8.8}$$

From (8.7) we deduce that the Minkowskian sine corresponding to two directions ψ_1, ψ_2 issuing from the origin may be written as[3]

$$\mathrm{sm}\,(\psi_1, \psi_2) = \sigma\,\frac{\sin\,(\psi_1 - \psi_2)}{G(\psi_1)\,G(\psi_2)}\,.$$

---

[1] Busemann [4], p. 161. The original definition given is not (8.7); rather (8.7) is a consequence of that definition. Although the former exhibits more clearly the fact that the Minkowskian sine-function is independent of the choice of the associated euclidean space, (8.7) permits a simple formulation. In a personal communication Prof. C. Y. Pauc suggests that one might profitably follow the method of Hilbert (as given for the case $m = 1$, $r = 1$, $n = 2$) in the general case: namely, by defining the angle of V_m and V_r as an ordered pair (V_m, V_r). The relative positions of V_m and V_r in the associated euclidean space may be described by a single number, namely the measure of the angle (V_m, V_r), the latter to be denoted by $\alpha(V_m, V_r)$, $0 \leq \alpha \leq \pi$, provided, of course, that condition (8.6) is fulfilled.

[2] Busemann [4], p. 162. It should be noted, however, that the cosine is defined only when the indicatrix is differentiable.

[3] Busemann [8].

Also, by means of (8.8) a measure of *angle* between these directions may be defined by writing

$$\Theta(\psi_1,\ \psi_2) = \sigma \int_{\psi_1}^{\psi_2} G^{-2}(\psi)\, d\psi\ . \tag{8.9}$$

From (8.8) we then deduce that this angle is equal to twice the Minkowskian area of the corresponding sector of the indicatrix.

So far we have restricted our attention to Minkowskian spaces. The extension to Finsler spaces F_n can be described most clearly if we introduce the latter as they appear in the work of BUSEMANN, CHOQUET, MENGER and PAUC[1]. An integrand $F(x,\dot{x})$ is defined for $x \in E_n$ and $\dot{x} \in E'_n$, E_n and E'_n denoting n-dimensional euclidean vector spaces. It is assumed that F is positive for $\dot{x} \neq 0$, continuous in $(x,\dot{x})$, symmetric, positively homogeneous and convex in $\dot{x}$. The tangent space $T_n(x)$ to F_n at x is the n-dimensional vector space endowed with the norm

$$||\dot{x}||_{F,\,x} = F(x,\dot{x})\ .$$

The associated euclidean metric of the tangent space $T_n(x)$ is the E_n-metric, where E_n and E'_n have been identified as a matter of convenience.

Thus for two-dimensional Finsler spaces the factor σ as defined by (8.3) may, in view of (8.8), be written as

$$\sigma(x) = \pi \left[\int_0^{\pi} G^{-2}(x,\psi)\, d\psi \right]^{-1}, \tag{8.10}$$

where $G(x,\psi) = F(x,\xi)$. Hence Choquet's area of a Lebesgue measurable subset M of F_2 is

$$\int_M \sigma(x)\, d\lambda_2(x) = \int_M \sigma(x)\, dx^1\, dx^2\ , \tag{8.10a}$$

the latter form being valid if we use "cartesian" coordinates.

A final remark of a more general nature may be relevant at this stage. For $r < n$ we have defined an r-dimensional measure only for subsets of r-dimensional linear subspaces of a Minkowskian space. This is called "Choquet measure" by PAUC. On the other hand, BUSEMANN[2] defines it for any BOREL set in the Minkowskian space as a Hausdorff measure (viz. as r-dimensional Hausdorff measure); this measure can aptly be named "Busemann measure". Only the n-dimensional measure in the n-dimensional Minkowskian space is a Haar measure. Certain complications arise in the transition to Finsler spaces: since a description of these is well beyond the scope of the present monograph, we refer the reader to a discussion of these matters by PAUC[3].

---

[1] See, for instance, PAUC [2], pp. 31—33, 37—40. Except for the symmetry condition A_1 this definition is slightly more general than that of § 1, the latter, however, being regarded as basic throughout this book.

[2] BUSEMANN [5].

[3] Compare, in particular, PAUC [3].

In conclusion we should mention that Busemann's measure may be derived from a somewhat different point of view, as was shown by BARTHEL[1]. This method can be summarised as follows: Consider the parallelepiped spanned in T_n by n vectors $A, A, \ldots, A$. It is required that the volume J of this parallelepiped possesses
${}_{(1)}{}_{(2)}{}_{(n)}$
the following properties:

(a) $J(A, \ldots, A) \geqq 0$,
${}_{(1)}{}_{(n)}$

(b) $J(A, \ldots, A + A, \ldots, A) = J(A, \ldots, A, \ldots, A)$, provided $i \neq k$,
${}_{(1)}{}_{(i)}{}_{(k)}{}_{(n)}{}_{(1)}{}_{(i)}{}_{(n)}$

(c) $J(A, \ldots, \lambda A, \ldots, A) = |\lambda|\, J(A, \ldots, A, \ldots, A)$.
${}_{(1)}{}_{(i)}{}_{(n)}{}_{(1)}{}_{(i)}{}_{(n)}$

It is well-known that these conditions determine J to within a positive factor $\sigma^{(n)}$ which is constant in T_n[2]:

$$J = \sigma^{(n)} \left| det\begin{pmatrix} A^i \\ (j) \end{pmatrix} \right| , \quad \begin{pmatrix} A^i = i^{th} \text{ component of } A \\ (j) \phantom{= i^{th} \text{ component of }} (j) \end{pmatrix} .$$

Thus, given a suitable orientation, the volume of a Lebesgue measurable set M of T_n is given by

$$J(M) = \sigma^{(n)} \lambda_n(M) .$$

The factor $\sigma^{(n)}$ may be determined by the following additional postulate:

(d) The Minkowskian volume of the indicatrix is the same for all n-dimensional Minkowskian spaces. This volume must therefore be given by (8.4), and hence we deduce that

$$\sigma^{(n)} \text{ (Lebesgue measure of the solid indicatrix)} = \omega^{(n)} ,$$

which determines $\sigma^{(n)}$[3].

In those theories of Finsler spaces (in the sense of the definition of § 1) in which any geometrical quantity may depend on a previously assigned direction (i. e. on an element of support) one can, of course, simply use an expression for the measure of volume which is a direct generalisation of the corresponding measure used in a locally euclidean geometry. Thus, given an arbitrary direction ξ^i in T_n, we may use the factor $\sqrt{g(x,\xi)}$ instead of $\sigma^{(n)}$, where $g(x,\xi) = \det(g_{ij}(x,\xi))$. For instance, if a continuous vector field $\xi^i(x^k)$ is defined over a Lebesgue measurable region R of the Finsler space F_n, R being bounded by a closed F_{n-1}, the definite integral

$$\int\limits_R \sqrt{g(x,\xi)}\, dx^1 \ldots dx^n \tag{8.11}$$

---

[1] BARTHEL [1]. The object of the alternative method appears to be the elimination of the notion of the associated euclidean spaces.

[2] See, for instance, SPERNER [1], pp. 118—124.

[3] BARTHEL [1], p. 360 et seq. Again the Minkowskian sine-function may be introduced. These notions are applied by BARTHEL to a discussion of the differential geometry of hypersurfaces in Minkowskian spaces.

extended over the region R would represent the volume of R with respect to the vector field $\xi^i(x^k)$ (on which it depends)[1].

In particular, it follows from (1.26) and (3.16) that

$$g(x, \xi) = F_1(x, \xi)\,[F(x, \xi)]^{n+1} \tag{8.12}$$

identically, so that for a two-dimensional Finsler space this type of area becomes

$$\iint\limits_{R} \sqrt{F^3(x, \xi)\,F_1(x, \xi)}\; dx^1\, dx^2 . \tag{8.13}$$

This is the form of a measure of area which was suggested by FUNK and BERWALD[2].

In conclusion we should mention a measure of area introduced by BLASCHKE[3]. This definition is formulated in terms of integral geometry and involves the area of the figuratrix (rather than that of the indicatrix) for the two-dimensional case.

---

[1] This procedure simply implies that the osculating indicatrices (ellipsoids corresponding to the vector field ξ^k) are used for purposes of measurement. There is still a close analogy with (8.3): for in the present case the choice of the factor σ of (8.3) is $\sqrt{g(x, \xi)}$, which implies that the n-dimensional volume of the ellipsoids (with respect to the osculating metric) is again precisely $\omega^{(n)}$.

[2] FUNK and BERWALD [1], p. 46. In this paper it is shown that this measure of area is closely related to the angle (7.12) of LANDSBERG. An alternative definition is given by GOLAB [7], for which it may be shown (under suitable differentiability assumptions) that it is independent of the vector field ξ^k if and only if the space is locally euclidean (GOLAB [8]). The area as defined by GOLAB is related to the angle (7.2) in the same manner as the area (8.13) is related to the angle (7.12) (GOLAB [10]).

[3] BLASCHKE [5].

Chapter II

Geodesics: Covariant Differentiation

Having described the more essential metric properties of the local tangent spaces, we now proceed to study the underlying manifold X_n, which we shall henceforth denote by F_n in order to stress the fact that a Finsler metric has been imposed upon it. Some of the most fundamental properties of F_n are described by the extremals of the problem in the calculus of variations which provides us with our metric. Thus we shall first derive the differential equations satisfied by the extremals — or "geodesics" — of F_n; this will be done not by means of the customary method involving the first variation of the length integral, but by means of a method specially adapted to illustrate clearly the geometrical background underlying this derivation. The basic problem in a geometry such as that of F_n is the investigation of the mutual relationship between tangent spaces attached to neighbouring points of F_n: more precisely, one seeks to establish geometrically meaningful mappings of one such tangent space onto another. In more elementary terms this problem may be formulated by posing the question as to the type of conditions which must be satisfied such that two vectors belonging to distinct but "neighbouring" tangent spaces may be described as being parallel. This question is by no means trivial: analytically it presents itself through the phenomenon that the ordinary derivative of a tensor is not in general also a tensor. It is from this point of view that we shall carry out a preliminary analysis of this problem in the present chapter.

§ 1. The Differential Equations Satisfied by the Geodesics

In this section we shall derive the differential equations satisfied by the geodesics of a Finsler space F_n. We shall present a direct geometrical approach which is based to some extent on the notions discussed in the previous chapter and which resembles the method of CARATHÉODORY[1]. The results of Chapter I concerning Minkowskian spaces (e. g. shortest distance of a point from a hyperplane) seem to indicate that the notion of normality provides the most natural approach to the problem of

---

[1] CARATHÉODORY [1], Chapter XII.

determining curves of minimum arc-length between two given points or between a given point and a given hypersurface. We shall be guided by this idea in the course of the construction of this section.

Suppose that we are given a family of hypersurfaces in F_n which are represented by the equation

$$S(x^i) = \Sigma , \qquad (1.1)$$

where Σ is the parameter of the family. We shall assume that the function $S(x^i)$ is at least of class C^2 and that the family (1.1) completely covers a finite region R of F_n simply. If δx^i represents a small displacement tangent to a hypersurface of the family [i. e. if the values x^i and $x^i + \delta x^i$ both satisfy equation (1.1) for the same value of Σ], we have

$$\frac{\partial S}{\partial x^i} \, \delta x^i = 0 . \qquad (1.2)$$

The $\dfrac{\partial S}{\partial x^i}$ form the components of a covariant vector; we construct a unit vector y_i in the same direction by putting

$$y_i = \varphi \, (x^i) \, \frac{\partial S}{\partial x^i} , \qquad (1.3)$$

where

$$(\varphi \, (x^j))^{-1} = H \left(x^j, \frac{\partial S}{\partial x^j} \right) \qquad (1.4)$$

represents the length of $\dfrac{\partial S}{\partial x^i}$. With y_i we may associate in the usual manner a unit contravariant vector

$$\xi^i = g^{ij}(x, y) \, y_j , \quad \text{or} \quad y_i = g_{ij}(x, \xi) \, \xi^j , \qquad (1.5)$$

so that in view of (1.3) equation (1.2) may be written in the form

$$g_{ij}(x, \xi) \, \xi^j \, \delta x^i = 0 . \qquad (1.6)$$

Since δx^i is an arbitrary displacement tangent to a hypersurface, we conclude in accordance with equation (1.6.2) that ξ^i is a unit normal vector to the hypersurface (all displacements tangent to the hypersurface being normal with respect to ξ^i).

By constructing the unit normal at each point of the family of hypersurfaces we thus find a vector-field $\xi^i(x^k)$ defined over the region R. By solving the differential equations

$$x'^i \equiv \frac{dx^i}{ds} = \xi^i(x^k) , \qquad (1.7)$$

we obtain a congruence G of curves whose tangent vectors coincide with the unit normals to the hypersurfaces. We shall call the latter family "transversal" to the congruence G. In order that the left-hand side of (1.7) shall also represent a unit vector it is necessary that the parameter s be the arc-length defined by the fundamental function $F(x, x')$

along the curves of G; for on dividing equation (1.3.9) by ds we immediately deduce that dx^i/ds is in fact a unit vector.

As regards notation, we may now introduce the following convention: whenever a directional argument such as dx^i/ds refers to a unit vector obtained by differentiation with respect to the arc-length s of a curve, this argument will be denoted by x'^i in contradistinction to $\dot{x}^i$.

Let Γ be a curve of the congruence G, intersecting two members of the family (1.1) corresponding to parameter values Σ_1 and Σ_2 in the points Q_1, Q_2 respectively. We may now construct an arbitrary curve C of class C^2 contained entirely in the region R, joining Q_1 and Q_2, which is nowhere tangent to a member of the family (1.1). If C is represented by the equations $x^i = x^i(\sigma)$, σ being the parameter of C, the latter condition is expressed by stipulating that Σ varies along C in such a manner that

$$\Sigma(\sigma) = S(x^i(\sigma))$$

is a strictly monotonic function of σ. Let P be an arbitrary point of C; the vector whose components are $dx^i/d\sigma$ is tangent to C and will be denoted by $\dot{x}^i$. We then have at P:

$$\frac{d\Sigma}{d\sigma} = \frac{\partial S}{\partial x^i}\,\dot{x}^i = \frac{y_i}{\varphi}\,\dot{x}^i$$

in view of (1.3); or, since y_i is a unit vector, we may use (1.5.13) so that

$$\varphi\,\frac{d\Sigma}{d\sigma} = F_{\dot{x}^i}(x, x')\,\dot{x}^i\,.$$

Here x'^i refers to the value at P of the vector field defined by (1.7). Introducing the expression (1.6.7) for the Weierstrass $\mathscr{E}$-function we see that the last equation may be written in the form

$$F(x, \dot{x}) - \varphi\,\frac{d\Sigma}{d\sigma} = \mathscr{E}(x, x', \dot{x})\,. \tag{1.8}$$

But since our convexity assumption implies $\mathscr{E} \geqq 0$ (§ 6, Ch. I), we find, on integrating along C from Q_1 to Q_2:

$$\int_{\sigma_1}^{\sigma_2}\left(F(x, \dot{x}) - \varphi\,\frac{d\Sigma}{d\sigma}\right) d\sigma \geqq 0\,,$$

where σ_1, σ_2 are the parameter values of σ corresponding to Q_1 and Q_2. But by definition (1.4) the function φ is positive and non-vanishing (§ 4, Ch. I), so that we have

$$\int_{C\ Q_1}^{\ \ Q_2}\frac{F(x, \dot{x})}{\varphi}\,d\sigma \geqq \Sigma_2 - \Sigma_1\,. \tag{1.9}$$

From (1.5) and (1.7) it follows that $y_i\, x'^i = 1$ (unit vector); thus using (1.3) we see that we have along Γ

$$1 = \varphi\, \frac{\partial S}{\partial x^i}\, x'^i = \varphi\, \frac{d\Sigma}{ds}\,, \tag{1.10}$$

where x'^i is tangent to Γ and $d\Sigma/ds$ measured along Γ. Hence we find by integration along Γ

$$\int_{\Gamma^{Q_1}}^{Q_2} \frac{F(x,x')}{\varphi}\, ds = \int_{\Gamma^{Q_1}}^{Q_2} \frac{d\Sigma}{ds}\, ds = \Sigma_2 - \Sigma_1\,. \tag{1.9a}$$

Comparing (1.9) and (1.9a) we deduce that curves such as Γ of the congruence G minimise the integral of F/φ, the factor φ^{-1} appearing in virtue of the fact that the family (1.1) of hypersurfaces was chosen arbitrarily. It is clear that this family must be specialised in a very definite manner in order that the curves of the resulting congruence G do in fact minimise the integral of F. In carrying out this specialisation we shall follow a method due to CARATHÉODORY[1].

Let us suppose then that the family (1.1) is such that $\varphi\,(x^i)$ is constant over each hypersurface of the family, i. e. φ^{-1} is a non-vanishing function $f(\Sigma)$ of Σ. From (1.10) we then have

$$\frac{d\Sigma}{ds} = f(\Sigma)\,, \tag{1.11}$$

$d\Sigma/ds$ being defined as before along curves of the congruence G. But if $\psi(t)$ is a monotonic function of its argument t, the family (1.1) may equally well be represented by the equations

$$\bar{S}(x^i) \equiv \psi(S(x^i)) = \psi(\Sigma) \equiv \bar{\Sigma}\,, \tag{1.12}$$

so that

$$\frac{d\bar{\Sigma}}{ds} = \frac{d\psi}{d\Sigma}\, \frac{d\Sigma}{ds} = f(\Sigma)\, \frac{d\psi}{d\Sigma} \tag{1.13}$$

in view of (1.11). Now we choose the function $\psi(\Sigma)$ such that

$$\psi(\Sigma) = \int_{\Sigma_1}^{\Sigma} f^{-1}(t)\, dt\,. \tag{1.14}$$

On substituting (1.14) in (1.13) we find that $\dfrac{d\bar{\Sigma}}{ds} = 1$ identically. It follows, therefore, that we may, without loss of generality, assume that $d\Sigma/ds$ of (1.11) is unity — this may always be achieved by "normalising" the equations (1.1) of the family in the sense defined by (1.12) and (1.14). Under these circumstances we deduce from (1.10) that $\varphi(x^i) = 1$ identically, and with the aid of equation (1.4) the special nature of the family

---

[1] FRANK and VON MISES [1], Ch. V.

of hypersurfaces is seen to be expressed by the partial differential equation

$$H\left(x^i, \frac{\partial S}{\partial x^i}\right) = 1 . \tag{1.15}$$

This is in fact the Hamilton-Jacobi equation for the simplest problem in the calculus of variations in parametric form[1]. It ensures that the curves of the congruence G minimise the integral of F; for in view of (1.15) and (1.4) a comparison of (1.9) and (1.9a) yields

$$\int_{C}^{Q_2} F(x^i, \dot{x}^i) \, d\sigma \geqq \int_{\Gamma}^{Q_2} F(x^i, x'^i) \, ds , \tag{1.16}$$

so that the curves of the congruence G are the geodesics of F_n. We may remark that this inequality is independent of the choice of the parameters σ and s appearing on either side since $F(x, \dot{x})$ is homogeneous of degree 1 in its directional arguments.

It is now a simple matter to derive the differential equations satisfied by the geodesics. From (1.15), (1.4) and (1.3) we now have

$$y_i = \frac{\partial S}{\partial x^i} . \tag{1.17}$$

Differentiating this equation with respect to s along Γ we find

$$\frac{d y_i}{ds} = \frac{\partial^2 S}{\partial x^i \partial x^j} x'^j = \frac{\partial^2 S}{\partial x^i \partial x^j} \frac{\partial H(x, y)}{\partial y_j} \tag{1.18}$$

in virtue of (1.5.13). Also, differentiating (1.15) partially with respect to x^i it follows from (1.17) that

$$\frac{\partial H(x, y)}{\partial x^i} + \frac{\partial H(x, y)}{\partial y_j} \frac{\partial^2 S}{\partial x^i \partial x^j} = 0 , \tag{1.19}$$

which together with (1.18) yields

$$\frac{d y_i}{ds} = - \frac{\partial H(x, y)}{\partial x^i} . \tag{1.20}$$

These are the required differential equations for the geodesics; the classical Euler-Lagrange equations may immediately be deduced from

---

[1] The Hamilton-Jacobi equation is not usually expressed in the form (1.15) which involves the unique Hamiltonian function H. For the standard treatment the reader should consult CARATHÉODORY [1], Ch. XIII, or BOLZA [1], Ch. V. Further properties of families of geodesics with special reference to contact transformations and Lagrange brackets are described by MAURIN [1], DOUGLAS [2], RUND [13]. The analogy of the above construction with geometrical optics (as well as with mechanics) should be immediately obvious to the reader. This analogy is discussed in some detail by CARATHÉODORY [5]. The approach of SYNGE [4] to geometrical optics may also be interpreted to some extent from this point of view.

them on substituting for y_i from (1.5.13) and on observing $(1.5.11)$[1]:

$$\frac{d}{ds}\left(\frac{\partial F(x, x')}{\partial x'^i}\right) - \frac{\partial F(x, x')}{\partial x^i} = 0 .\qquad(1.21)$$

Any reader who is familiar with analytical mechanics will not fail to recognise the resemblance between equations (1.20), (1.21) and the equations of motion of a dynamical system. Indeed, we had already remarked in § 5 of Chapter I that the transition from the variables $(x^i, \dot{x}^i)$ to the variables (x^i, y_i) corresponds to the transition in mechanics from the generalised components of velocity to the canonical momenta. Equations (1.20) therefore represent the first set of the equations of motion in canonical form, the second set being represented by (1.5.7). These two sets constitute $2n$ differential equations of the first order, while the set (1.21) consists of n second order differential equations.

The extremals of the problem in the calculus of variations defined by the integral (1.1.7) thus satisfy the equivalent differential equations (1.20) and (1.21): the arcs of these curves consequently represent the "shortest distances" in F_n (provided the arcs concerned are contained in a sufficiently restricted region of F_n), and therefore we shall henceforth call such arcs *geodesic arcs*, or simply *geodesics*. We remark that it may be shown that the differential equations (1.21) always possess a unique solution corresponding to a given initial line-element, which means that through each line-element of a Finsler space we can construct a unique geodesic. Further, a given solution of the differential equations (1.21) may always be imbedded in a family of solutions. The proofs of these observations depend on the theory of differential equations and the reader is therefore referred to standard treatises on this subject[2].

Again, if we integrate $dS = (\partial S/\partial x^i)\, dx^i$ along an arbitrary curve C joining two points corresponding to parameter values Σ_1 and Σ_2 of S, it follows from (1.17), (1.15) and (1.5.13) that we have

$$\Sigma_2 - \Sigma_1 = \int_{C}^{Q_2}{}_{Q_1} y_i\, dx^i = \int_{C}^{Q_2}{}_{Q_1} \frac{\partial F(x, x')}{\partial x'^i}\frac{dx^i}{d\sigma}\, d\sigma ,\qquad(1.22)$$

where the argument vector x'^i in the second integral refers to the tangent vector of the congruence G of geodesics. The value of this integral is seen to be independent of the choice of the curve C, depending only on the end-points Q_1 and Q_2. This integral is, in fact, the well-known *independent integral of* HILBERT[3].

---

[1] The Euler-Lagrange equations may be deduced under much weaker differentiability assumptions with respect to the function $F(x, \dot{x})$; in fact, it is shown by CARATHÉODORY [4] that it is sufficient to assume that $F(x, \dot{x})$ is of class C^2.

[2] CARATHÉODORY [1], pp. 240—245. See also Ch. III, § 6.

[3] BOLZA [1], p. 258.

Further, if the curve C happens to be closed, we deduce from (1.22) that

$$\oint y_i \, dx^i = 0 \,. \tag{1.22a}$$

Recalling once more that the variables y_i may be interpreted as the canonical momenta of a dynamical system, we recognise in the left-hand side of (1.22a) the fundamental *integral invariant* of such a system.

In conclusion we may verify that the geodesics of a Minkowskian space are the straight-line segments. In fact, a Minkowskian space is characterised by the existence of coordinate systems in which the metric function F is independent of the x^i. In such a system $F_{x^i} = -H_{x^i} = 0$ [eqn. (1.5.11)]. According to (1.20) the geodesics of a Minkowskian space then satisfy the equations $y_i = $ const., and hence, in virtue of (1.5.8), $dx^i/ds = $ const., which gives $d^2 x^i/ds^2 = 0$[1].

§ 2. The Explicit Expression for the Second Derivatives in the Differential Equations of the Geodesics

We shall now endeavour to solve equations (1.21) algebraically for the second derivatives x''^i. Before doing so, we remark that equations (1.21) are valid also when we replace the parameter s in these equations by an arbitrary parameter t which is a monotonically increasing function $t(s)$ of s (by suitable adjustment of the sign of t if necessary). For if we now denote derivatives of x^i with respect to t along some curve by $\dot{x}^i$, we have $x'^i = \dot{x}^i \dfrac{dt}{ds}$, and since F is positively homogeneous of degree 1 in its directional arguments, we have $F(x^i, \dot{x}^i) = F(x^i, x'^i) \, (ds/dt)$. On differentiating the latter equation with respect to x^i and $\dot{x}^i$ we find

$$F_{x^i}(x, \dot{x}) = F_{x^i}(x, x') \frac{ds}{dt}, \quad F_{x'^i}(x, x') = F_{\dot{x}^i}(x, \dot{x}) \,.$$

Thus equations (1.21) may be written in the form

$$\frac{d}{dt}\left(\frac{\partial F(x, \dot{x})}{\partial \dot{x}^i} \right) - \frac{\partial F(x, \dot{x})}{\partial x^i} = 0 \,. \tag{2.1}$$

---

[1] Throughout this discussion we have assumed that the family of curves cutting the hypersurfaces (1.1) orthogonally cover the region R simply. It should be pointed out, however, that this assumption may be satisfied only within a restricted region of R. In fact, according to a theorem due to MORSE and LITTAUER [1], the presence of focal points (MORSE [1], p. 51) is sufficient to destroy this property of a family of extremals. More precisely, a necessary and sufficient condition that the point p on an extremal Γ normal to a hypersurface σ be a focal point of σ is that the family of extremals cut transversally by σ near Γ shall fail to cover the neighbourhood of p simply. MORSE and LITTAUER proved this theorem on the assumption that the Finsler space F_n as well as the hypersurface σ are analytic; it was shown later by SAVAGE [1] that it is possible to prove the same theorem on the weaker hypothesis that F_n and σ are of class C^3. Further global properties of families of extremals are discussed by REEB [1, 2, 3].

We shall now derive a different form for the expression on the left-hand side of this equation.

From (1.20) and (1.5.11) together with (1.3.1) we deduce that the geodesics satisfy the differential equations

$$\frac{\partial}{\partial x^i}\left(g_{hk}(x,\, x')\, x'^h\, x'^k\right)^{1/2} = \frac{1}{2}\,\frac{\partial g_{hk}(x,\, x')}{\partial x^i}\, x'^h\, x'^k = \frac{dy_i}{ds}\,, \tag{2.2}$$

[noting that due to our choice of arc-length as parameter $F(x,\, x') = 1$]. At this stage it is necessary to introduce the Christoffel symbols of the first kind which are defined as in Riemannian geometry by the equations

$$\gamma_{ihk}(x,\, \dot{x}) = \frac{1}{2}\left(\frac{\partial g_{ih}(x,\, \dot{x})}{\partial x^k} + \frac{\partial g_{hk}(x,\, \dot{x})}{\partial x^i} - \frac{\partial g_{ki}(x,\, \dot{x})}{\partial x^h}\right). \tag{2.3}$$

As a result of the symmetry properties of the g_{ij} we have

$$\gamma_{ihk}(x,\, \dot{x})\, \dot{x}^h\, \dot{x}^k = \frac{1}{2}\,\frac{\partial g_{hk}(x,\, \dot{x})}{\partial x^i}\, \dot{x}^h\, \dot{x}^k\,, \tag{2.4}$$

so that equations (2.2) become

$$\frac{dy_i}{ds} - \gamma_{ihk}(x,\, x')\, x'^h\, x'^k = 0\,. \tag{2.5}$$

This is the tensor form of the differential equations which the covariant components of the tangent vector of a geodesic have to satisfy[1].

In order to find the corresponding equations for the derivatives of the contravariant components x'^i, we replace y_i in (2.2) in accordance with (1.4.1); and on carrying out the differentiation on the left-hand side [noting (1.3.5) as we carry out this operation], we find

$$g_{ij}(x,\, x')\, x''^j + \left(\frac{\partial g_{ih}(x,\, x')}{\partial x^k} - \frac{1}{2}\,\frac{\partial g_{hk}(x,\, x')}{\partial x^i}\right) x'^h\, x'^k = 0\,.$$

Again, due to the symmetry properties of the g_{ij} we then have in virtue of (2.3):

$$g_{ij}(x,\, x')\, x''^j + \gamma_{hik}(x,\, x')\, x'^h\, x'^k = 0\,. \tag{2.6}$$

If we denote the so-called Christoffel symbols of the second kind by

$$\gamma^{j}_{hk}(x,\, x') = g^{ij}(x,\, y)\, \gamma_{hik}(x,\, x')\,, \tag{2.7}$$

---

[1] The reader may verify that the expressions on the left-hand side of (2.1) transform like the components of a covariant vector under the transformation (1.1.1). Thus equations (2.1) and hence also (2.5) are invariant (which is also obvious from our construction). However, the tensor character of (2.5) may also be established by direct transformation. It is to be noted that although the Christoffel symbols (2.3) by themselves do not possess the same transformation properties as in Riemannian geometry, the combination (2.4) is such as to cause the left-hand side of (2.5) to represent the components of a covariant vector [see equation (3.11)]. This question will be fully dealt with in the next section.

[where the y_i correspond to the x'^i by (1.4.1)] we see that equation (2.6) is equivalent to the equation

$$x''^j + \gamma^j_{hk}(x, x')\, x'^h\, x'^k = 0 \,. \tag{2.8}$$

These are the desired differential equations of second order, whose contravariant vector characteristics follow from the tensor properties of equations (2.5). The parameter of differentiation in (2.2) is the arclength s as indicated by our notation. It is easily seen that when we perform a parameter transformation $t = t(s)$ (with $dt/ds \neq 0$) equations (2.8) become

$$\ddot{x}^j + \gamma^j_{hk}(x, \dot{x})\, \dot{x}^h\, \dot{x}^k - \dot{x}^j \left(\frac{d^2 s}{dt^2}\right) \bigg/ \frac{ds}{dt} = 0 \,. \tag{2.8'}$$

In conclusion it might be useful to write down the identities satisfied by the Christoffel symbols:

$$\gamma_{hik}(x, \dot{x}) = \gamma_{kih}(x, \dot{x}) \;; \quad \gamma_h{}^i{}_k(x, \dot{x}) = \gamma_k{}^i{}_h(x, \dot{x}) \;;$$
$$g_{hk}(x, \dot{x})\, \gamma_i{}^k{}_j(x, \dot{x}) = \gamma_{ihj}(x, \dot{x}) \;; \tag{2.9}$$

together with

$$\frac{\partial g_{ij}(x, \dot{x})}{\partial x^k} = \gamma_{ijk}(x, \dot{x}) + \gamma_{jik}(x, \dot{x}) \;;$$
$$\frac{\partial g^{ik}(x, y)}{\partial x^j} = - g^{hk}(x, y)\, \gamma_h{}^i{}_j(x, \dot{x}) - g^{hi}(x, y)\, \gamma_h{}^k{}_j(x, \dot{x}) \,. \tag{2.10}$$

In these formulae — as well as in (2.3) and (2.7) — care must be taken that the directional arguments $\dot{x}^i$ and y_i correspond to each other according to (1.4.1) and its inverse.

§ 3. The Differential of a Vector

At this stage it is convenient to begin the discussion of the possible ways according to which a vector may be differentiated such that this process yields once more a vector or a tensor. We shall see that various approaches to this problem exist: hence we shall start with elementary analytical considerations in this section, postponing the relevant geometrical discussion until later.

Let C: $x^i = x^i(t)$ be a curve of class C^2 in the Finsler space F_n, and suppose that there is defined a continuous and continuously differentiable vector field $X^i(t)$ along C. In a new coordinate system $(x^{i'})$ obtained from the (x^i) coordinate system by (1.1.1) [subject, of course, to (1.1.2)] this vector field will be given by the equation

$$X^i = A^i_{i'}\, X^{i'} \,, \tag{3.1}$$

where we have put

$$A^i_{i'} = \frac{\partial x^i}{\partial x^{i'}} \,. \tag{3.2}$$

We note for future reference that if we write $A_i^{i'} = \partial x^{i'}/\partial x^i$, we have the identities:

$$A_{i'}^i A_j^{i'} = \delta_j^i. \tag{3.3}$$

Let us differentiate (3.1) with respect to t, the parameter of the curve C. We find

$$\frac{dX^i}{dt} = A_{i'}^i \frac{dX^{i'}}{dt} + (\partial_j A_{i'}^i) X^{i'} \dot{x}^{j'}, \tag{3.4}$$

where we have written

$$\partial_{j'} A_{i'}^i = \frac{\partial A_{i'}^i}{\partial x^{j'}} = \frac{\partial^2 x^i}{\partial x^{i'} \partial x^{j'}}. \tag{3.5}$$

Equation (3.4) clearly indicates that the dX^i/dt do not form the components of a vector in view of the presence of the term involving $\partial_{j'} A_{i'}^i$. Geometrically it is obvious that this should be so: for the vectors X^i and $X^i + dX^i$ are elements of two *distinct* tangent spaces, namely $T_n(P)$ and $T_n(Q)$ respectively, where P and Q are neighbouring points with coordinates x^i and $x^i + dx^i$ on C, the displacement dx^i along C corresponding to the increment dt of t in (3.4). Since we are dealing with a metric space, a natural approach would be to attempt to express this difference between the neighbouring tangent spaces in terms of the metric tensor and its derivatives. In studying the change of the vector field $X^i(t)$ along C it appears, therefore, that we have to consider two factors as we pass from P to Q, namely (a) the change $dX^i = (dX^i/dt)\, dt$ in $X^i(t)$ which depends solely on the definition of the field X^i and is naturally independent of the metric of the space, and (b) the difference in metric between the tangent spaces $T_n(P)$ and $T_n(Q)$. One would then surmise that a tensorial differential, i. e. an expression which has the correct invariance properties, for the change in the vector field X^i would consist of the sum of two terms, each of which corresponds to one of the factors (a) and (b).

A naive, analytical attempt to determine the latter factor is to examine in some detail the second term on the right-hand side of equation (3.4): we shall now show that this term can be completely expressed as a function of two expressions of which one is written in terms of the (x^i) coordinate system, while the other is written in terms of the $(x^{i'})$ system, such that each of these expressions involves the metric tensor and its derivatives in a similar way in the two systems. This process will yield the required vector differential.

Let $g_{ij}(x, \dot{x})$ represent the metric tensor along C: its law of transformation under (1.1.1) is given by

$$g_{i'j'}(x^{k'}, \dot{x}^{k'}) = g_{ij}(x^k, \dot{x}^k)\, A_{i'}^i A_{j'}^j. \tag{3.6}$$

If we differentiate this equation with respect to $x^{k'}$, we have [1]

$$\frac{\partial g_{i'j'}}{\partial x^{k'}} = \frac{\partial g_{ij}}{\partial x^{k}} A_{i'}^{i} A_{j'}^{j} A_{k'}^{k} + 2 C_{ijh} A_{i'}^{i} A_{j'}^{j} (\partial_{k'} A_{h'}^{h}) \dot{x}^{h'} + \\ + g_{ij} (A_{i'}^{i} \partial_{k'} A_{j'}^{j} + A_{j'}^{j} \partial_{k'} A_{i'}^{i}) , \tag{3.7}$$

where we have made use of (1.3.4) and the relations (3.1) as applied to the vector $\dot{x}^{i}$, viz.

$$\dot{x}^{i'} = A_{i}^{i'} \dot{x}^{i} . \tag{3.8}$$

From equation (3.7) we obtain two similar equations by means of a cyclic interchange of the indices i', j', k'. From the sum of the latter two equations we subtract (3.7) and divide by 2. The resulting expression simplifies considerably provided we take into account equation (2.3) and the fact that $\partial_{j'} A_{i'}^{i} = \partial_{i'} A_{j'}^{i}$ in view of (3.5). In fact, we obtain

$$\gamma_{i'k'j'} = A_{i'}^{i} A_{j'}^{j} A_{k'}^{k} \gamma_{ikj} + g_{ik} A_{k'}^{k} (\partial_{j'} A_{i'}^{i}) + C_{ijh} \{ A_{j'}^{j} A_{k'}^{k} \partial_{i'} A_{h'}^{h} + \\ + A_{k'}^{i} A_{i'}^{j} \partial_{j'} A_{h'}^{h} - A_{i'}^{i} A_{j'}^{j} \partial_{k'} A_{h'}^{h} \} \dot{x}^{h'} . \tag{3.9}$$

We note that as a result of (3.6) we have $g_{ik} A_{k'}^{k} = g_{r'k'} A_{i}^{r'}$. Thus on solving for the second expression on the right-hand side of (3.9), we find after multiplication with $\dot{x}^{j'}$,

$$A_{i}^{r'} (\partial_{j'} A_{i'}^{i}) \dot{x}^{j'} = \gamma_{i'j'}^{r'} \dot{x}^{j'} - \gamma_{ij}^{r} A_{r}^{r'} A_{i'}^{i} \dot{x}^{j} - \\ - g^{r'k'} (C_{ijh} A_{k'}^{i} A_{i'}^{j} (\partial_{j'} A_{h'}^{h}) \dot{x}^{h'} \dot{x}^{j'}) , \tag{3.10}$$

where we have made use of (1.5.9), (1.3.5), (2.7) and (3.8). This equation still involves the quantity $\partial_{j'} A_{i'}^{i}$, which is the object of our search, on both sides: but we may easily overcome this difficulty by multiplying (3.10) by $\dot{x}^{i'}$ and noting that as a result of (3.8) and (1.3.5) the last term on the right-hand side vanishes. In this manner we obtain, after suitable interchange of indices,

$$(\partial_{j'} A_{h'}^{h}) \dot{x}^{j'} \dot{x}^{h'} = A_{h'}^{h} \gamma_{p'j'}^{h'} \dot{x}^{p'} \dot{x}^{j'} - \gamma_{pj}^{h} \dot{x}^{p} \dot{x}^{j} . \tag{3.11}$$

We substitute this result in the last term on the right-hand side of (3.10): this term thus becomes

$$- g^{r'k'} C_{ijh} A_{k'}^{i} A_{i'}^{j} (A_{h'}^{h} \gamma_{p'j'}^{h'} \dot{x}^{p'} \dot{x}^{j'} - \gamma_{pj}^{h} \dot{x}^{p} \dot{x}^{j}) .$$

But we have seen (Ch. I) that partial differentiation of a tensor with respect to directional arguments again leads to a tensor. It therefore follows from (3.6) that we have

$$C_{ijh} A_{k'}^{i} A_{i'}^{j} A_{h'}^{h} = C_{k'i'h'} .$$

As a result of this equation and (1.5.9) the term in question reduces to

$$- g^{r'k'} C_{k'i'h'} \gamma_{p'j'}^{h'} \dot{x}^{p'} \dot{x}^{j'} + A_{r}^{r'} [g^{jr} C_{ijh} A_{i'}^{i} \gamma_{pj}^{h} \dot{x}^{p} \dot{x}^{j}] .$$

---

[1] For the rest of this section it will be evident that the directional argument of the g_{ij} will be the tangent vector $\dot{x}^{i} = dx^{i}/dt$ of C: we may therefore omit an explicit representation of the directional argument without danger of confusion.

When the last term of the right-hand side of (3.10) is replaced by this expression, we find after some further interchange of indices and use of (3.2),

$$(\partial_{j'} A^{i}_{i'})\, \dot{x}^{j'} = A^{i}_{r'} \{ \gamma_{i'}{}^{r'}{}_{j'} - g^{r'h'} C_{h'i'l'}\, \gamma_{p'}{}^{l'}\, \dot{x}^{p'} \}\, \dot{x}^{j'} - A^{k}_{i'} \{ \gamma_{k}{}^{i}{}_{j} - g^{ih} C_{hkl}\, \gamma_{p}{}^{l}{}_{j}\, \dot{x}^{p} \}\, \dot{x}^{j} \,. \tag{3.12}$$

This is the desired result: it will be noted that the two expressions in curly brackets have identical structures but refer to different coordinate systems. Putting

$$g^{ih} C_{hkl} = g^{ih} C_{khl} = C^{i}_{kl} \,, \tag{3.13}$$

we shall now write

$$P^{i}_{kj}(x, \dot{x}) = \gamma_{k}{}^{i}{}_{j}(x, \dot{x}) - C^{i}_{kl}(x, \dot{x})\, \gamma_{p}{}^{l}{}_{j}(x, \dot{x})\, \dot{x}^{p} \,, \tag{3.14}$$

with a similar equation in the $(x^{i'})$-system. Thus equation (3.12) reduces to

$$(\partial_{j'} A^{i}_{i'})\, \dot{x}^{j'} = A^{i}_{r'} P^{r'}_{i'j'}\, \dot{x}^{j'} - A^{k}_{i'} P^{i}_{kj}\, \dot{x}^{j} \,. \tag{3.15}$$

Let us substitute this expression for the last term on the right-hand side of (3.4). Using (3.1) we obtain

$$\frac{dX^{i}}{dt} = A^{i}_{i'}\, \frac{dX^{i'}}{dt} + A^{i}_{r'} P^{r'}_{i'j'} X^{i'}\, \dot{x}^{j'} - P^{i}_{hj} X^{h}\, \dot{x}^{j} \,,$$

or, on rearrangement,

$$\left(\frac{dX^{i}}{dt} + P^{i}_{h}\, X^{h}\, \dot{x}^{j} \right) = A^{i}_{i'} \left(\frac{dX^{i'}}{dt} + P^{i'}_{h'j'} X^{h'}\, \dot{x}^{j'} \right) \,. \tag{3.16}$$

It follows, therefore, that the expressions defined by

$$\frac{\delta X^{i}}{\delta t} = \frac{dX^{i}}{dt} + P^{i}_{hj}(x, \dot{x})\, X^{h}\, \dot{x}^{j} \tag{3.17}$$

form the components of a contravariant vector. We shall regard $\delta X^{i}/\delta t$ as the first of several plausible differentials[1]. The process of differentiation as exemplified by (3.17) will be designated as "δ-differentiation".

In particular, we observe that this process gives rise to a well-defined *parallel displacement*: the vector $X^{i} + dX^{i}$ of $T_{n}(x^{i} + dx^{i})$ is said to result from the vector X^{i} of $T_{n}(x^{i})$ by parallel displacement if $\delta X^{i} = 0$, i. e. if[2]

$$dX^{i} = - P^{i}_{hj}(x, dx)\, X^{h}\, dx^{j} \,. \tag{3.18}$$

---

[1] The $P^{i}_{hk}(x, \dot{x})$ were first introduced in the form (3.14) by RUND [3, 4], but as we shall see later, these quantities bear a close relationship to similar coefficients introduced by CARTAN [1].

[2] We may replace the argument $(x^{k}, \dot{x}^{k})$ in $P^{i}_{hj}(x^{k}, \dot{x}^{k})$ by (x^{k}, dx^{k}), for from the definition (3.14) we deduce that the P^{i}_{hj} are homogeneous of degree zero in their directional arguments as a result of (1.3.4) and (3.13). Furthermore, it should be noted that the P^{i}_{hj} are not symmetric in their lower indices [unlike the Christoffel symbols (2.7)].

Clearly this parallel displacement depends only on the vector X^i and the displacement dx^i, in contradistinction to the alternative forms of parallelism that will be discussed later.

The following geometrical interpretation clearly establishes the fundamental significance of this definition of parallelism. Let us suppose, for the moment, that the Finsler space is Minkowskian. In other words, we suppose that there exists a coordinate system in which $\partial F/\partial x^i = 0$. However, if we apply a transformation of the type (1.1.1) to these derivatives, we find by differentiation that

$$\frac{\partial F}{\partial x^i} = \frac{\partial F}{\partial x^i}\, A^i_{i'} + \frac{\partial F}{\partial \dot x^i}\, (\partial_{i'} A^i_{j'})\, \dot x^{j'},$$

so that in the (x'^i)-system $\partial F/\partial x^{i'}$ does not in general vanish. In fact, it vanishes only if the transformation happens to be linear. Hence we shall call those coordinate systems in the Minkowskian space for which $\partial F/\partial x^i$ vanishes "linear coordinates", while the others will be regarded as being "curvilinear". It is evident from (1.3.1) and (2.3) that in a linear system $\gamma_{h\ k}^{\ i} = 0$, and thus in view of (3.14) also $P^i_{hk} = 0$. Hence in such a system the condition (3.18) for parallelism reduces to the form $dX^i = 0$, while this is not the case for curvilinear systems. Now in a linear system two vectors of equal length are parallel if their components are identical (Ch. I, § 2), which implies that such a field of parallel vectors satisfies the differential equations $dX^i = 0$. It follows, therefore, *that the general definition* (3.18) *of parallelism reduces to the elementary vectorial definition if the space is Minkowskian.* Further, it is clear that when we transform the components dX^i from a linear system into an arbitrary curvilinear system in the Minkowskian space, we shall obtain the components δX^i as defined by (3.17) in the latter system. Summarising, we may assert: *In order that a field of vectors results from the parallel displacement of a given vector in a Minkowskian space, it is necessary and sufficient that it satisfies the differential equations* (3.18) *in an arbitrary curvilinear coordinate system*[1].

---

[1] In fact, it was this property which originally led the present author to *define* the P^i_{hk}; for a construction very similar to the one outlined in this section gives rise to (3.14) if this property is postulated. This method is directly analogous to one frequently used in euclidean curvilinear coordinates; for if we transform the differential equations of straight lines: $d^2x^i/ds^2 = 0$ ($s =$ arc-length) from linear coordinates to curvilinear coordinates, we obtain the Christoffel symbols of Riemannian geometry. The analogy goes still further: for the principal difference between a Minkowskian geometry in curvilinear coordinates and Finsler geometry lies in the fact that in the latter equations (3.18) are, in general, not integrable in contradistinction to the Minkowskian case. VARGA [1] also considers curvilinear coordinates in a Minkowskian space, but with the help of a euclidean metric associated with each direction, i. e. with the system of osculating indicatrices (Ch. I, § 3). His subsequent discussion of the parallel displacement will become more relevant in the light of the next chapter.

The parallel displacement (3.18) enjoys a further, most significant property. A curve is called *autoparallel*, if its tangent vectors result from each other by successive, infinitesimal parallel displacements of the type (3.18). We may state the following theorem: *The geodesics* (2.8) *of the Finsler space are autoparallel curves*[1].

This result follows immediately from (2.8), (3.14) and (3.18) if we observe that in view of (1.3.5) we have identically

$$P_{hj}^{i}(x, \dot{x})\, \dot{x}^{h} = \gamma_{h}{}^{i}{}_{j}(x, \dot{x})\, \dot{x}^{h} \,. \tag{3.19}$$

Thus the equations of the geodesics of the Finsler space may be written in the form

$$\frac{\delta x'^{i}}{\delta s} = 0 \,, \quad \text{where} \quad x'^{i} = \frac{dx^{i}}{ds} \,. \tag{3.20}$$

At this stage we could, should we wish to do so, discuss certain geometrical properties of δ-differentiation as well as its application to tensors in general and sums or products of tensors. However, it will be found convenient to deal with these questions at a later stage for reasons to be explained in the next section.

§ 4. Partial Differentiation of Vectors

Suppose that instead of being given a vector field as a function of a single parameter t (defined along a curve), we are now given a vector field $X^{i}(x^{k})$, defined over a finite region of the space F_{n}, as functions of the coordinates $x^{1}, \ldots, x^{n}$. We assume that the X^{i} are continuous and continuously differentiable functions of these n variables; and on forming the partial derivatives, we deduce from (3.1) and (3.5) that these transform according to the equation

$$\frac{\partial X^{i}}{\partial x^{j}} = A^{\,i}_{\,i'} A^{j'}_{j} \frac{\partial X^{i'}}{\partial x^{j'}} + A^{j'}_{j}\, \partial_{j'} A^{i}_{i'} X^{i'} \,. \tag{4.1}$$

Again, the presence of the term $\partial_{j'} A^{i}_{i'}$ indicates that these partial derivatives do not form the components of a tensor. But it would be erroneous to assume that we would obtain a tensor if we were to add to the partial derivative a term $P_{hj}^{i} X^{h}$ as equation (3.17) would suggest. For it is easily seen that while $P_{hj}^{i}(x, \dot{x})\, \dot{x}^{j}$ possesses the correct transformation properties, this is not so for the term $P_{hj}^{i}(x, \dot{x})$ without the $\dot{x}^{j}$. However, on the basis of the formulae of the previous section it is a relatively simple matter to deduce a suitable term which has to be added to the partial derivatives $\partial X^{i}/\partial x^{k}$ in order to obtain the relevant tensor[2].

---

[1] Note that this is a property enjoyed by straight lines in euclidean geometry.

[2] Once more this additional expression will involve the metric tensor and its derivatives: but any function of the metric tensor must have some directional argument. Thus we must choose some line-element $(x^{i}, \dot{x}^{i})$, to which we shall adhere throughout this section, so that we may omit directional arguments; $\dot{x}^{i}$ refers to a displacement $d x^{i}$ when we write $d X^{i} = (\partial X^{i}/\partial x^{k})\, d x^{k}$.

From equation (4.1) it is clear that we must endeavour to separate the last term on the right-hand side into two expressions which are similar but refer to different coordinate systems. This term is given explicitly by equation (3.9), provided we eliminate expressions of the type $(\partial_{j'} A^h_{h'})\, \dot{x}^{h'}$ which appear inside the curly brackets of (3.9). We may do so by means of (3.15), having taken into account the symmetry in (3.5). In this way we obtain

$$g_{ik} A^k_{k'} (\partial_{j'} A^i_{i'}) = \gamma_{i'k'j'} - A^i_{i'} A^j_{j'} A^k_{k'}\, \gamma_{ikj} -$$
$$- C_{ijh}(A^i_{j'} A^i_{k'} A^h_{h'} P^{h'}_{i'l'} + A^i_{k'} A^j_{i'} A^h_{h'} P^{h'}_{j'l'} - A^i_{i'} A^j_{j'} A^h_{h'} P^{h'}_{k'l'})\, \dot{x}^{l'} + \tag{4.2}$$
$$+ C_{ijh}(A^i_{j'} A^j_{k'} A^r_{i'} P^h_{rl} + A^i_{k'} A^j_{i'} A^r_{j'} P^h_{rl} - A^i_{i'} A^j_{j'} A^k_{k'} P^h_{kl})\, \dot{x}^l .$$

Applying (3.13) and interchanging indices, we see that this becomes

$$g_{ik} A^k_{k'} (\partial_{j'} A^i_{i'}) = \gamma_{i'k'j'} - (C_{j'k'h'} P^{h'}_{i'l'} + C_{k'i'h'} P^{h'}_{j'l'} - C_{i'j'h'} P^{h'}_{k'l'})\, \dot{x}^{l'} -$$
$$- A^i_{i'} A^j_{j'} A^k_{k'} \{\gamma_{ikj} - (C_{jkh} P^h_{il} + C_{kih} P^h_{jl} - C_{ijh} P^h_{kl})\, \dot{x}^l\}. \tag{4.3}$$

We therefore define new coefficients as follows:

$$P^*_{ikj}(x, \dot{x}) = \gamma_{ikj}(x, \dot{x}) - \{C_{jkh}(x, \dot{x})\, P^h_i(x, \dot{x}) +$$
$$+ C_{kih}(x, \dot{x})\, P^h_{jl}(x, \dot{x}) - C_{ijh}(x, \dot{x})\, P^h_{kl}(x, \dot{x})\}\, \dot{x}^l , \tag{4.4}$$

together with

$$P^{*h}_{ij}(x, \dot{x}) = g^{hk}(x, y)\, P^*_{ikj}(x, \dot{x}) , \quad (y_i \equiv g_{ij}(x, \dot{x})\, \dot{x}^j) . \tag{4.5}$$

Equation (4.3) then reduces to the form

$$g_{ik} A^k_{k'} (\partial_{j'} A^i_{i'}) = P^*_{i'k'j'} - A^i_{i'} A^j_{j'} A^k_{k'} P^*_{ikj} . \tag{4.6}$$

This leads to the correct transformation law: for when we multiply this equation by the expression $g^{k'h'} A^r_{h'} A^{j'}_{h}$, noting (4.5), (3.2), (3.5) and (3.6), we find after suitable interchange of indices:

$$A^{j'}_j (\partial_{j'} A^i_{i'}) = A^i_{h'} A^{j'}_j P^{*h'}_{i'j'} - A^h_{i'} P^{*i}_{hj} ; \tag{4.6a}$$

and on substituting this expression in (4.1), we finally have in virtue of (3.1):

$$\left(\frac{\partial X^i}{\partial x^j} + P^{*i}_{hj} X^h\right) = A^i_{i'} A^{j'}_j \left(\frac{\partial X^{i'}}{\partial x^{j'}} + P^{*i'}_{h'j'} X^{h'}\right). \tag{4.7}$$

Thus the quantities $X^i_{;j}$ defined by

$$X^i_{;j}(x, \dot{x}) = \frac{\partial X^i}{\partial x^j} + P^{*i}_{hj}(x, \dot{x})\, X^h , \tag{4.8}$$

do in fact form the components of a mixed tensor of contravariant and covariant valency one. We shall call the process (4.8) partial δ-differentiation[1]: for the intimate relation between the processes (4.8) and (3.17) is best expressed by the relation

$$\frac{\delta X^i}{\delta t} = X^i_{;j}(x,\dot{x})\,\dot{x}^j = X^i_{;j}(x,\dot{x})\,\frac{dx^j}{dt}\,, \qquad (4.9)$$

which is analogous to the usual formula $d\Phi/dt = (\partial\Phi/\partial x^i)\,\dot{x}^i$. Equation (4.9) is easily deduced as follows: from (4.4), (4.5) and (1.3.5) we have

$$P^{*\,h}_{ij}\dot{x}^j = g^{hk}(\gamma_{ikj} - C_{kir}P^r_{jl}\dot{x}^l)\,\dot{x}^j\,,$$

or, using (2.7), (3.13) and (3.19),

$$P^{*\,h}_{ij}\,\dot{x}^j = (\gamma_i{}^h{}_j - C^h_{ir}\gamma_j{}^r{}_l\,\dot{x}^l)\,\dot{x}^j\,.$$

From the definition (3.14) we thus have the result

$$P^{*\,h}_{ij}(x,\dot{x})\,\dot{x}^j = P^h_{ij}(x,\dot{x})\,\dot{x}^j\,, \qquad (4.10)$$

from which (4.9) follows. It is clear, also, that we may replace the P^i_{hj} by the $P^{*\,i}_{hj}$ in the process (3.17) without effecting any change: nevertheless, it is often preferable not to do so, especially when the calculations involve the precise expression of these coefficients, of which the former have a much simpler form. We may observe also that in view of (3.19) and (4.10) the equations of the geodesics (2.8) may be written in the form:

$$\frac{d^2 x^i}{ds^2} + P^{*\,i}_{h\,k}\left(x,\frac{dx}{ds}\right)\frac{dx^h}{ds}\frac{dx^k}{ds} = 0\,. \qquad (4.11)$$

In conclusion we note that due to the symmetry of the Christoffel symbols a closer inspection of (4.4) indicates the symmetry of the $P^{*\,i}_{hj}$ in their lower indices:

$$P^{*\,i}_{hj}(x,\dot{x}) = P^{*\,i}_{jh}(x,\dot{x})\,, \qquad (4.12)$$

while the difference between the two coefficients is given by the formula

$$P^*_{ihj} - P_{ihj} = (C_{ijk}P^k_{hl} - C_{jhk}P^k_{il})\,\dot{x}^l + C_{hik}C^k_{jr}\gamma_p{}^r{}_l\,\dot{x}^p\,\dot{x}^l\,. \qquad (4.13)$$

This relation is a direct consequence of equations (3.14) and (4.4).

§ 5. Elementary Properties of δ-differentiation

It is clear that the differentiation processes as defined by equations (3.17) and (4.8) provide us with a basic method by means of which we may proceed to construct a general theory of Finsler spaces. However,

---

[1] The definition of the $P^{*\,h}_{ij}$ in the form (4.4), (4.5) was given by RUND [5]. However, it was later shown by E. T. Davies that these coefficients are identical to a set of coefficients (denoted by $\Gamma^{*\,h}_{ij}$) introduced much earlier by E. CARTAN [1], whose method we shall describe presently. We have purposely retained the above notation in the present context in the hope that this will help the reader to distinguish clearly between the various forms of covariant differentiation which will be described below. The definition (4.4) may be written in a different form, involving a finite recursion process (RUND [6]) which may be useful with a view to further generalisations. Alternative derivations of the $P^{*\,h}_{ij}$ will be considered in Ch. III.

we do not propose to do so directly, since it is necessary to give a full discussion of an alternative treatment of covariant derivatives, due to E. CARTAN, which has practically dominated the literature on Finsler spaces. We shall thus restrict ourselves in the present section to an investigation of the most elementary properties of the δ-derivative, since this will enable us to draw the relevant comparisons more readily at a later stage.

Clearly we may extend the process of δ-differentiation of contravariant vectors to arbitrary tensors by means of the P_{hk}^{*i}, simply following the example of Riemannian or Non-Riemannian geometry. Yet a certain difficulty arises which was not apparent in the previous sections, where we had assumed that the vector field X^i depends only on position or on a single parameter. It may happen, however, that a vector field X^i depends on line-elements (x^k, ξ^k) instead of just on the position: $X^i = X^i(x^k, \xi^k)$. Then it is clear that we must replace the first terms on the right-hand sides of (3.17) and (4.8) by the expressions

$$\frac{\partial X^i}{\partial x^j}\frac{dx^j}{dt} + \frac{\partial X^i}{\partial \dot{x}^j}\cdot\frac{d\xi^j}{dt} \quad\text{and}\quad \frac{\partial X^i}{\partial x^j} + \frac{\partial X^i}{\partial \dot{x}^k}\cdot\frac{\partial \xi^k}{\partial x^j}$$

respectively. Although this does not change the tensor-character of (3.17) and (4.8) in any way[1], the addition of the extra terms is awkward since they involve the derivatives of the directional part ξ^k of the line-elements (x^k, ξ^k) on which the vector field X^i depends. Nevertheless, this cannot be avoided, and we shall see in later applications that in general the geometrical problem under consideration clearly indicates — or rather, forces — a suitable choice for the derivatives of the directional argument[2]. However, this does not invalidate the statement made in connection with equation (3.18): given some vector X^i in $T_n(P)$, this equation uniquely determines the "parallel" vector $X^i + dX^i$ in the neighbouring tangent space.

Accordingly, we define the partial δ-derivative with respect to x^k in the direction $\dot{x}^i$ of the arbitrary tensor $T^{i_1\cdots i_r}{}_{j_1\ldots j_s}(x, \xi)$ by the formula

$$T^{i_1\cdots i_r}{}_{j_1\ldots j_s;k} = \frac{\partial T^{i_1\cdots i_r}{}_{j_1\ldots j_s}}{\partial x^k} + \frac{\partial T^{i_1\cdots i_r}{}_{j_1\ldots j_s}}{\partial \dot{x}^h}\frac{\partial \xi^h}{\partial x^k} +$$

$$+ \sum_{\mu=1}^{r} T^{i_1\cdots i_{\mu-1}\,h\,i_{\mu+1}\cdots i_r}{}_{j_1\ldots j_s}\, P_{hk}^{*i_\mu}(x, \dot{x}) - \qquad (5.1)$$

$$- \sum_{\nu=1}^{s} T^{i_1\cdots i_r}{}_{j_1\ldots j_{\nu-1}\,h\,j_{\nu+1}\ldots j_s}\, P_{j_\nu k}^{*h}(x, \dot{x}).$$

---

[1] This is immediately obvious if (3.1) is differentiated, while the variation of the directional arguments is being taken into account.

[2] We shall amplify this remark in Ch. III in connection with the covariant derivative of E. CARTAN.

The tensor character of (5.1) follows directly from the transformation law (4.6a).

The following rules are direct consequences of this definition:

(1) The δ-derivative (or differential) of a sum of two tensors is the sum of the δ-derivatives (or differentials) of these tensors.

(2) The δ-derivative (or differential) obeys the same product rules as the ordinary derivative (or differential).

Of course these rules only apply when the tensors under consideration depend on the same field ξ^k of directions (if at all), i. e. the terms $\partial \xi^k/\partial x^h$ must be the same for all. Finally, one may also deduce from (5.1) that

(3) The δ-derivative of a scalar function is its ordinary derivative.

As an immediate application of definition (5.1) let us consider the δ-derivative in the direction $\dot{x}^i$ of the metric tensor $g_{ij}(x, \xi)$, corresponding to some line-element (x, ξ). From (5.1) and (1.3.4) we have

$$g_{ij;k}(x, \xi) = \frac{\partial g_{ij}(x, \xi)}{\partial x^k} + 2C_{ijh}(x, \xi)\,\frac{\partial \xi^h}{\partial x^k} -$$
$$- g_{hj}(x, \xi)\,P^{*h}_{ik}(x, \dot{x}) - g_{ih}(x, \xi)\,P^{*h}_{jk}(x, \dot{x})\,. \tag{5.2}$$

The following observations are immediately evident. Firstly, the presence of the term $\partial \xi^h/\partial x^k$ indicates clearly that the δ-derivative (5.2) has no meaning until this term is specified. Again, the geometrical reason for this is clear, for the metric tensor must be defined also with respect to direction as well as position when its value at a neighbouring point is specified. Secondly, no further simplification is possible unless the direction $\dot{x}^i$ in which we differentiate is identical with the direction ξ^i for which g_{ij} is defined. Let us therefore put $\dot{x}^i = \xi^i$; in view of (4.5) the last two terms on the right of (5.2) reduce to

$$- (P^{*}_{ijk}(x, \xi) + P^{*}_{jik}(x, \xi))\,.$$

But in virtue of the symmetry of the C_{ijh} and equation (2.10) the following identities result from the definition (4.4):

$$P^{*}_{ijk} + P^{*}_{jik} = \frac{\partial g_{ij}}{\partial x^k} - 2C_{ijh}P^{h}_{kl}(x, \dot{x})\,\dot{x}^l\,, \tag{5.3}$$

or

$$P^{*}_{ijk} + P^{*}_{jik} = \frac{\partial g_{ij}}{\partial x^k} - 2C_{ijh}P^{*h}_{lk}\,\dot{x}^l\,, \tag{5.3'}$$

where we have used (4.10) and (4.12). Thus equation (5.2) (with $\dot{x}^i = \xi^i$) becomes

$$g_{ij;k}(x, \xi) = 2C_{ijh}(x, \xi)\left[\frac{\partial \xi^h}{\partial x^k} + P^{*h}_{lk}(x, \xi)\,\xi^l\right]. \tag{5.4}$$

In view of (4.8) this reduces to

$$g_{ij;k}(x, \xi) = 2C_{ijh}(x, \xi)\,\xi^h_{;k}\,. \tag{5.5}$$

This result represents the generalisation of Ricci's Lemma of Riemannian geometry: for if our space were Riemannian, the tensor C_{ijh} would vanish identically[1]. The general form (5.5) clearly indicates how the δ-derivative of the g_{ij} must depend on the variation of the directional argument ξ^k. In some cases, however, the fact that the right-hand side of (5.5) does not vanish causes no further analytical complications, for by (1.3.5) we have the identities

$$g_{ij;k}(x, \xi)\, \xi^i = g_{ij;k}(x, \xi)\, \xi^j = 0 , \tag{5.6}$$

irrespective of the values of $\partial \xi^h / \partial x^k$.

An interesting special case arises when $\xi^i = \dot{x}^i = dx^i/dt$ represents the tangent vector to a curve C with arc-length s, so that the process of δ-differentiation actually takes place along the curve. On multiplying (5.5) by dx^k/dt we then have from (4.9),

$$\frac{\delta g_{ij}(x, \dot{x})}{\delta t} = 2 C_{ijh}(x, \dot{x})\, \frac{\delta \dot{x}^h}{\delta t} , \tag{5.7}$$

and corresponding to (5.6) the useful identities

$$\frac{\delta g_{ij}(x, \dot{x})}{\delta t}\, \dot{x}^i = \frac{\delta g_{ij}(x, \dot{x})}{\delta t}\, \dot{x}^j = 0 . \tag{5.8}$$

Equations (5.7) and (5.8) hold for arbitrary curves C of class C^2; but from (3.20), (2.8') and (5.8) we deduce that *in general*

$$\frac{\delta g_{ij}(x, \dot{x})}{\delta t} = 0 \tag{5.8a}$$

if we differentiate along a geodesic of the Finsler space[2].

An immediate consequence of the fact that the covariant derivative of the metric tensor with arbitrary argument does not in general vanish is the fact that under parallel displacement (as defined in § 3 with respect to δ-differentiation) of a vector X^i the length of the latter does not in general remain invariant, which is in direct contrast to Riemannian geometry. The change of the length of X^i when displaced by parallelism from the point $P(x^i)$ to the point $Q(x^i + dx^i)$ is given by the relation

$$\frac{\delta}{\delta s}\, (g_{ij}(x, X)\, X^i X^j)\, ds = \frac{\delta g_{ij}(x, X)}{\delta s}\, X^i X^j\, ds , \tag{5.9}$$

---

[1] RUND [5, 6]. The reader is referred to a further discussion of this question in chapter III, § 2.

[2] In view of (1.3.5) the left-hand side of (5.7) vanishes if $\delta \dot{x}^i/\delta t = \lambda \dot{x}^i$ along the curve C for some function $\lambda(t)$. But if the curve satisfies this differential equation it must be a geodesic; for we have $g_{ij}(x, \dot{x})\, \dot{x}^i \dot{x}^j = (ds/dt)^2 = \dot{s}^2$ in view of (1.3.9), and if we apply the process of δ-differentiation to this equation we find $g_{ij}(x, \dot{x})\, \dot{x}^i(\delta \dot{x}^j/\delta t) = \dot{s}\ddot{s}$, which, together with the two previous equations yields $\lambda = \ddot{s}/\dot{s}$. Thus the differential equations satisfied by C are $\delta \dot{x}^i/\delta t - (\ddot{s}/\dot{s})\, \dot{x}^i = 0$, which are equivalent to the equations (2.8') for the geodesics in view of (3.19). We may remark also that if $\lambda(t) = 0$ along C, then $t = as + b$, where a, b are constants.

which follows from (3.18), where ds is the length of the displacement dx^i. However, if Γ is the unique geodesic of F_n joining the points P and Q, it is easily seen that the scalar product of X^i with the tangent vector x'^i to the geodesic at P, namely the expression $g_{ij}(x, x')\, x'^i X^j$, remains invariant in consequence of equation (5.8). In particular, therefore, the length of a vector remains invariant under parallel displacement if the displacement is taken in the direction of the vector.

This condition, together with two others, namely, that the geodesics should be autoparallel curves and that the dX^i should be linear in X^i, was postulated by Rund [4] in an attempt to give a geometrical treatment of the problem of parallel displacement in a Finsler space. In this manner the P^i_{hk} of § 3 may be deduced. However, it is clear that these conditions are not sufficient to determine the P^i_{hk} uniquely; at best one may say that they are the simplest coefficients whose properties ensure that the corresponding parallelism enjoys those attributes. This is pointed out again explicitly by Laugwitz [2]; in fact it is shown that if a term of the type $B^i_{hj}(x, dx)\, X^h\, dx^j$ is added to the right-hand side of the equation (3.18) of parallelism, the above three conditions are still satisfied, provided that the tensor $a_{ij} = B_{ijk}\, dx^k = g_{ih}\, B^h_{jk}\, dx^k$ is skew-symmetric and satisfies the condition $a_{ij}(x, dx)\, dx^j = 0$. The latter relations hold if the tensor B_{ijk} is itself skew-symmetric in all its indices, and non-vanishing tensors of this kind can exist only if the number of dimensions exceeds 2. However the given conditions determine our parallelism uniquely for $n = 2$, but not generally. A further detailed study of similar conditions is made by Ohkubo [1, 2], who introduces a new metric connection whose coefficients are not evaluated explicitly.

The following formulae will be found useful in the sequel. One frequently has to deal with quantities involving the directional derivatives of the coefficients P^i_{hk} and P^{*i}_{hk}. We have from (4.10) and (3.19) the relation

$$P^{*i}_{hk}(x, \dot{x})\, \dot{x}^h\, \dot{x}^k = \gamma^i_{hk}(x, \dot{x})\, \dot{x}^h\, \dot{x}^k \,. \tag{5.10}$$

Differentiating the right-hand side of this equation with respect to $\dot{x}^l$, we obtain in view of (2.3), (2.7) and (1.3.6) the expression

$$\frac{\partial g^{ir}}{\partial \dot{x}^l}\, g_{rs}\, \gamma^s_{hk}\, \dot{x}^h\, \dot{x}^k + 2\gamma^i_{hl}\, \dot{x}^h,$$

which, in consequence of (1.5.9), is equivalent to

$$2\left[\gamma^i_{lh} - C^i_{lr}\, \gamma^r_{hk}\, \dot{x}^k\right] \dot{x}^h \,.$$

Thus it follows from (3.14) that

$$\frac{\partial}{\partial \dot{x}^l}\left(\gamma^i_{hk}\, \dot{x}^h\, \dot{x}^k\right) = 2 P^i_{lh}\, \dot{x}^h \,, \tag{5.11}$$

and on differentiating equation (5.10) with respect to $\dot{x}^l$, we have

$$\frac{\partial P^{*i}_{hk}}{\partial \dot{x}^l}\, \dot{x}^h\, \dot{x}^k + 2 P^{*i}_{lh}\, \dot{x}^h = 2 P^i_{lh}\, \dot{x}^h \,,$$

and hence, in virtue of (4.10), we obtain the useful identity

$$\frac{\partial P^{*i}_{hk}}{\partial \dot{x}^l}\, \dot{x}^h\, \dot{x}^k = 0 \,. \tag{5.12}$$

Similarly, on differentiating (3.14) and making use of similar reductions, we find

$$\frac{\partial P_{ihk}}{\partial \dot{x}^j}\, \dot{x}^k = \frac{\partial C_{ihj}}{\partial x^k}\, \dot{x}^k - C_{ihl}\, P^l_{jr}\, \dot{x}^r - \left(\frac{\partial C_{ihr}}{\partial \dot{x}^j} - C_{hik}\, C^k_{jr}\right)\gamma^{\,r}_{p\,l}\, \dot{x}^p\, \dot{x}^l. \qquad (5.13)$$

This equation is often more useful if the last term on the right-hand side is eliminated by means of (4.13). A short calculation yields the formula

$$\frac{\partial P_{ihk}}{\partial \dot{x}^j}\, \dot{x}^k = (P^*_{ihj} - P_{ihj}) + 2\, C_{khj}\, P^k_{ir}\, \dot{x}^r + C_{ihj|k}\, \dot{x}^k\,, \qquad (5.14)$$

where the tensor $C_{ihj|k}$ is derived from the covariant derivative $C_{ihj;k}$ by replacing the derivatives $\partial \xi^h / \partial x^k$ in the corresponding expression (5.1) by the terms $-P^h_{kr}\, \dot{x}^r$ (see Ch. III, § 2). It is easily verified that the relation (5.12) may also be derived less directly from formula (5.14).

In a manner analogous to that leading up to (5.5) and (5.6) we may also show that

$$\xi^i\, C_{ijh;k}(x,\, \xi) = -C_{ljh}(x,\, \xi)\, \xi^l_{;k}\,. \qquad (5.15)$$

On differentiating (5.5) with respect to x^h and on interchanging the indices $h,\, k$ in the equation thus obtained, we find

$$\tfrac{1}{2}\,(g_{ij;hk} - g_{ij;kh}) = C_{ijl;k}\, \xi^l_{;h} - C_{ijl;h}\, \xi^l_{;k} + C_{ijl}(\xi^l_{;hk} - \xi^l_{;kh})\,, \qquad (5.16)$$

and hence, in view of (5.15) and (1.3.5),

$$\begin{aligned}
\tfrac{1}{2}\,(g_{ij;hk}(x,\, \xi) &- g_{ij;kh}(x,\, \xi))\, \xi^i = \\
&= -C_{jlm}(x,\, \xi)\, \xi^m_{;k}\, \xi^l_{;h} + C_{jlm}(x,\, \xi)\, \xi^m_{;h}\, \xi^l_{;k} = 0\,,
\end{aligned} \qquad (5.17)$$

as a result of the symmetry of C_{jlm} in all its indices.

Chapter III

The "Euclidean Connection" of E. Cartan

A distinguishing feature of the process of δ-differentiation as developed in the previous chapter is the fact that the covariant derivative of the metric tensor does not in general vanish. Consequently, further developments of the theory of Finsler spaces will differ radically from the established body of theorems of Riemannian geometry, for in the latter the lemma of Ricci (according to which the covariant derivative of the metric tensor vanishes) plays a most decisive role. This divergence cannot be avoided if one continues to regard Finsler spaces as locally Minkowskian spaces, only linear connections being considered. On the other hand, a basically different point of view may be adopted if one introduces the so-called element of support: if this device is accepted, one can indeed construct covariant derivatives for which the analogue of Ricci's lemma is generally valid. This is achieved by the "euclidean connection" of Cartan, to which the first part of the present chapter is devoted. In view of the great influence which this approach has exerted upon the general development of our subject, the theory of Cartan will be developed *ab initio*. In the latter half of this chapter we shall introduce a generalisation of Finsler spaces by defining a general space of paths: this will lead to alternative forms of covariant derivatives which will be compared with each other in the light of results of this and the previous chapter.

§ 1. The Fundamental Postulates of Cartan

In the previous chapter we have outlined the analytical basis on which one may construct a theory of Finsler spaces, regarding the latter as locally Minkowskian spaces. The theory of E. Cartan[1] which treats Finsler spaces from an entirely different point of view, has played the most predominant role in the development of the subject, and in order to do full justice to the methods of Cartan, it is essential that we should return, momentarily, to the point of departure, namely the elementary definitions of Chapter I. In this section we shall endeavour to present the point of view taken by Cartan in his monograph, and we shall discuss the postulates by means of which he defined his covariant derivatives, while the formal calculations of Chapter II will enable us to dispense with some fairly complicated analysis in connection with the

---

[1] Cartan [1, 2].

actual determination of the form of the coefficients of these covariant derivatives[1].

In order to be able to endow the Finsler space F_n with a so-called "euclidean connection", Cartan considers the manifold X_{2n-1} of the line-elements $(x^i, \dot{x}^i)$, which is $(2n-1)$-dimensional since only the ratios of the $\dot{x}^i$ are necessary to define a direction in the tangent space $T_n(x^i)$, the coordinate x^i referring to the centre of the line-element $(x^i, \dot{x}^i)$. All quantities, such as tensors, are to be defined as functions of line-elements. In the space F_n a metric is defined by means of a function $F(x^i, \dot{x}^i)$ satisfying the conditions of Ch. I, but the manifold X_{2n-1} is said to be endowed with a *euclidean connection* if the following construction is imposed on X_{2n-1}:

(I) A metric tensor with symmetric components $g_{ij}(x, \dot{x})$ is given, such that the square of the distance between the centres x^i and $x^i + dx^i$ of the two neighbouring line-elements $(x^i, \dot{x}^i)$ and $(x^i + dx^i, \dot{x}^i + d\dot{x}^i)$ is given by the expression[2]

$$g_{ij}(x, \dot{x})\, dx^i\, dx^j \, . \tag{1.1}$$

Since the dx^i form the components of a contravariant vector, it follows that the square of the length of an arbitrary contravariant vector is to be defined by the quadratic form

$$g_{ij}(x, \dot{x})\, X^i X^j \, . \tag{1.1'}$$

This expression is no longer a function of the position x^i and the vector X^i alone, but depends on the so-called *element of support* $(x^i, \dot{x}^i)$ of the vector (which has to be chosen in advance)[3].

---

[1] Shorter methods for the determination of Cartan's coefficients have been given at various times, in particular by Varga [2], Laugwitz [3] and Sulanke [1]. However, since these authors use special devices such as the osculating Riemannian space (see § 4) or the general geometry of paths (in the sense of Douglas [1]), their methods might not exhibit the basic ideas of Cartan as clearly as the original method of the latter. It might, therefore, be more advantageous to follow the historical approach and to postpone the discussion of these more recent constructions. Also, it should be pointed out that Schouten and Haantjes [1] gave a more general theory for the determination of the connection coefficients of spaces in which the fundamental metric function depends on co- and contravariant vector densities. In this connection the reader should also consult Schouten and Hlavatý [1].

[2] Note that in this expression there need not be any relation between the differential dx^i and the directional argument $\dot{x}^i$. No relation is assumed between the tensor g_{ij} and the metric function $F(x^i, \dot{x}^i)$ of F_n at the present stage: this relation is to be deduced from later postulates.

[3] The reader will recognise immediately that the measure of length $(1.1')$ corresponds to the measure of length with respect to an osculating indicatrix [Ch. I, § 3, equation (3.10)] constructed in $T_n(x^i)$, the coefficients of this quadric hypersurface being given by $g_{ij}(x^k, \dot{x}^k)$, where the $x^k, \dot{x}^k$ are kept fixed. In this manner it is ensured that the expression $(1.1')$ is similar to the measure of length in locally euclidean geometry.

(II) An analytical expression is given which represents the variation of the vector X^i when its element of support $(x^i, \dot{x}^i)$ undergoes an infinitesimal change, becoming $(x^i + d x^i, \dot{x}^i + d \dot{x}^i)$. This variation of X^i is to be represented by means of a *covariant* (or *absolute*) differential:

$$D X^i = d X^i + C^i_{kh}(x, \dot{x}) X^k d \dot{x}^h + \Gamma^i_{kh}(x, \dot{x}) X^k d x^h , \qquad (1.2)$$

where the coefficients C^i_{kh} and Γ^i_{kh} are functions of the element of support. These coefficients are by no means arbitrary; in fact, the first condition to be imposed upon them is the following: If the vector X^i is transported from $(x^i, \dot{x}^i)$ to $(x^i + d x^i, \dot{x}^i + d \dot{x}^i)$ by *parallel displacement*, i. e. if the actual change $d X^i$ in X^i is in accordance with the equation

$$D X^i = 0 \quad \text{or} \quad d X^i = -C^i_{kh} X^k d \dot{x}^h - \Gamma^i_{kh} X^k d x^h , \qquad (1.3)$$

then the length of X^i as given by $(1.1')$ *is to remain invariant.* It is easy to deduce the restriction on the C^i_{kh} and the Γ^i_{kh} as implied by this postulate. Firstly, let us introduce the notation

$$\omega_k{}^i = C^i_{kh} d \dot{x}^h + \Gamma^i_{kh} d x^h , \qquad (1.4)$$

so that (1.2) becomes

$$D X^i = d X^i + \omega_k{}^i X^k . \qquad (1.2')$$

On differentiating $(1.1')$ it follows that under the parallel displacement (1.3) the variation of the square of the length of X^i becomes

$$(d g_{ij} - g_{ki} \omega_j{}^k - g_{jk} \omega_i{}^k) X^i X^j ;$$

so that if we take symmetric parts of the expression inside the bracket, we see that a sufficient condition for the required invariance of the length of X^i under parallel displacement reads

$$d g_{ij} = \omega_{ji} + \omega_{ij} , \qquad (1.5)$$

where we have put

$$\omega_{ji} = g_{ik} \omega_j{}^k . \qquad (1.6)$$

In view of (1.4), since the $d x^i$ and $d \dot{x}^i$ are independent, the conditions (1.5) imply that the Γ^i_{jh} and the C^i_{jh} must satisfy the equations

$$\frac{\partial g_{ij}}{\partial x^h} = \Gamma_{ijh} + \Gamma_{jih} , \qquad (1.7)$$

and

$$\frac{\partial g_{ij}}{\partial \dot{x}^h} = C_{ijh} + C_{jih} , \qquad (1.7\,\text{a})$$

where we have written

$$\Gamma_{ijh} = g_{jk} \Gamma^k_{ih} , \qquad (1.8)$$

together with

$$C_{ijh} = g_{jk} C^k_{ih} . \qquad (1.8\,\text{a})$$

Clearly the relations (1.7) and (1.7a) are not in themselves sufficient to determine the Γ^k_{ih} and the C^k_{ih} in terms of the tensor g_{ij} and its derivatives; furthermore, no mention has yet been made of the given metric function $F(x^i, \dot{x}^i)$ of F_n. In order to link the latter with the metric tensor and in order to establish unique coefficients for the covariant derivative (1.2), the following additional postulates have to be introduced[1].

(III) (a) If the direction of a vector X^i coincides with that of its element of support $(x^i, \dot{x}^i)$, its length is to be equal to $F(x^i, X^i)$.

(b) Let X^i, Y^i represent two vectors with a common element of support $(x^k, \dot{x}^k)$. When the latter performs an infinitesimal rotation about its own centre x^k, thus becoming $(x^k, \dot{x}^k + d\dot{x}^k)$, while the components X^i and Y^i remain fixed, let the corresponding covariant differentials (1.2) be DX^i and DY^i. The following symmetry condition is required:

$$g_{ij}(x, \dot{x}) X^i D Y^j = g_{ij}(x, \dot{x}) Y^i D X^j .$$

(c) If the direction of a vector with fixed components X^i coincides with that of its element of support, then its covariant differential (1.2) corresponding to an infinitesimal rotation of its element of support about its own centre vanishes identically.

(d) The coefficients — to be denoted by Γ^{*i}_{hk} — which appear in the covariant differential (1.2) when the displacement is such that the element of support is transported parallel to itself from x^k to $x^k + dx^k$ in accordance with (1.3), are to be symmetric in their lower indices k, h [2].

Let us now draw some analytical conclusions from these conditions. Obviously, in view of (1.1'), condition (a) yields

$$F^2(x, \dot{x}) = g_{ij}(x, \dot{x}) \dot{x}^i \dot{x}^j . \tag{1.9}$$

From (b) we have, under the given circumstances, in view of (1.2) and (1.8a)

$$C_{kih} Y^k X^i d\dot{x}^h = C_{kih} X^k Y^i d\dot{x}^h ,$$

---

[1] In point of fact, Cartan gives a further postulate in addition to conditions (a) to (d) ([1], p. 10); however, this postulate is superfluous since it may be derived from the others (see Cartan [3]). It merely involves the relation between perpendicularity and transversality with respect to the element of support.

[2] A most interesting geometrical interpretation of these postulates may be derived by a method employed by Varga [1] for the deduction of the coefficients of the covariant differential in Minkowskian spaces. By means of the osculating indicatrix a euclidean metric is attached to each direction $\dot{x}^i$ of the space. This leads to the construction of "cartesian" coordinate systems, whose n generating vectors $e^k_{(i)}$ $(i, k = 1, \ldots, n)$ satisfy the equations $e_{(i)} e_{(k)} = g_{ik}(\dot{x})$, such that the vectors $e^k_{(i)}$ are continuous and continuously differentiable functions of $\dot{x}^i$. Again an analytical expression is sought which is to represent the change in a given vector field depending on $\dot{x}^i$ when the latter undergoes a change $d\dot{x}^j$. Because of the special coordinate system the corresponding analysis is greatly simplified. Finally, the introduction of curvilinear coordinates in the Minkowskian space practically reduces the problem of finding the corresponding coefficients in these coordinates to the problem of the present section.

which is satisfied for all possible sets of values of X^i, Y^i and $d\dot{x}^i$, only if we have the symmetry relation

$$C_{kih} = C_{ikh}. \tag{1.10}$$

But from (1.7a) we then deduce immediately,

$$C_{ijh} = \frac{1}{2} \frac{\partial g_{ij}}{\partial \dot{x}^h}. \tag{1.11}$$

Also, condition (c) may be written in the form

$$C^i_{kh}\, \dot{x}^k\, d\dot{x}^h = 0\,,$$

as a result of (1.2); again, since this is to hold for all possible sets of values of $d\dot{x}^h$, we must have

$$C_{kih}\, \dot{x}^k = 0\,, \tag{1.12}$$

where we have used (1.8a) once more. Differentiating equation (1.9) in succession with respect to $\dot{x}^k$ and $\dot{x}^h$, and taking into account (1.10), (1.11) and (1.12), we find

$$F(x, \dot{x})\, F_{\dot{x}^k}(x, \dot{x}) = g_{ik}(x, \dot{x})\, \dot{x}^i\,, \tag{1.13}$$

and

$$\tfrac{1}{2} F^2_{\dot{x}^h \dot{x}^k} \equiv F\, F_{\dot{x}^h \dot{x}^k} + F_{\dot{x}^h} F_{\dot{x}^k} = g_{hk}\,. \tag{1.14}$$

Thus it follows from (1.9), (1.14) and (1.11) that *the tensors g_{ij}, C_{ijh} as introduced in this section are identical with those denoted similarly and defined directly in Chapter I* [§ 3, equations (3.1), (3.2) and (3.4)], enjoying, of course, the same properties. In particular, the condition C of Chapter I imposed on the function $F(x, \dot{x})$ ensures that the metric (1.1′) is positive definite[1].

As a result of (1.9) we see that the unit vector l^i in the direction of the element of support $(x^i, \dot{x}^i)$ is given by

$$l^i = \frac{\dot{x}^i}{F(x, \dot{x})}\,, \tag{1.15}$$

while its covariant counterpart results from (1.13) and (1.15):

$$l_i = F_{\dot{x}^i}(x, \dot{x}) = g_{ik}(x, \dot{x})\, l^k\,. \tag{1.15′}$$

In order to exploit the final condition (d) we have to evaluate the covariant differential of l^i. In virtue of (1.12) the corresponding equation (1.2) reduces to

$$Dl^i = dl^i + \Gamma^i_{kh}\, l^k\, dx^h\,, \tag{1.16}$$

and when l^i is displaced parallel to itself, i. e. if $Dl^i = 0$, we have from (1.15)

$$d\left(\frac{\dot{x}^i}{F(x, \dot{x})}\right) = -\Gamma^i_{kh}\, l^k\, dx^h\,,$$

---

[1] It should be stressed, however, that the notion of length as defined in Chapter I is in general not identical to that defined by (1.1′), the identity holding only if the direction of the vector X^i coincides with its own element of support.

or

$$d\,\dot{x}^i = \dot{x}^i \left(\frac{dF}{F}\right) - \Gamma^i_{kh}\,\dot{x}^k d\,x^h\,. \tag{1.16'}$$

Thus when the element of support is displaced in this manner, the $d\dot{x}^i$ are given as functions of $(x^k,\dot{x}^k)$, and on substituting these values in the expression (1.2) for the covariant differential, we find, on taking (1.12) into account once more, that under these circumstances

$$D X^i = d X^i + \Gamma^{*i}_{kj} X^k\,d\,x^j\,, \tag{1.17}$$

where we have put

$$\Gamma^{*i}_{kj} = \Gamma^i_{kj} - C^i_{kh}\Gamma^h_{rj}\,\dot{x}^r\,. \tag{1.18}$$

Condition (d) now states that

$$\Gamma^{*i}_{kj} = \Gamma^{*i}_{jk}\,. \tag{1.19}$$

On interchanging indices in (1.18) and observing (1.8), (1.8a) and (1.19) we obtain

$$\Gamma_{kij} - \Gamma_{jik} = (C_{kih}\Gamma^h_{rj} - C_{jih}\Gamma^h_{rk})\,\dot{x}^r\,. \tag{1.20}$$

Noting that the coefficients C_{ijh} have already been determined by equations (1.14), it remains to find the coefficients Γ_{kij}. But equations (1.20) together with (1.7) comprise a totality of n^3 equations for the n^3 unknowns Γ_{kij}, and thus these equations will determine the Γ_{kij} uniquely[1], or, equivalently, these equations will determine the Γ^{*i}_{kj} uniquely in virtue of (1.18).

We could now follow the treatment of CARTAN and solve the equations in question for these coefficients. However, the analysis of our previous chapter will enable us to avoid this somewhat lengthy calculation. Firstly, we note that in view of (1.12) we have from (1.18)

$$\Gamma^{*i}_{kj}\,\dot{x}^k = \Gamma^i_{kj}\,\dot{x}^k\,, \tag{1.21}$$

so that (1.18) may be written in the form

$$\Gamma^{*i}_{kj} = \Gamma^i_{kj} - C^i_{kh}\Gamma^{*h}_{rj}\,\dot{x}^r\,. \tag{1.22}$$

Thus writing equations (1.7) in terms of the Γ^*_{ijh} rather than in terms of the Γ_{ijh} we have

$$\frac{\partial g_{ij}}{\partial x^h} = \Gamma^*_{ijh} + \Gamma^*_{jih} + 2C_{ijk}\Gamma^{*k}_{rh}\,\dot{x}^r\,, \tag{1.23}$$

where, of course,

$$\Gamma^*_{ijh} = g_{kj}\Gamma^{*k}_{ih}\,. \tag{1.24}$$

The set of equations (1.23) is one set of the relations which have to be satisfied by the Γ^*_{ijh}: but fortunately we already know a solution. For if we glance back at equation (5.3') of § 5, Ch. II, we see that the P^*_{ijh} defined in § 4 of Ch. II satisfy just this same equation. Furthermore, if

---

[1] CARTAN [1], p. 15.

we substitute P^{*}_{ijh} for Γ^{*}_{ijh} in (1.18) we obtain a set of values Γ_{ijh} which satisfy equations (1.20), again as a result of the symmetry relation $P^{*j}_{ih} = P^{*j}_{hi}$ [equation (2.4.12)]. Thus we have a unique solution of the equations (1.7) and (1.20), given by

$$\Gamma_{kij} = \Gamma^{*}_{kij} + C_{kih} \Gamma^{*h}_{rj} \dot{x}^{r}, \tag{1.25}$$

where the Γ^{*}_{kij} are given by equation (2.4.4)[1].

We shall now express these coefficients in the form in which they were originally derived by CARTAN. In order to do so, let us write

$$2 G^{i}(x, \dot{x}) = \gamma^{i}_{hk}(x, \dot{x}) \dot{x}^{h} \dot{x}^{k}. \tag{1.26}$$

Differentiating this equation with respect to $\dot{x}^{r}$, we find by (2.5.11)

$$\frac{\partial G^{i}}{\partial \dot{x}^{r}} = P^{i}_{rh}(x, \dot{x}) \dot{x}^{h}. \tag{1.27}$$

Hence in view of (1.25), (1.12) and (2.4.10)

$$\Gamma^{i}_{kj} \dot{x}^{k} = \Gamma^{*i}_{kj} \dot{x}^{k} = P^{*i}_{kj} \dot{x}^{k} = P^{i}_{jk} \dot{x}^{k} = \frac{\partial G^{i}}{\partial \dot{x}^{j}}. \tag{1.27'}$$

Thus equation (2.4.4) may be written as

$$\Gamma^{*}_{kij} = \gamma_{kij} - C_{jih} \frac{\partial G^{h}}{\partial \dot{x}^{k}} - C_{kih} \frac{\partial G^{h}}{\partial \dot{x}^{j}} + C_{kjh} \frac{\partial G^{h}}{\partial \dot{x}^{i}}, \tag{1.28}$$

which is the form given by CARTAN[2]. Also, using (1.27') and (1.27) we see that (1.25) becomes

$$\Gamma_{kij} = \Gamma^{*}_{kij} + C_{kih} \frac{\partial G^{h}}{\partial \dot{x}^{j}}, \tag{1.29}$$

so that we have from (1.28)

$$\Gamma_{kij} = \gamma_{kij} - C_{jih} \frac{\partial G^{h}}{\partial \dot{x}^{k}} + C_{kjh} \frac{\partial G^{h}}{\partial \dot{x}^{i}}. \tag{1.30}$$

We conclude that the coefficients specifying the covariant differential (1.2) are thus completely determined by the equations (1.11) and (1.30)[3].

---

[1] Except under special circumstances we shall, in the sequel, adhere to Cartan's notation for the Γ^{*i}_{hk} instead of writing P^{*i}_{hk}.

[2] CARTAN [1], p. 16.

[3] The application of Cartan's derivative to problems in the calculus of variations is studied by STOKES [1]. VARGA [3] considers covariant derivatives of the type (1.2), but the symmetry conditions usually imposed on the coefficients are not used, the only restriction being $C^{i}_{kh}(x, \dot{x}) \dot{x}^{k} = 0$. OHKUBO [1, 3, 5] introduces conditions similar to those of CARTAN, but in his theory the relations (1.14) need not hold necessarily. A more general method, based on the use of exterior differential forms, which leads to a *class* of euclidean connections, is due to CHERN [1, 2]. The connection (1.30) is a special member of this class. Some implications of Chern's method are described in Ch. VI, § 3.

§ 2. Properties of the Covariant Derivative

Having established a covariant differential (1.2) of an arbitrary contravariant vector X^i we may now extend this process of differentiation to tensors of arbitrary rank. We define

$$
\begin{aligned}
D T^{i_1 \cdots i_r}{}_{j_1 \ldots j_s} =\; & d T^{i_1 \cdots i_r}{}_{j_1 \ldots j_s} + \\
& + \sum_{\mu=1}^{r} T^{i_1 \cdots i_{\mu-1} k i_{\mu+1} \cdots i_r}{}_{j_1 \ldots j_s} \left(C^{i_\mu}_{kh} d\dot{x}^h + \Gamma^{i_\mu}_{kh} d x^h \right) - \\
& - \sum_{\nu=1}^{s} T^{i_1 \cdots i_r}{}_{j_1 \ldots j_{\nu-1} k j_{\nu+1} \ldots j_s} \left(C^{k}_{j_\nu h} d\dot{x}^h + \Gamma^{k}_{j_\nu h} d x^h \right) .
\end{aligned}
\tag{2.1}
$$

Here, of course, it is to be noted that the tensor T is a function of the element of support $(x, \dot{x})$, so that the term dT involves the variation of the latter. If this fact is taken into account, it may easily be shown that this differentiation process again obeys the usual laws of covariant differentiation, namely that the differential of a sum is the sum of the differentials, while the product rule of ordinary differentiation holds. The covariant differential of a scalar is its ordinary differential.

The definition (2.1) may easily be re-written in a form which leads directly to the covariant (partial) derivatives with respect to a coordinate, say x^h. However, this is most clearly illustrated if we take a tensor of fairly low rank, the corresponding generalisation to tensors of arbitrary rank being obvious. Consider a tensor $T_{ij} = T_{ij}(x, \dot{x})$, for which equation (2.1) reads

$$
\begin{aligned}
D T_{ij} =\; & d T_{ij} - T_{kj} (C^{k}_{ih} d\dot{x}^h + \Gamma^{k}_{ih} d x^h) - \\
& - T_{ik} (C^{k}_{jh} d\dot{x}^h + \Gamma^{k}_{jh} d x^h) .
\end{aligned}
\tag{2.1$'$}
$$

In this expression we may replace the $d\dot{x}^h$ by the covariant differential of the unit vector l^i in the direction of the element of support. Thus we solve equations (1.16) for $d\dot{x}^h$, obtaining

$$
d\dot{x}^h = F\, D l^h + \dot{x}^h \left(\frac{dF}{F} \right) - \Gamma^{h}_{rs} \dot{x}^r d x^s ,
\tag{2.2}
$$

where we have made use of (1.15). When these expressions are substituted in (2.1$'$) we find that as a result of (1.12) the coefficient of $D l^h$ becomes

$$
- \{ T_{kj} (F C^{k}_{ih}) + T_{ik} (F C^{k}_{jh}) \}.
\tag{2.3}
$$

Thus, following Cartan, we shall define a new tensor A^{k}_{ih} as follows:

$$
A^{k}_{ih}(x, \dot{x}) = F(x, \dot{x})\, C^{k}_{ih}(x, \dot{x}) ,
\tag{2.4}
$$

so that in view of (1.11) and (1.14) we have[1]

$$A_{ijh} = g_{jk} A_{ih}^{k} = \tfrac{1}{4} F \frac{\partial^3 F^2}{\partial \dot{x}^i \partial \dot{x}^j \partial \dot{x}^h} . \tag{2.4'}$$

On the basis of this notation equation (2.1') in terms of the expression (2.3) reduces to

$$\begin{aligned} D T_{ij} &= d T_{ij} - (T_{kj} A_{ih}^{k} + T_{ik} A_{jh}^{k}) D l^h - \\ &\quad - (T_{kj} \Gamma_{ih}^{*k} + T_{ik} \Gamma_{jh}^{*k}) d x^h , \end{aligned} \tag{2.5}$$

where we have made use of equations (1.18).

However, if we wish to express the variation of the element of support in T_{ij} more explicitly, we substitute the identity

$$d T_{ij} = \frac{\partial T_{ij}}{\partial x^h} d x^h + \frac{\partial T_{ij}}{\partial \dot{x}^h} d \dot{x}^h$$

in (2.5). On applying (2.2) once more, we find after some further simplification[2]

$$D T_{ij} = \left(F \frac{\partial T_{ij}}{\partial \dot{x}^h} - T_{kj} A_{ih}^{k} - T_{ik} A_{jh}^{k} \right) D l^h + T_{ij|h} d x^h , \tag{2.5'}$$

where we have put

$$T_{ij|h} = \frac{\partial T_{ij}}{\partial x^h} - \frac{\partial T_{ij}}{\partial \dot{x}^k} \Gamma_{rh}^{k} \dot{x}^r - T_{kj} \Gamma_{ih}^{*k} - T_{ik} \Gamma_{jh}^{*k} . \tag{2.5''}$$

In view of (1.27') this may be written in the form

$$T_{ij|h} = \frac{\partial T_{ij}}{\partial x^h} - \frac{\partial T_{ij}}{\partial \dot{x}^k} \frac{\partial G^k}{\partial \dot{x}^h} - T_{kj} \Gamma_{ih}^{*k} - T_{ik} \Gamma_{jh}^{*k} . \tag{2.6}$$

Clearly these quantities are the components of a tensor, namely the covariant derivative of T_{ij} with respect to x^h. While the total variation of T_{ij} is given by (2.5'), the coefficients of $D l^h$ also forming the com-

---

[1] From this tensor we may construct a contracted tensor

$$A_i = A_{ik}^{k} = \tfrac{1}{2} F g^{hk} \frac{\partial g_{hk}}{\partial \dot{x}^i} , \tag{2.4a}$$

or

$$A_i = \tfrac{1}{2} F \frac{\partial}{\partial \dot{x}^i} (\log g) , \tag{2.4b}$$

where $g = \det|g_{ij}|$. Considerable attention had been paid to such Finsler spaces for which $A_i = 0$; for when g is a function of position only, an invariant measure of volume $\sqrt{g}\, d x^1 \ldots d x^n$ may be defined precisely as in Riemannian geometry [eqn. (1.8.11)]. While it was known that the only *two-dimensional Finsler spaces with $A_i = 0$ were in fact Riemannian spaces* (BERWALD [10, IV], p. 769), it was only recently that DEICKE [1], using the methods of relative affine geometry, showed that this result is *true for Finsler spaces of arbitrary dimension.*

[2] CARTAN [1], p. 17. It must be assumed, however, that $T_{ij}(x, \dot{x})$ is homogeneous of degree zero in the $\dot{x}^i$ so that $\dot{x}^k (\partial T_{ij}/\partial \dot{x}^k) = 0$. We also note that CARTAN denotes the expression $F (\partial T_{ij}/\partial \dot{x}^k)$ by $T_{ij||k}$; however, in order to avoid introducing too many abbreviations of this type which might confuse the reader we shall not follow his example.

ponents of a tensor, the expression $T_{ij|h}\,dx^h$ would represent the variation of T_{ij} if the element of support were transported by parallel displacement from the point x^i to the point $x^i + dx^i$.

In particular, we see that if $T_{ij} \equiv g_{ij}$, then both Dg_{ij} and $g_{ij|h}$ vanish identically. This, of course, follows directly from the construction of § 1, but may be verified directly by substitution in (2.5) and (2.6)[1].

Clearly the above construction of a covariant differential of a tensor in terms of the covariant differential of the unit vector in the direction of the element of support is applicable to tensors of arbitrary rank. For a contravariant vector $X^i = X^i(x, \dot{x})$ (again homogeneous of degree zero with respect to $\dot{x}^k$) we find similarly:

$$DX^i = \left(F\,\frac{\partial X^i}{\partial \dot{x}^h} + A^i_{kh}X^k\right)Dl^h + X^i_{|h}\,dx^h\,, \tag{2.7}$$

where we have written[2]

$$X^i_{|h} = \frac{\partial X^i}{\partial x^h} - \frac{\partial X^i}{\partial \dot{x}^k}\,\frac{\partial G^k}{\partial \dot{x}^h} + \Gamma^{*i}_{kh}X^k\,. \tag{2.8}$$

If we apply this formula to the unit vector l^i in the direction of the element of support, as given by (1.15), we observe first that

$$\frac{\partial l^i}{\partial \dot{x}^k} = [F(x, \dot{x})]^{-1}\left(\delta^i_k - \frac{\dot{x}^i}{F(x, \dot{x})}\,l_k\right)\,, \tag{2.9}$$

in view of (1.15′), so that

$$Dl^i = (\delta^i_k - l^i l_k)\,Dl^k + l^i_{|h}\,dx^h\,, \tag{2.10}$$

where we have taken into account (1.12) and (2.4). But since l^i is of unit length, we find on differentiation of the identity $g_{hk}(x, \dot{x})\,l^h l^k = 1$, that

$$l_k Dl^k = 0 \tag{2.11}$$

in consequence of $Dg_{hk}(x, \dot{x}) = 0$; on substituting this result in (2.10) we therefore have

$$l^i_{|h} = 0\,. \tag{2.12}$$

Furthermore, if we write

$$F_{|k} = \frac{\partial F}{\partial x^k} - \frac{\partial F}{\partial \dot{x}^l}\,\frac{\partial G^l}{\partial \dot{x}^k}\,,$$

a simple reduction based on equations (1.27′), (2.3.14) and (1.3.5) yields

$$F_{|k} = 0 \tag{2.13}$$

---

[1] This could be expected also from a purely geometrical point of view, since the element of support defines two osculating (i. e. quadric) indicatrices for the two neighbouring points in analogy with Riemannian geometry.

[2] If the vector $X^i(x, \dot{x})$ is positively homogeneous of degree p in the $\dot{x}^k$ the formula (2.7) must be replaced by

$$DX^i = \left(F\,\frac{\partial X^i}{\partial \dot{x}^h} + A^i_{kh}X^k\right)Dl^h + X^i_{|h}\,dx^h + pX^i\,\frac{dF}{F}\,. \tag{2.7a}$$

identically. Hence, for future reference, we deduce from (2.4) that

$$A_{ihj|k} = F C_{ihj|k} =$$
$$= F \left\{ \frac{\partial C_{ihj}}{\partial x^k} - \frac{\partial C_{ihj}}{\partial \dot{x}^l} \frac{\partial G^l}{\partial \dot{x}^k} - C_{ljh} \Gamma^{*l}_{ik} - C_{ilj} \Gamma^{*l}_{hk} - C_{ihl} \Gamma^{*l}_{jk} \right\}. \tag{2.14}$$

Also, using (1.3.5) and (1.3.6) we note that

$$A_{ihj|k} \dot{x}^i = -F \left(\frac{\partial C_{ihj}}{\partial \dot{x}^l} \frac{\partial G^l}{\partial \dot{x}^k} \dot{x}^i + C_{ljh} \Gamma^{*l}_{ik} \dot{x}^i \right).$$

But from the definition (1.3.4) we see that $\partial C_{ihj}/\partial \dot{x}^l = \partial C_{ljh}/\partial \dot{x}^i$, while the C_{ihj} are homogeneous of degree -1 in the $\dot{x}^k$. Thus from Euler's theorem and equation (1.27') it follows that

$$A_{ihj|k} \dot{x}^i = 0, \qquad C_{ijh|k} \dot{x}^i = 0. \tag{2.14a}$$

At this stage it may be instructive to compare this process of covariant differentiation with the δ-process described in the previous chapter. Again, for the purpose of comparison let us take a covariant tensor T_{ij}. Suppose, for the moment, that this tensor is a function of position only, so that $\partial T_{ij}/\partial \dot{x}^k = 0$. Then, since $P^{*i}_{hk} = \Gamma^{*i}_{hk}$, it follows immediately from the definitions (2.5.1) and (2.6) that $T_{ij|h} = T_{ij;h}$, *the significant difference being that in (2.5') the differential $D T_{ij}$ still depends on the variation $D l^h$ of the element of support, while the corresponding differential δT_{ij} depends only on the direction $\dot{x}^k$ in which the process of differentiation takes place.* On the other hand, if T_{ij} depends on a vector field ξ^k, $T_{ij} = T_{ij}(x, \xi)$, we may put $\xi^i = \dot{x}^i$ (element of support) at the point considered. Then we see from (2.5.1) that $T_{ij;h}$ does depend on the variation $\partial \xi^k/\partial x^h$ of this vector field, while $T_{ij|h}$ results from $T_{ij;h}$ if we put $\partial \xi^k/\partial x^h = -\partial G^k/\partial \dot{x}^h$ [1]. Thus in general we have

$$T_{ij;h} - T_{ij|h} = \frac{\partial T_{ij}}{\partial \dot{x}^k} \left(\frac{\partial \xi^k}{\partial x^h} + \frac{\partial G^k}{\partial \dot{x}^h} \right),$$

and since $\xi^i = \dot{x}^i$, we find by means of (1.27') and (2.4.8) that

$$T_{ij;h} - T_{ij|h} = \frac{\partial T_{ij}}{\partial \dot{x}^k} \xi^k_{;h}. \tag{2.15}$$

A further important special case arises when the direction of the displacement $d x^k$ from the point x^k to the point $x^k + d x^k$ coincides with the direction $\dot{x}^k$ of the element of support. In view of (1.30) and (1.12)

---

[1] This construction was in fact carried out independently of these considerations by RUND [5] on the basis of certain geometrical considerations in terms of non-integrable vector fields. However, it was pointed out in this paper that by assigning pre-arranged values to $\partial \xi^k/\partial x^h$ the derivative thus obtained would not, in general, give a correct description of the variation of the tensor field under consideration. This difficulty does not arise in the case of Cartan's differential (2.5) since the first expression on the right-hand side of (2.5) involving the variation $D l^h$ of the element of support contains the necessary compensating terms.

we then have

$$\Gamma_{kih}(x, \dot{x})\, d x^h = \gamma_{kih}(x, \dot{x})\, d x^h\,, \tag{2.16}$$

so that the covariant differential (1.2) of the vector X^i reduces to

$$D_c X^i = d X^i + C^i_{kh} X^k\, d \dot{x}^h + \gamma_{k\,h}^{\;i} X^k\, d x^h\,. \tag{2.17}$$

This is the type of covariant differential which one would obtain if one were to differentiate a vector field $X^i(t)$ along a given curve $C: x^i = x^i(t)$, the element of support being taken as $\dot{x}^i = d x^i/dt$ (hence the notation D_c). The expression (2.17) was found independently and almost simultaneously by Synge and Taylor[1]. Again, it is easy to compare the differential (2.17) with the δ-differential of Chapter II. From (2.3.14), (2.3.17) and (2.17) we have

$$D_c X^i - \delta X^i = C^i_{k\,h} X^k (d \dot{x}^h + \gamma_{j\,l}^{\;h} \dot{x}^j\, d x^l)\,,$$

or, noting (2.3.19),

$$D_c X^i - \delta X^i = C^i_{kh} X^k\, \delta \dot{x}^h\,. \tag{2.18}$$

This relation expresses the fact that when the transition between two neighbouring points x^i and $x^i + d x^i$ on C is considered, the corresponding differential $D_c X^i$ depends on the second derivative $\ddot{x}^i$ of the tangent vector of C, while this is not the case for δX^i. From the point of view of defining parallelism between the two tangent spaces $T_n(x^k)$ and $T_n(x^k + d x^k)$ such a dependence on the curve C may be regarded as being undesirable. Furthermore, in view of (1.12) *the differentials $D_c X^i$, δX^i will coincide if $\delta \dot{x}^i$ is of the form $\lambda \dot{x}^i$, i. e. if C is a geodesic*[2].

§ 3. The General Geometry of Paths: The Connection of Berwald

A well-known generalisation of Riemannian geometry is obtained by the transition to the so-called "geometry of paths" (sometimes referred to as the geometry of "non-Riemannian" spaces)[3]. This transition may

---

[1] Synge [1], Taylor [1]. The corresponding formula for covariant vectors was given by Subramanian [1].

Barthel [2] introduced a more complicated process of "partial" covariant differentiation: the symmetry condition (1.19) being replaced by the relation

$$\Gamma^*_{ikh} - \Gamma^*_{hki} = l_i A_{k|h} - l_h A_{k|i}\,,$$

where the latter derivatives refer to a process such as (2.8) with the new Γ^*_{ihk}. The explicit expression of these coefficients is not given. The purpose of the introduction of this derivative is the fact that under these circumstances hypersurfaces with vanishing second fundamental form (provided that these exist) are then minimal surfaces, the "area" being defined as in Riemannian geometry [eqn. (1.8.11)] on the basis of an element of support identical with the normal vector of the hypersurface.

[2] See the footnote to equation (2.5.8a), where an analogous problem is treated.

[3] See, for instance, T. Y. Thomas [1] or Eisenhart [2].

be described as follows: in a Riemannian space endowed with a metric tensor $g_{ij}(x^k)$ the geodesics are given by the differential equations

$$\frac{d^2 x^i}{ds^2} + \gamma_h{}^i{}_k(x^j)\,\frac{d x^h}{ds}\,\frac{d x^k}{ds} = 0 \, ,$$

the Christoffel symbols being functions of position only, these functions having been derived according to (2.2.3) and (2.2.7) from the $g_{ij}(x^k)$. Conversely, however, one may suppose that there is given a set of functions $h^i_{hk}(x^j)$ — called connection coefficients — depending on the positional coordinates x^j of a manifold X_n in which *no* metric is defined. It is assumed that these functions satisfy the transformation law

$$h^{i'}_{h'k'} = h^i_{hk} A^{i'}_i A^h_{h'} A^k_{k'} + (\partial_{h'} A^i_{k'})\, A^{i'}_i \, ,$$

so that the differential equations

$$\frac{d^2 x^i}{dt^2} + h^i_{hk}(x^j)\,\frac{d x^h}{dt}\,\frac{d x^k}{dt} = 0$$

are invariant, the left-hand side representing a contravariant vector. The curves $x^i = x^i(t)$ which satisfy these equations are called the "paths" of X_n, and the investigation of their properties is the object of the study of the "geometry of paths". The underlying manifold X_n endowed with these paths constitutes what is often called a "Non-Riemannian" space. If there exists a tensor $g_{ij}(x^k)$ from which the $h^i_{hk}(x^j)$ may be derived according to (2.2.3) and (2.2.7), the paths (with $t = s$) are simply the geodesics of the Riemannian metric defined by the $g_{ij}(x^k)$. However, in general such a tensor does not exist.

From the point of view of the type of generalisation which is suggested by the theory of Finsler spaces it is important to note that in the differential equations of the paths the $d^2 x^i/dt^2$ are given as polynomials of the second degree in the $d x^h/dt$. Hence the corresponding transition which suggests itself in the case of Finsler spaces involves the introduction of a system of differential equations of the type

$$\frac{d^2 x^i}{dt^2} + 2 H^i\left(x^j, \frac{d x^j}{dt}\right) = 0 \, , \tag{3.1}$$

where the functions $H^i(x^j, \dot{x}^j)$ are given functions of the line-elements of an underlying manifold X_n, these functions being positively homogeneous of the second degree in the $\dot{x}^j$. Naturally it is also assumed that the transformation properties of the H^i are such as to render the equations (3.1) invariant. The curves $x^i = x^i(t)$ which satisfy equations (3.1) are the "paths" of the manifold, the manifold endowed with these paths constituting a *general space of paths*. These spaces, which had originally been introduced by Douglas[1], represent a direct generalisation of

---

[1] Douglas [1], also Knebelman [1].

Finsler spaces, for if there exists a metric function $F(x, \dot{x})$ and hence a metric tensor $g_{ij}(x, \dot{x})$ such that the functions $H^i(x, \dot{x})$ are equivalent to the functions $G^i(x, \dot{x})$ [as defined by (1.26)], it follows from (2.2.8) that the equations (3.1) with $t = s$ represent the geodesics of the metric $F(x, \dot{x})$. In general, of course, this reduction is not possible.

As a result of the homogeneity assumption concerning the functions $H^i(x, \dot{x})$ it is evident that the equations (3.1) may be written in the form

$$\frac{d^2 x^i}{dt^2} + H_{hk}\left(x^j, \frac{dx^j}{dt}\right) \frac{dx^h}{dt} \frac{dx^k}{dt} = 0 , \tag{3.2}$$

where we have put

$$H^i_{hk}(x, \dot{x}) = \frac{\partial^2 H^i(x, \dot{x})}{\partial \dot{x}^h \, \partial \dot{x}^k} , \tag{3.3}$$

for, by Euler's theorem, we have

$$2 H^i(x, \dot{x}) = H^i_{hk}(x, \dot{x}) \, \dot{x}^h \, \dot{x}^k .$$

We note that the H^i_{hk} are homogeneous of degree zero in the $\dot{x}^i$ and symmetric in their lower indices.

In order that the equations (3.1) be invariant, we have to stipulate that the functions $H^i(x, \dot{x})$ transform according to

$$2 H^{i'} = 2 H^i A^{i'}_i - (\partial_i A^{i'}_j) \, \dot{x}^i \, \dot{x}^j . \tag{3.4}$$

If this condition is satisfied, it is easily verified[1] that the "connection coefficients" (3.3) satisfy the desired transformation rule:

$$H^{i'}_{h'k'} = H^i_{hk} A^{i'}_i A^h_{h'} A^k_{k'} + (\partial_{h'} A^i_{k'}) A^{i'}_i . \tag{3.5}$$

The H^i_{hk} thus satisfy the same transformation law as the P^{*i}_{hk} (or Γ^{*i}_{hk}), as is evident from (2.4.6a). This is to be expected. Thus in the general geometry of paths the H^i_{hk} play the same role as the P^{*i}_{hk} (or Γ^{*i}_{hk}) of the metric theory.

---

[1] This follows by direct differentiation of (3.4) with respect to $\dot{x}^{h'}$, $\dot{x}^{k'}$, the second term on the right-hand side being a consequence of (2.3.3). Alternatively, equation (3.5) is a direct consequence of the following useful lemma which is due to Knebelman ([1], p. 529): Given a set of functions $f^{(\alpha)}_{(\beta) i_1 i_2 \ldots i_r}(\dot{x})$ which are homogeneous of degree zero in the n variables $\dot{x}^1, \dot{x}^2, \ldots, \dot{x}^n$, symmetric in the indices $i_1, i_2, \ldots, i_r$, while (α) and (β) represent any additional sets of indices. Suppose further that

$$f^{(\alpha)}_{(\beta) i_1 i_2 \ldots i_r}(\dot{x}) \, \dot{x}^{i_1} \dot{x}^{i_2} \ldots \dot{x}^{i_r} = 0 ,$$

$$\frac{\partial f^{(\alpha)}_{(\beta) i_1 i_2 \ldots i_r}}{\partial \dot{x}^p} = \frac{\partial f^{(\alpha)}_{(\beta) p i_2 \ldots i_r}}{\partial \dot{x}^{i_1}} ;$$

then

$$f^{(\alpha)}_{(\beta) i_1 i_2 \ldots i_r}(\dot{x}) = 0 .$$

This lemma is easily established by repeated differentiation and the application of Euler's theorem.

Should the H^i_{hk} be independent of direction, we are confronted with the so-called "restricted geometry of paths", i. e. a non-Riemannian geometry.

The concept of the general geometry of paths has proved to be a most fruitful one, and in subsequent chapters we shall frequently return to this idea. But even in the metric case the fundamental approach as exemplified by (3.3) is of special significance, as it leads us directly to the connection coefficients used by BERWALD.

Returning, then, to a Finsler space F_n, we note that in virtue of equations (2.2.8) and (1.26) the differential equations of the geodesics may be written in the form

$$\frac{d^2 x^i}{ds^2} + 2 G^i\left(x^j, \frac{dx^j}{ds}\right) = 0 ,$$

the parameter s representing the arc-length. Thus equations (3.3) suggest that we define new connection coefficients by putting

$$G^i_{hk}(x, \dot{x}) = \frac{\partial^2 G^i(x, \dot{x})}{\partial \dot{x}^h \, \partial \dot{x}^k} . \tag{3.6}$$

In fact, these are the coefficients which were used in the earliest treatments of Finsler spaces (regarded as spaces of line-elements) from the point of view of differential invariants by E. NOETHER[1]. The geometrical theory based on the connection (3.6) was developed in great detail by BERWALD[2], especially the two-dimensional case[3].

It is a simple matter to establish the relationship between the coefficients (3.6) and the coefficients Γ^{*i}_{hk}. For from (3.6) and (1.27) we have

$$g_{hk} G^k_{ij} = g_{hk} \frac{\partial}{\partial \dot{x}^j} (P^k_{ir} \dot{x}^r)$$

$$= g_{hk}\left\{\frac{\partial}{\partial \dot{x}^j} (g^{ks} P_{isr}) \dot{x}^r + P^k_{ij}\right\},$$

or, using (1.5.9),

$$G_{ihj} = P_{ihj} + \frac{\partial}{\partial \dot{x}^j} (P_{ihr}) \dot{x}^r - 2 C_{khj} P^k_{ir} \dot{x}^r . \tag{3.7}$$

This relation may be simplified by means of (2.5.14); noting once more that $\Gamma^{*h}_{ij} = P^{*h}_{ij}$, we thus find

$$G_{ihj} = \Gamma^*_{ihj} + C_{ihj|k} \dot{x}^k , \tag{3.8}$$

or, alternatively, if we use (2.14) and (1.15),

$$G_{ihj} = \Gamma^*_{ihj} + A_{ihj|k} l^k . \tag{3.8a}$$

---

[1] NOETHER [1].

[2] BERWALD [1, 2, 4]. Expository articles dealing with this development are due to KOSCHMIEDER [1], WINTERNITZ [1], BERWALD [13]. Certain aspects of the metric theory of BERWALD in relation to the work of DOUGLAS [1] are discussed by SLEBODZINSKI in an article [1] based on the "kinematic group" (WUNDHEILER [2]).

[3] BERWALD [5, 6]. It should be stressed that Berwald's theory is also based on the use of the element of support.

This equation was first given by Cartan[1]; as was pointed out by him, the connection coefficients (3.6) involve 4th order derivatives of the metric function while the Γ^{*}_{ihj} involve only derivatives up to the third order.

Also, from (2.14a) and (3.8a) we see that

$$G_{ihj}\,\dot{x}^i = \Gamma^{*}_{ihj}\,\dot{x}^i\,, \tag{3.8b}$$

and therefore the law of parallel displacement of the element of support is the same for both connections.

Berwald defines partial covariant derivatives in a manner analogous to that of Cartan as exemplified by equation (2.6), the only difference being that the $\Gamma^{*\,i}_{h\,k}$ are replaced by the $G^i_{h\,k}$. In this case, too, the partial covariant derivatives of F and of the unit vector l^i in the direction of the element of support vanish identically[2] [as in equations (2.12) and (2.13)]. In order to avoid confusion with the notation used above, we denote such covariant derivatives with respect to, say x^k, by a suffix k in brackets, so that the above statement reads

$$F_{(k)} = 0\,,\qquad l^i_{(k)} = 0\,. \tag{3.9}$$

Furthermore, Berwald's covariant derivative of the metric tensor is defined by

$$g_{ij(k)} = \frac{\partial g_{ij}}{\partial x^k} - \frac{\partial g_{ij}}{\partial \dot{x}^l}\frac{\partial G^l}{\partial \dot{x}^k} - G_{ijk} - G_{jik}\,, \tag{3.10}$$

and on comparing this result with equation (3.8a) and noting that $g_{ij|h} = 0$, it follows from the symmetry of the A_{ihk} that[3]

$$g_{ij(k)} = -2A_{ijk|h}\,l^h\,. \tag{3.11}$$

---

[1] Cartan [1], formula XV, p. 19. Further comparisons between the two connections are made by Davies [2], p. 262, Hosokawa [3], Laptew [2] and Berwald [7, 9, 10], who also derives relations between the curvature tensors resulting from the two connections. (See Ch. IV). Sasaki [1] investigates the relationship between the coefficients (3.6) associated with special metric functions in connection with his generalisation of projective differential geometry of curved spaces as discussed by Veblen [1].

[2] Berwald [2], equations (25) and (35).

[3] Berwald [9], equation (5). See also Schouten and Haantjes [1], § 5. Equation (3.11) is actually a special case of the following lemma. Suppose we are given two connection coefficients such that

$$\overset{(1)}{\Gamma^i_{hk}} = \overset{(2)}{\Gamma^i_{hk}} + T^i_{hk},$$

where T^i_{hk} is some tensor. The difference between the corresponding covariant derivatives of the metric tensor g_{ij} with respect to x^h [defined as for (3.10)] is then given by $T_{ijh} + T_{jih}$. Thus if one of the connections is "metric" [i. e. the corresponding expression (3.10) vanishes identically] the other will be metric if and only if T_{ijh} is skew-symmetric in the first two indices. For Riemannian spaces this result was pointed out by Nalli [1] and generalised later by Kawaguchi [1].

Those Finsler spaces for which the G^i_{hk} are independent of the directional arguments $\dot{x}^i$ are called by BERWALD "*affinely-connected spaces*"[1]. In order to find necessary and sufficient conditions which must be satisfied in order that this requirement be fulfilled we have to evaluate the derivatives of the G^h_{ij} or Γ^{*h}_{ij} with respect to direction.

Let us write

$$Q_{ikj} = -\frac{1}{2}\frac{\partial g_{ij}}{\partial x^k} + C_{ijh}\frac{\partial G^h}{\partial \dot{x}^k}. \tag{3.12}$$

On differentiating this expression with respect to $\dot{x}^r$, we find that in view of (1.3.4), (2.14), (3.6) and (3.8)

$$\frac{\partial Q_{ikj}}{\partial \dot{x}^r} + C_{ijr|k} = C_{ijh}C^h_{kr|l}\,\dot{x}^l - C_{hjr}\,\Gamma^{*h}_{ik} - C_{ihr}\,\Gamma^{*h}_{jk}.$$

By cyclic interchange of the indices i, k, j we obtain two similar equations. These latter we add and subtract the result from the first equation; and on observing (3.12), (2.4.4) (with $P^{*i}_{kh} = \Gamma^{*i}_{kh}$) we thus obtain

$$\begin{aligned}
C_{jkr|i} + C_{kir|j} - C_{ijr|k} - \frac{\partial \Gamma^*_{ikj}}{\partial \dot{x}^r} &= \\
&= (C_{jkh}C^h_{ir|l} + C_{kih}C^h_{jr|l} - C_{ijh}C^h_{kr|l})\,\dot{x}^l - 2C_{hkr}\,\Gamma^{*h}_{ij}.
\end{aligned} \tag{3.13}$$

But if we differentiate the identity (1.24) with respect to $\dot{x}^r$, noting (1.3.4) once more, equation (3.13) reduces to

$$\begin{aligned}
\frac{\partial \Gamma^{*h}_{ij}}{\partial \dot{x}^r} &= C^h_{jr|i} + C^h_{ir|j} - g^{hk}C_{ijr|k} \\
&\quad - (C^h_{jl}C^l_{ir|k} + C^h_{il}C^l_{jr|k} - C^l_{ij}C^h_{lr|k})\,\dot{x}^k.
\end{aligned} \tag{3.14}$$

In particular, we deduce from this equation and (1.3.5), (2.14) with (2.14a) that

$$\frac{\partial \Gamma^{*h}_{ij}}{\partial \dot{x}^r}\,\dot{x}^j = C^h_{ir|j}\,\dot{x}^j = \frac{\partial \Gamma^{*h}_{rj}}{\partial \dot{x}^i}\,\dot{x}^j, \tag{3.15}$$

so that we may write (3.8a) in the form

$$G^k_{ij} = \Gamma^{*k}_{ij} + \frac{\partial \Gamma^{*k}_{ir}}{\partial \dot{x}^j}\,\dot{x}^r. \tag{3.16}$$

Thus if the metric of the space is such that $C_{ijk|h} = 0$ identically, it follows from (3.8) that $\Gamma^{*i}_{hk} = G^i_{hk}$, while the directional derivatives of the G^i_{hk} vanish in view of (3.14). *Thus the "affinely-connected spaces" are characterised by the condition* $C_{ijk|h} = 0$. Conversely, one may

---

[1] BERWALD [2], p. 47. In contrast to this VAGNER [1, 4] calls such spaces "Berwald spaces" and shows that the determination of such spaces is related to the problem of determining all hypersurfaces which admit a transitive subgroup of centrally affine transformations. In the paper VAGNER [1] special cases are discussed.

show[1] that the equations $\partial G^i_{hk}/\partial \dot{x}^l = 0$ imply this condition and hence $\partial \Gamma^{*i}_{hk}/\partial \dot{x}^l = 0$. *The geometrical significance of these special spaces is due to the fact that in virtue of equations (1.17) the parallel displacement of a vector (in the sense of* Cartan*) is independent of its initial element of support, provided the latter is displaced parallel to itself*[2].

§ 4. Further Connections Arising from the General Geometry of Paths

Let us return to the non-metric case for the moment, namely to the manifold X_n whose paths are given by the differential equations (3.1). By means of the coefficients (3.3) a parallel displacement of a vector X^i from the point x^i to the point $x^i + dx^i$ may be defined as usual by the vanishing of the corresponding covariant derivative, i. e. we regard the equations

$$dX^i = -H^i_{hk}(x, \dot{x}) X^h dx^k \qquad (4.1)$$

as the conditions for such a displacement. Here the $\dot{x}^i$ (the directional argument in the H^i_{hk}) represent the element of support. This would, in fact, be equivalent to the parallel displacement according to the covariant derivative of Berwald (§ 3) in the metric case. However, following a method due to Laugwitz, we may also define — in various ways — a parallel displacement independent of the element of support. If we write

$$H^i_k(x, \dot{x}) = \frac{\partial H^i(x, \dot{x})}{\partial \dot{x}^k}, \qquad (4.2)$$

these functions are positively homogeneous of degree one in the $\dot{x}^k$ in view of the homogeneity of the functions H^i, so that we have by (3.3)

$$H^i_{kj}(x, \dot{x}) \dot{x}^j = H^i_k(x, \dot{x}) . \qquad (4.3)$$

We may now define the parallel displacement of a vector X^i from the point x^i to the point $x^i + dx^i$ — without taking recourse to an element of support — by putting either[3]

$$dX^i = -H^i_k(x, dx) X^k , \qquad (4.4a)$$

---

[1] Berwald [9], § 1. The proof of this statement is not difficult but somewhat lengthy, depending on the basic fact that it is possible to represent the tensor $C_{ihj|k}$ in terms of linear combinations of the 3rd and 4th derivatives of the G^i with respect to the directional arguments. The same theorem (but in a slightly different form) is developed by Vagner [13], p. 123, the proof being based on the affine differential geometry of the indicatrix.

[2] Cartan [1], p. 38.

[3] Laugwitz [3], p. 23. The basic idea of this method is similar to one described briefly by Bompiani [1]. A general geometry of paths based on the non-linear displacement (4.4b) had already been discussed by Mikami [1] in an extension of the work of Bortolotti [1, 3] and Friesecke [1]. See also Schouten [2] and Kawaguchi [3].

or, alternatively,

$$dX^i = -H^i_k(x, X)\, dx^k .\qquad(4.4\,\mathrm{b})$$

If we introduce a Finsler metric in X_n such that the paths (3.1) correspond to the geodesics of this metric, we may put $H^i(x, \dot{x}) = G^i(x, \dot{x})$. In view of (4.2), (1.27) and (2.3.18) it then follows that the parallel displacement (4.4a) is simply the parallel displacement corresponding to the process of δ-differentiation described in § 3 of Chapter II.

On the other hand, we may deduce in the same manner that the alternative parallelism defined by (4.4b) leads to

$$dX^i = -P^i_{hj}(x, X)\, X^j\, dx^h ,\qquad(4.5)$$

where, by (2.3.14),

$$P^i_{hj}(x, X) = \gamma_h{}^i{}_j(x, X) - C^i_{hl}(x, X)\, \gamma_p{}^l{}_j(x, X)\, X^p .\qquad(4.5\,\mathrm{a})$$

These coefficients correspond to a process of covariant differentiation introduced by BARTHEL[1]. It is seen that the corresponding covariant differential of X^i is non-linear in X^i, although it is homogeneous of degree 1 in the latter. An important feature of this derivative — in contrast to the δ-process — is the fact that the length of a vector remains invariant under parallel displacement (as in Riemannian geometry). This follows directly from the fact that in virtue of (1.3.5), (2.3.19) and (2.2.10) we have

$$X^i X^j\left[dg_{ij}(x, X) - g_{ih}(x, X)\, P^h_{kj}(x, X)\, dx^k - g_{hj}(x, X)\, P^h_{ki}(x, X)\, dx^k\right] =$$

$$= X^i X^j\left[\frac{\partial g_{ij}(x, X)}{\partial x^k} - \gamma_{jik}(x, X) - \gamma_{ijk}(x, X)\right] dx^k = 0 .$$

It was, in fact, this fundamental property which led BARTHEL to consider such non-linear connections. We shall return to this point in the next section[2].

Again, the Γ^{*i}_{hk} (or P^{*i}_{hk}) may also be derived in a very elegant manner due to SULANKE [1, 2] from the equations (3.2) of the paths. In this method, however, we do not assume that the coefficients H^i_{hk} result from differentiation of a given function H^i, but we suppose that the H^i_{hk} have been given initially. In X_n we define a covariant derivative in the usual manner with respect to the H^i_{hk}. For instance, if $a_{ij}(x, \dot{x})$ is a covariant tensor of rank 2, homogeneous of degree zero in the $\dot{x}^i$, we would write[3]

$$\Delta_j a_{ik} = \frac{\partial a_{ik}}{\partial x^j} - \frac{\partial a_{ik}}{\partial \dot{x}^l}\, H^l_{pj}\, \dot{x}^p - a_{hk}H^h_{ij} - a_{ih}H^h_{kj} .\qquad(4.6)$$

---

[1] BARTHEL [3, 4],

[2] For a more general discussion of non-linear connections see Ch. VI, § 4. A general theory of such connections is developed by VAGNER [13].

[3] Compare, for instance, equation (2.6).

If the tensor a_{ij} as well as the coefficients H^h_{ik} are symmetric in their lower indices, it follows from (4.6) that[1]

$$H^h_{ki} = a^{jh} \left\{ \frac{\partial a_{ij}}{\partial x^k} + \frac{\partial a_{kj}}{\partial x^i} - \frac{\partial a_{ik}}{\partial x^j} - \right.$$
$$\left. - \frac{1}{2} \left(\frac{\partial a_{ij}}{\partial \dot{x}^l} H^l_{pk} + \frac{\partial a_{kj}}{\partial \dot{x}^l} H^l_{pi} - \frac{\partial a_{ik}}{\partial \dot{x}^l} H^l_{pj} \right) \dot{x}^p \right\} - \qquad (4.7)$$
$$- \tfrac{1}{2} a^{jh} \{ \Delta_k a_{ij} + \Delta_i a_{kj} - \Delta_j a_{ik} \} .$$

Let us now impose a Finsler metric once more, and let us determine the — thus far arbitrary — coefficients H^i_{kh} according to the following conditions: (a) The H^h_{ki} must be symmetric in k, i; (b) the covariant derivative of the metric tensor g_{ik} defined with respect to the H^h_{ki} according to (4.6) must vanish identically.

Applying equation (4.7) to the tensor g_{ij} [which is permissible in view of condition (a)], it follows from (b) that under these circumstances we have

$$H^h_{ki} = g^{jh} \{ \gamma_{ijk} - (C_{ijl} H^l_{pk} + C_{kjl} H^l_{pi} - C_{ikl} H^l_{pj}) \dot{x}^p \} , \qquad (4.8)$$

where we have made use of (1.3.4) and (2.2.3). In view of (2.2.7) we also deduce from (4.8) and (1.3.5)

$$H^i_{hk} \dot{x}^h \dot{x}^k = \gamma_h{}^i{}_k \dot{x}^h \dot{x}^k . \qquad (4.9)$$

Thus on multiplying (4.8) by $\dot{x}^k$ and noting (1.3.5), we have

$$H^h_{ki} \dot{x}^k = \gamma_i{}^h{}_k \dot{x}^k - C^h_{il} \gamma_p{}^l{}_k \dot{x}^p \dot{x}^k = P^h_{ik} \dot{x}^k$$

in virtue of (2.3.14). Hence from (2.4.4) and (4.8) it follows that $H^h_{ki} = P^{*h}_{ki} = \Gamma^{*h}_{ki}$. We therefore see that the conditions (a) and (b) imposed on the paths are sufficient to specify completely suitable connection coefficients, namely those which define covariant δ-differentiation[2].

§ 5. The Osculating Riemannian Space

We shall devote this section to a discussion of a method of great simplicity which may be applied to various topics in Finsler geometry and which has, in particular, been used for the purpose of deriving and comparing the various covariant derivatives. The basic idea of the osculating Riemannian space originated in the thesis of Nazim[3]. The latter author did not derive explicit expressions for the coefficients of the connection used by him, but it is clear from his developments that he was dealing with the covariant derivative of Taylor and Synge, which, as we have seen in § 2, is a special case of the covariant derivative of Cartan. It was Varga who undertook a systematic investigation of the osculating Riemannian spaces, from which he then derived the connection of Cartan in a most elegant manner. A similar method was

---

[1] Compare Norden [1], p. 135. In fact, (4.7) may be deduced by cyclic interchange of the indices i, k, j in (4.6), which gives two new equations from the sum of which (4.6) may be subtracted. On simplifying the result we find (4.7).

[2] It should be noted, however, that the derivative (4.6) is not in general identical with the corresponding δ-derivative, the difference between the two processes being due to different coefficients of the terms involving $\partial a_{ij}/\partial \dot{x}^l$.

[3] Nazim [1].

then devised by LAUGWITZ [2] in a comparative study of the more recent definitions of covariant derivatives.

Let C be a curve of the Finsler space F_n, referred to its arc-length s as parameter: $x^i = x^i(s)$, and let us denote the unit tangent vectors $x'^i = dx^i/ds$ by $\xi^i(s)$. We shall suppose these vectors to be imbedded locally in a vector field $\xi^i(x^k)$ defined over a region R of F_n. Then a Riemannian metric $\gamma_{ij}(x)$ is defined over R by means of the relations

$$\gamma_{ij}(x) = g_{ij}(x, \xi(x)) . \tag{5.1}$$

This region is called the "osculating Riemannian manifold" along the curve C. But in R we may use the ordinary covariant derivative of Riemannian geometry[1] arising out of the parallelism of LEVI-CIVITA; if $X^i(x^k)$ is a vectorfield defined over R, we may write

$$D_R X^i = dX^i + \{_k{}^i{}_j\} X^k dx^j , \tag{5.2}$$

where the $\{_k{}^i{}_j\}$ are the Christoffel symbols of the second kind defined with respect to the γ_{ij} in the usual manner, i. e.

$$\{_k{}^i{}_j\} = \tfrac{1}{2} \gamma^{ih} \left(\frac{\partial \gamma_{jh}}{\partial x^k} + \frac{\partial \gamma_{kh}}{\partial x^j} - \frac{\partial \gamma_{kj}}{\partial x^h} \right) . \tag{5.2a}$$

The tensor character of the expression (5.2) is assured as a consequence of the transformation law of these symbols. If we substitute in (5.2a) from (5.1), we find in view of (2.2.3) and (1.3.4)

$$\begin{aligned} \{_k{}^i{}_j\} = {}&\gamma_k{}^i{}_j(x, \xi) + C^i_{jl}(x, \xi) \frac{\partial \xi^l}{\partial x^k} + C^i_{kl}(x, \xi) \frac{\partial \xi^l}{\partial x^j} - \\ &- g^{ih}(x, \xi) C_{kjl}(x, \xi) \frac{\partial \xi^l}{\partial x^h} . \end{aligned} \tag{5.3}$$

Thus if we use (5.2) together with (5.3) in order to define a covariant derivative in F_n, we see that the latter will depend on the choice of the vector field $\xi^h(x^k)$, since the partial derivatives of the field enter the expression (5.3). On the other hand, if we differentiate along the curve C itself, we may apply the homogeneity relation (1.3.5) to the second term on the right-hand side of (5.2), thus obtaining

$$D_R X^i = dX^i + \gamma_k{}^i{}_j(x, \xi) X^k dx^j + C^i_{kl}(x, \xi) \frac{d^2 x^l}{ds^2} X^k ds . \tag{5.4}$$

We observe that this differential depends on the second derivatives along the curve C, although it is otherwise independent of the choice of the vector field $\xi^i(x^k)$.

As is customary in the "locally Minkowskian" theory (viz. the theory independent of the notion of the element of support), one would demand that the differential of X^i corresponding to a displacement from the point x^i to the point $x^i + dx^i$ should depend only on X^i and dx^i (apart

---

[1] EISENHART [1], p. 62.

from the metric tensor and its derivatives which must of necessity enter any such expression). This may easily be achieved by specialisation of (5.2) and (5.3) as was shown by Laugwitz[1]. In fact, if C is chosen to be the geodesic through the point x^i whose tangent vector at x^i coincides with ξ^i, we obtain from (5.4) the δ-process (§ 3, Ch. II). On the other hand, if we imbed the vector X^i in a vector field $X^i(x^k)$ and replace the directional arguments ξ^i in (5.1) by the X^i, we construct another osculating Riemannian space corresponding to this vector field. From the equations corresponding to (5.3) and the relations (1.3.5) we see that the new covariant derivative resulting from this construction is given by

$$dX^i + \gamma_k{}^i{}_j(x, X)\, X^k\, d x^j + C^i_{jl}(x, X)\, \frac{\partial X^l}{\partial x^k}\, X^k\, d x^j. \tag{5.5}$$

Clearly this expression is still dependent on the vector field $X^i(x^k)$ in view of the derivatives $\partial X^i/\partial x^k$. But we may replace the latter by means of the (in general non-integrable) equations

$$\frac{\partial X^l}{\partial x^k}\, X^k = -\, \gamma_k{}^l{}_j(x, X)\, X^k X^j. \tag{5.6}$$

When this has been done, the expression (5.5) is uniquely determined and does, in fact, represent the covariant differential of Barthel (§ 4). From a geometrical point of view equation (5.6) indicates that the field vector X^k at the point with coordinates $x^i + X^i\, ds$ results from the field vector X^k at x^i by parallel displacement according to (5.5).

Furthermore, the Γ^{*i}_{hk} of Cartan may be obtained directly from (5.3) as follows. Let us form the partial covariant derivative of a vector field $X^i(x^k)$ with respect to the symbols (5.3) and let us suppose — following Varga[2] — that the vector field ξ^i is so chosen that the corresponding covariant derivatives of the ξ^i vanish identically at the point under consideration, i. e. that

$$\frac{\partial \xi^i}{\partial x^j} + \{_k{}^i{}_j\}\, \xi^k = 0. \tag{5.7}$$

In view of (1.3.5) and (5.3) this is equivalent to

$$\frac{\partial \xi^i}{\partial x^j} = -\, \gamma_k{}^i{}_j(x, \xi)\, \xi^k - C^i_{jl}(x, \xi)\, \frac{\partial \xi^l}{\partial x^k}\, \xi^k. \tag{5.8}$$

But from (5.7) and (1.3.5) we also have

$$\frac{\partial \xi^l}{\partial x^k}\, \xi^k = -\, \{_k{}^l{}_h\}\, \xi^k\, \xi^h = -\, \gamma_k{}^l{}_h\, \xi^k\, \xi^h.$$

On substituting these expressions in (5.8) we obtain

$$\frac{\partial \xi^i}{\partial x^j} = -\, P^i_{jk}(x, \xi)\, \xi^k, \tag{5.9}$$

---

[1] [2], p. 450—452.
[2] Varga [2], p. 172. See also Varga [6].

in consequence of definition (2.3.14). We may now replace the derivatives according to relation (5.9) in (5.3), and on observing the definition (2.4.4) we see that we finally have

$$\{_j{}^i{}_k\} = P^{*i}_{jk} = \Gamma^{*i}_{jk} \tag{5.10}$$

for the initial line-element (x, ξ).

In essence the above construction is the one used by VARGA [2] in his derivation of the Γ^{*i}_{jk}. We refer the reader to this paper for a more precise description of the properties of osculating Riemannian manifolds. Similarly, it is not difficult to derive the coefficients G^i_{jk} of BERWALD (§ 3) in this manner[1].

In conclusion, we may summarise the results of the last three sections as follows: The covariant derivatives of BERWALD, BARTHEL and of the author may be derived by means of the method originating from the general geometry of paths, while this is not true of the covariant derivatives of TAYLOR, SYNGE and CARTAN. However, all these covariant derivatives may be constructed by means of the method of the osculating Riemannian space, although in some cases such a construction may appear to be somewhat artificial. The reader is referred to a lucid tabulated description of these relationships as summarised by LAUGWITZ[2].

§ 6. Normal coordinates

It is well-known that in Riemannian geometry and in the "restricted" geometry of paths [i. e. the geometry of paths for which the coefficients (3.3) are independent of direction] the use of normal coordinates has played an important part in the development of the subject. We recall that these coordinates are defined as follows: Let O be a fixed point, called the origin, and consider the family of all paths through O. The normal coordinates relative to the given coordinate system (x^i) and origin O are coordinates z^i, which are analytic functions of the x^i, such that the paths through O are given by linear equations of the type

$$z^i = a^i\, t + b^i, \qquad (a^i, b^i \text{ constant}),$$

where $z^i = 0$ at O. These conditions determine z^i up to homogeneous linear transformations.

It is natural to inquire as to whether or not it is possible to introduce normal coordinates in the general geometry of paths and hence in Finsler spaces. We shall, in the present section, concern ourselves with this problem, and, in particular, we shall deal with it from the more general point of view of the general geometry of paths, for clearly our results will be directly applicable to Finsler spaces. We shall see that strictly

---

[1] LAUGWITZ [3], p. 26.

[2] LAUGWITZ [3], p. 25.

speaking the answer to the above question is negative, but that it will be possible to construct coordinate systems which possess some of the properties usually required of normal coordinates.

Let us consider then a path Γ, given by (3.1), through a fixed origin O with coordinates $x^i_{(0)}$, its tangent at O being denoted by $\dot{x}^i_{(0)} = [d x^i/d t]_{t=t_0}$, where t_0 is the value of the parameter t of Γ at O[1]. Differentiating (3.1) with respect to t, we find

$$\frac{d^3 x^i}{d t^3} + 2 H^i_3\left(x, \frac{d x}{d t}\right) = 0 \, ,$$

where

$$H^i_3\left(x, \frac{d x}{d t}\right) = \frac{\partial H^i}{\partial x^k} \frac{d x^k}{d t} - 2 \frac{\partial H^i}{\partial \dot{x}^k} H^k \, ,$$

the last term resulting from an application of (3.1). Clearly $H^i_3(x^k, \dot{x}^k)$ is positively homogeneous of degree 3 in the $\dot{x}^k$, and by induction we conclude that generally[2]

$$\frac{d^m x^i}{d t^m} + 2 H^i_m\left(x, \frac{d x}{d t}\right) = 0 \tag{6.1}$$

along Γ, where $H^i_m\left(x^k, \frac{d x^k}{d t}\right)$ is positively homogeneous of degree m in the $\frac{d x^k}{d t}$.

The parametric equations $x^i = x^i(t)$ of Γ are determined by the sequence (6.1), which results from assigning the values 2, 3, ... to m, in conjunction with the given initial values at $t = t_0$. We have, in fact

$$x^i(t) = x^i_{(0)} + \dot{x}^i_{(0)}(t - t_0) - \sum_{m=2}^{\infty} \frac{2}{m!} H^i_m(x^j_{(0)}, \dot{x}^j_{(0)})(t - t_0)^m \, . \tag{6.2}$$

We remark that this power series is convergent for sufficiently small values of $|t - t_0|$. Thus in a given region containing O the path is uniquely determined by the direction $\dot{x}^i_{(0)}$ through O. Putting

$$z^i = (t - t_0)\, \dot{x}^i_{(0)} \, , \tag{6.3}$$

it follows from the homogeneity properties of the functions H^i_m that the series (6.2) may be written in the form

$$x^i(t) = x^i_{(0)} + z^i - \sum_{m=2}^{\infty} \frac{2}{m!} H^i_m(x^j_{(0)}, z^j) \, . \tag{6.4}$$

---

[1] Throughout this section we shall suppose that the function $H^i(x, \dot{x})$ of (3.1) is analytic in all its arguments.

[2] For the sake of uniformity of notation, we shall write H^i_2 for H^i.

We may regard this equation as a transformation defining the new variables z^i. Its Jacobian at O is

$$\det \left| \frac{\partial x^i}{\partial z^j} \right|_{t=t_0} = |\delta^i_j| = 1 \,. \tag{6.5}$$

But in general the transformation is not analytic at O. For if we evaluate the second derivatives, we find

$$\frac{\partial^2 x^i(t)}{\partial z^j \, \partial z^k} = - \frac{\partial^2 \underset{2}{H^i}(x, z)}{\partial z^j \, \partial z^k} - \frac{1}{3} \frac{\partial^2 \underset{3}{H^i}(x, z)}{\partial z^j \, \partial z^k} - \cdots \,,$$

and, when $t \to t_0$, each term on the right-hand side, with the exception of the first, tends to zero, since these terms are positively homogeneous of degree 1, 2, 3, ... in the z^i. But the limit of the first term is indeterminate, for in view of its dependence on the z^i its limiting value will in general depend on the direction of approach. Obviously the only case in which this difficulty does not occur arises when the second derivatives of the H^i are independent of direction. But by (3.3) this implies that the theory is then reduced to the "restricted" geometry of paths. We may summarise this state of affairs as follows:

Let O denote an arbitrary fixed point in a space of paths referred to the coordinate system (x^i). If and only if the space of paths is of the restricted type does there exist a system of coordinates (z^i) such that (1) the transformation from x^i to z^i is analytic at O and (2) the paths through O have the linear equations (6.3)[1].

Regarding the line-element $(x^i_{(0)}, \dot{x}^i_{(0)})$ as an element of support, one can, however, construct a system of coordinates which satisfies property (1), but not (2). This construction is easily carried out as follows. With respect to our element of support one may define a sequence of parameters

$$H^i_{j_1 j_2 \ldots j_m} = \frac{2}{m!} \cdot \frac{\partial^m \underset{m}{H^i}(x_{(0)}, \dot{x}_{(0)})}{\partial \dot{x}^{j_1} \, \partial \dot{x}^{j_2} \ldots \partial \dot{x}^{j_m}} \,, \qquad (m = 2, 3, \ldots) \,. \tag{6.6}$$

---

[1] This theorem is due to Douglas [1], p. 163. A similar result has been stated by Busemann [11]. In this context we should remark that the paths (and hence the geodesics of a Finsler space) naturally continue to possess the following property: for any point O of a region there exists a domain R containing O such that any point P of R is joined to O by one and only one path lying in R. In fact, an even more general result is proved by Whitehead [1, 2]: For any point O there exists a domain D containing O such that any two points P and Q of D can be joined by one and only one path lying in the domain D. See also Whitehead [4].

A discussion of normal coordinates determined by a family of geodesics is given by Noether [1]. Further properties of normal coordinates in F_n are derived by Cairns [1], where the existence of a type of normal coordinates under weak assumptions, namely that $F(x, \dot{x})$ and $F_{\dot{x}^i}(x, \dot{x})$ be of class C^1, is proved. See also Duschek and Mayer [1].

From these we form the expressions

$$P^i_m \left(x^k_{(0)}, \dot{x}^k_{(0)}, z^k \right) = H_{j_1 j_2 \ldots j_m} z^{j_1} z^{j_2} \cdots z^{j_m} \, . \tag{6.7}$$

The *osculating analytic function* in the direction $\dot{x}^i_{(0)}$ to the generally non-analytic function (6.4) is obtained by the replacement of the $H^i_m \left(x^k_{(0)}, z^k \right)$ in (6.4) by the corresponding functions $P^i_m \left(x^k_{(0)}, \dot{x}^k_{(0)}, z^k \right)$:

$$x^i(t) = x^i_{(0)} + z^i - \sum_{m=2}^{\infty} \frac{1}{m!} H^i_{j_1 j_2 \ldots j_m} z^{j_1} z^{j_2} \ldots z^{j_m} \, . \tag{6.8}$$

Again, this multiple power series is convergent for sufficiently small values of the $|z^i|$. For if we put $z^i = \dot{x}^i_{(0)} (t - t_0)$, the series (6.8) reduces to (6.4), since in view of the homogeneity properties of the functions H^i we have from (6.6) and (6.7)

$$P^i_m \left(x^k_{(0)}, \dot{x}^k_{(0)}, \dot{x}^j_{(0)} \right) = H^i_{j_1 j_2 \ldots j_m} \dot{x}^{j_1}_{(0)} \dot{x}^{j_2}_{(0)} \ldots \dot{x}^{j_m}_{(0)} = 2 H^i_m \left(x_{(0)}, \dot{x}_{(0)} \right) \, , \tag{6.9}$$

and, as (6.4) is convergent for sufficiently small values of the $|z^i|$ for which the z^i are proportional to the $\dot{x}^i_{(0)}$, it follows that (6.8) is convergent for all sufficiently small values of the $|z^i|$.

The coordinates z^i defined by (6.8) are *the normal coordinates relative to the system (x^i) with respect to the element of support $(x^k_{(0)}, \dot{x}^k_{(0)})$*. As in the restricted geometry of paths, it is easily shown that when the coordinates x^i undergo an arbitrary transformation, the corresponding normal coordinates z^i undergo a homogeneous linear transformation[1].

Furthermore, the process of "extension" as well as "normal" tensors may be introduced in the usual manner. We shall not, however, pursue this subject any further, since the relevant generalisations differ but slightly from their original counterparts[2].

An alternative definition of normal coordinates in manifolds of line-elements which are endowed with a different type of connection is due to Varga[3]. We shall briefly describe these coordinates, because they, too, may be applied directly to Finsler spaces. The manifolds in question are endowed with a connection in the sense that a tensor $C^i_{hk}(x, \dot{x})$ and symmetric connection parameters $\Gamma^{*i}_{hk}(x, \dot{x})$ are defined for each line-element $(x, \dot{x})$[4]. With respect to a given family of line-elements

$$x^i = x^i(t) \, , \qquad v^i = v^i(t) \, , \tag{6.10}$$

---

[1] Douglas [1], p. 164 et seq.

[2] Douglas [1], §§ 10—11. For the original definitions of extension and normal tensors see T. Y. Thomas [1], pp. 14 and 102.

[3] Varga [13]. For the general theory of the spaces referred to, the so-called „Affinzusammenhängende Mannigfaltigkeiten", see Varga [11].

[4] The C^i_{hk} and Γ^{*i}_{hk} play a similar role in the present theory to that played by the quantities denoted by the same symbols in the theory of § 1, but it should be remembered that the latter are derivable from a metric function, which is not defined at present. The C^i_{hk} and Γ^{*i}_{hk} are here assumed to be analytic.

the parallel displacement of a vector field $X^i(x^k, v^k)$ is defined by the formula

$$\frac{dX^i}{dt} + \left\{ C^i_{hk} \left[\frac{dv^h}{dt} + \Lambda^h_j(x, v) \frac{dx^j}{dt} \right] + \Gamma^{*i}_{hk}(x, v) \frac{dx^h}{dt} \right\} X^k = 0 , \quad (6.11)$$

where we have put

$$\Lambda^h_j(x, v) = \Gamma^{*h}_{jl}(x, v) \, v^l. \quad (6.12)$$

If the family (6.10) is a "parallel" family, i. e. if we assume that this family satisfies the differential equations

$$\frac{dv^h}{dt} + \Lambda^h_j(x, v) \frac{dx^j}{dt} = 0 , \quad (6.13)$$

the condition (6.11) reduces to

$$\frac{dX^i}{dt} + \Gamma^{*i}_{hk}(x, v) \frac{dx^h}{dt} X^k = 0 . \quad (6.14)$$

In order to define a normal coordinate system let us consider a fixed line-element $(x_{(0)}, v_{(0)})$, and a vector $\ddot{x}^i_{(0)}$ defined at the point $x^i_{(0)}$ as a function of this line-element. We shall now endeavour to construct a curve C passing through $x^i_{(0)}$ and tangent to $\ddot{x}^i_{(0)}$ which possesses the following property: if $v^i_{(0)}$ is transported by parallelism along C, then the field of tangent vectors $\dot{x}^i$ of C is parallel with respect to the field v^i. The line-elements (x^i, v^i) must therefore satisfy (6.13), while the tangent vectors satisfy (6.14), or, more precisely,

$$\frac{d^2x^i}{dt^2} + \Gamma^{*i}_{hk}(x, v) \frac{dx^h}{dt} \frac{dx^k}{dt} = 0 . \quad (6.15)$$

The curve C together with the requisite field $v^i(t)$ is then uniquely determined by the differential equations (6.13) and (6.15) in conjunction with the given initial conditions at the point $x^i_{(0)}$, where $t = t_0$. The following expansion will then represent the solution of (6.15):

$$x^i(t) = x^i_{(0)} + \dot{x}^i_{(0)}(t - t_0) -$$
$$- \sum_{m=2}^{\infty} \frac{1}{m!} \Gamma^{*i}_{j_1 j_2 \ldots j_m}(x_{(0)}, v_{(0)}) \, \dot{x}^{j_1}_{(0)} \, \dot{x}^{j_2}_{(0)} \ldots \dot{x}^{j_m}_{(0)}(t - t_0)^m , \quad (6.16)$$

while for the solution of (6.13) we have similarly

$$v^i(t) = v^i_{(0)} - \Lambda^i_j(x_{(0)}, v_{(0)}) \, \dot{x}^j_{(0)}(t - t_0) -$$
$$- \sum_{m=2}^{\infty} \frac{1}{m!} \Lambda^i_{j_1 j_2 \ldots j_m}(x_{(0)}, v_{(0)}) \, \dot{x}^{j_1}_{(0)} \, \dot{x}^{j_2}_{(0)} \ldots \dot{x}^{j_m}_{(0)}(t - t_0)^m . \quad (6.17)$$

The coefficients appearing in these expansions are found to be given by the following reduction formulae:

$$\Gamma^{*i}_{j_1 j_2 \ldots j_m} = \frac{1}{m} P \left\{ \frac{\partial \Gamma^{*i}_{j_1 j_2 \ldots j_{m-1}}}{\partial x^{j_m}} - (m-1) \Gamma^{*i}_{s j_2 \ldots j_{m-1}} \Gamma^{*s}_{j_1 j_m} - \frac{\partial \Gamma^{*i}_{j_1 j_2 \ldots j_{m-1}}}{\partial \dot{x}^s} \Lambda^s_{j_m} \right\} \quad (6.18)$$

together with

$$\Lambda^i_{j_1 j_2 \ldots j_m} = \frac{1}{m} P \left\{ \frac{\partial \Lambda^i_{j_1 j_2 \ldots j_{m-1}}}{\partial x^{j_m}} - (m-1) \Lambda^i_{s j_2 \ldots j_{m-1}} \Gamma^{*s}_{j_1 j_m} - \frac{\partial \Lambda^i_{j_1 j_2 \ldots j_{m-1}}}{\partial \dot{x}^s} \Lambda^s_{j m} \right\} , \tag{6.19}$$

where we have used the notation of Eisenhart[1] in the sense that the symbol P before an expression indicates the sum of the terms obtained by permuting the subscripts $j_1, j_2, \ldots, j_m$ cyclically, m denoting the number of subscripts.

At the point $x^i_{(0)}$ we may now vary the vector $\dot{x}^i_{(0)}$, keeping $v^i_{(0)}$ fixed. In this manner we obtain a family of curves C, which, as we shall see, covers a region containing $x^i_{(0)}$ simply. If we introduce the substitution

$$\bar{z}^i = \dot{x}^i_{(0)} (t - t_0) , \tag{6.20}$$

equations (6.16) yield a representation of the new coordinates $\bar{z}^i$:

$$x^i(t) = x^i_{(0)} + \bar{z}^i - \sum_{m=2}^{\infty} \frac{1}{m!} \Gamma^{*i}_{j_1 j_2 \ldots j_m} (x_{(0)}, v_{(0)}) \bar{z}^{j_1} \bar{z}^{j_2} \ldots \bar{z}^{j_m} , \tag{6.21}$$

the Jacobian of which also satisfies (6.5) at $x^i_{(0)}$. Similarly we find from (6.17)

$$v^i(t) = v^i_{(0)} - \Lambda^i_j (x_{(0)}, v_{(0)}) \bar{z}^j - \sum_{m=2}^{\infty} \frac{1}{m!} \Lambda^i_{j_1 j_2 \ldots j_m} (x_{(0)}, v_{(0)}) \bar{z}^{j_1} \bar{z}^{j_2} \ldots \bar{z}^{j_m}. \tag{6.22}$$

The inverse of (6.21) will be of the form

$$\bar{z}^i = \bar{z}^i (x^k_{(0)}, v^k_{(0)}, x^k - x^k_{(0)}) , \tag{6.23}$$

which, when substituted in (6.22) gives a unique representation of the field as a function of position:

$$v^i = v^i (x^k_{(0)}, v^k_{(0)}, x^k) . \tag{6.24}$$

The coordinates $\bar{z}^i$ as defined by (6.21), being uniquely determined by the initial line-element $(x^k_{(0)}, v^k_{(0)})$ in a region containing $x^k_{(0)}$, may be regarded as *normal coordinates with respect to the line-element* $(x^k_{(0)}, v^k_{(0)})$[2].

As before, we see that these normal coordinates do not possess all the properties which are usually required of normal coordinates in the classical sense. The family of curves C is represented by linear equations (6.20), *but these curves are not, in general, the autoparallel curves* (i. e. the paths of the present theory). In fact, it follows from (6.13) and (6.15) that amongst the curves of this family only that one is autoparallel for which the additional conditions

$$v^i_{(0)} = \dot{x}^i_{(0)} = \left(\frac{d x^i}{d t} \right)_{t=t_0}$$

---

[1] Eisenhart [2], p. 58.
[2] Varga [13], § 1.

are satisfied. However, from the point of view of the applicability of these coordinates to the problem of the determination of invariants this is not too serious, since an arbitrary transformation of the (x^i) again gives rise to a linear homogeneous transformation of the $(\bar{z}^i)$. Thus once more the process of extension and the notion of normal tensors may be introduced with respect to the normal coordinates $\bar{z}^i$. Furthermore, it is possible to derive generalisations of the well-known replacement theorem of T. Y. Thomas[1].

Normal coordinates of the type defined by (6.4) may be used in connection with the study of rectifiable arcs, as is shown in the work of Myers [1]. In this investigation it is assumed that the underlying manifold M is of class C^4, the metric function $F(x, \dot{x})$ being of class C^3. The so-called "integral" arc-length may be defined by a Lebesgue integral of F along an absolutely continuous representation of a curve. A Finsler space becomes a metric manifold (in the usual topological sense) if the distance $\varrho(P, Q)$ between the two points P, Q of M is defined as the greatest lower bound of the integral arc-length of arcs joining P to Q. Now, as is well-known, in a euclidean space a necessary and sufficient condition that an arc $x^i = x^i(t)$, $t_0 \leq t \leq t_1$, be rectifiable is that the functions $x^i(t)$ be of bounded variation in the interval $t_0 \leq t \leq t_1$. In order to formulate the corresponding result for Finsler spaces, we shall, following Myers, say that a manifold M possesses the property I, if corresponding to an arbitrary compact subset S of M and an arbitrary admissible coordinate system (x^i) covering it, there exist positive constants A and B such that for any two points $P(x^i_{(0)})$, $Q(x^i_{(1)})$ of S the inequalities

$$A \left| x^i_{(1)} - x^i_{(0)} \right| \leq \varrho(P, Q) \leq B \sum_{j=1}^{n} \left| x^j_{(1)} - x^j_{(0)} \right|, \qquad (i = 1, \ldots, n),$$

are satisfied. *Using normal coordinates it may be shown that every Finsler manifold possesses the property I*[2]. Furthermore, the following theorem holds: On a metric manifold of class C^1 with property I, a necessary and sufficient condition that an arc g be rectifiable is that every sub-arc be given by functions $x^i = x^i(t)$, $t_0 \leq t \leq t_1$, of bounded variation in the interval $t_0 \leq t \leq t_1$. Hence, in a Finsler space a necessary and sufficient condition that an arc be rectifiable is that every sub-arc be given by functions $x^i = x^i(t)$, $t_0 \leq t \leq t_1$, of bounded variation in the interval $t_0 \leq t \leq t_1$.

---

[1] Varga [13], p. 155 et seq.; for the classical replacement theorem see T. Y. Thomas [1], § 39. The normal coordinates of Varga are used by Rapcsák [2] to derive an invariant Taylor expansion for tensors of a Finsler space. For a modification of Varga's definitions see Rapcsák [3].

[2] For the proof of this and of the following statements see Myers [1], § 3. Cf. also Busemann and Mayer [1].

Chapter IV

The Theory of Curvature

The theory of curvature describes the essential differences between the underlying manifold and its tangent spaces. This cannot be done solely in terms of the connection coefficients, since these do not possess suitable transformation properties. We are therefore compelled to seek a set of tensors for this purpose. The simplest analytical approach to this problem is indicated by the fact that covariant differentiation is not in general commutative. Thus, in the first section of this chapter, we shall study the relevant commutation formulae which give rise to the required curvature tensors. It will be seen that the latter satisfy a large number of identities, the most important of which we shall derive. Unfortunately the analysis is occasionally apt to prove somewhat complicated. We have therefore endeavoured to break the monotony of mere analysis as soon as possible, namely by inserting a section on geodesic deviation, by means of which some of the most striking geometrical properties of the curvature and related tensors may be illustrated. We shall also discuss the second variation of the length integral, thus re-establishing the fundamental link with the calculus of variations, and in particular with the accessory problem. Also, a further set of curvature tensors plays a fundamental role in the general geometry of paths: furthermore, formally similar tensors were used by BERWALD in his very extensive theory of curvature. It is essential, therefore, that these tensors be also examined, especially in view of their close affinity to the projective curvature tensors. By means of the analytical apparatus thus developed we shall be in a position to study special types of Finsler spaces, such as spaces of constant curvature or zero projective curvature.

§ 1. The Commutation Formulae

In the two preceding chapters we have studied the various forms of covariant derivatives which may be used in Finsler geometry, and we noted that these differentiation-processes have one feature in common: namely, that the elementary rules for the differentiation of a sum or product of two or more tensors hold in their usual form, while the order of differentiation with respect to two distinct parameters or positional coordinates may not, in general, be interchanged. If we interpret the covariant derivative in the customary manner by means of the notion of parallel displacement, this means, geometrically, that when a vector is transported by parallel displacement around some closed circuit, the final values of its components do not generally coincide with the initial values: in fact, these values depend on the choice of the circuit. This phenomenon is frequently referred to as the "non-integrability" of the parallel displacement and is well-known in Riemannian geometry. Often this is taken as the starting point of the theory of curvature of Riemannian spaces, and it is natural for us to begin our discussion of the curvature properties of Finsler spaces in a similar manner.

In the present section we shall therefore derive the commutation formulae for repeated differentiation corresponding to the various forms of covariant derivatives and thus we shall be led to define several new tensors, the so-called curvature tensors, whose analytical and geometrical properties will be studied in this chapter.

We shall begin with the commutation formulae corresponding to the δ-derivative discussed in Chapter II since these appear to possess the simplest structure. A study of Cartan's covariant derivative will yield three distinct curvature tensors, one of which is intimately connected with the tensor arising from the δ-process, the knowledge of which will enable us to avoid further lengthy calculations. It should be pointed out, however, that CARTAN[1] arrived at the three above-mentioned curvature tensors by means of his method of exterior differential forms which we shall briefly outline towards the end of this section for the benefit of those readers who are familiar with these concepts.

1°. Commutation Formulae Resulting from δ-differentiation

As in § 5 of Chapter II let us consider a vector field $X^i(x^k, \xi^k)$, depending on position as well as on a direction field: $\xi^k = \xi^k(x^k)$. According to definition (2.5.1) its covariant δ-derivative at the point x^k in the direction ξ^k is given by

$$X^i_{;h} = \frac{\partial X^i}{\partial x^h} + \frac{\partial X^i}{\partial \dot{x}^l}\frac{\partial \xi^l}{\partial x^h} + \Gamma^{*i}_{rh}(x, \xi)\, X^r\,, \tag{1.1}$$

while

$$X^i_{;hk} = \frac{\partial}{\partial x^k}(X^i_{;h}) + \frac{\partial}{\partial \dot{x}^l}(X^i_{;h})\frac{\partial \xi^l}{\partial x^k} + \Gamma^{*i}_{jk}X^j_{;h} - X^i_{;j}\Gamma^{*i}_{hk}. \tag{1.2}$$

---

[1] CARTAN [1], Ch. XIII.

We substitute from (1.1) in (1.2) and carry out the operations indicated in (1.2), and on collecting terms we obtain

$$
\begin{aligned}
X^i_{;hk} = \Big\{ & \frac{\partial^2 X^i}{\partial x^k\,\partial x^h} + \Big(\frac{\partial^2 X^i}{\partial x^k\,\partial \dot{x}^l}\,\frac{\partial \xi^l}{\partial x^h} + \frac{\partial^2 X^i}{\partial \dot{x}^l\,\partial x^h}\,\frac{\partial \xi^l}{\partial x^k} \Big) \\
& + \frac{\partial^2 X^i}{\partial \dot{x}^l\,\partial \dot{x}^m}\,\frac{\partial \xi^m}{\partial x^h}\,\frac{\partial \xi^l}{\partial x^k} + \frac{\partial X^i}{\partial \dot{x}^l}\,\frac{\partial^2 \xi^l}{\partial x^k\,\partial x^h} + \Big(\frac{\partial X^j}{\partial x^h}\,\Gamma^{*\,i}_{jk} + \frac{\partial X^j}{\partial x^k}\,\Gamma^{*\,i}_{jh} \Big) \\
& + \frac{\partial X^j}{\partial \dot{x}^l}\Big(\Gamma^{*\,i}_{jh}\,\frac{\partial \xi^l}{\partial x^k} + \Gamma^{*\,i}_{jk}\,\frac{\partial \xi^l}{\partial x^h} \Big) - X^i_{;j}\,\Gamma^{*\,j}_{hk} \Big\} \\
& + X^j\Big[\frac{\partial \Gamma^{*\,i}_{jh}}{\partial x^k} + \frac{\partial \Gamma^{*\,i}_{jh}}{\partial \dot{x}^l}\,\frac{\partial \xi^l}{\partial x^k} + \Gamma^{*\,i}_{mk}\,\Gamma^{*\,m}_{jh} \Big].
\end{aligned}
\tag{1.3}
$$

Inspection of this equation shows that the expression contained in the curly brackets on the right-hand side of (1.3) is symmetric in the indices h, k. Thus, if we form the derivative $X^i_{;kh}$ by interchanging h,k and subtract the equation thus obtained from (1.3), we find

$$
X^i_{;hk} - X^i_{;kh} = \tilde{K}^i_{jhk}(x,\,\xi)\,X^j\,,
\tag{1.4}
$$

where we have put

$$
\begin{aligned}
\tilde{K}^i_{jhk} = {} & \Big(\frac{\partial \Gamma^{*\,i}_{jh}}{\partial x^k} + \frac{\partial \Gamma^{*\,i}_{jh}}{\partial \dot{x}^l}\,\frac{\partial \xi^l}{\partial x^k} \Big) - \Big(\frac{\partial \Gamma^{*\,i}_{jk}}{\partial x^h} + \frac{\partial \Gamma^{*\,i}_{jk}}{\partial \dot{x}^l}\,\frac{\partial \xi^l}{\partial x^h} \Big) \\
& + \Gamma^{*\,i}_{mk}\,\Gamma^{*\,m}_{jh} - \Gamma^{*\,i}_{mh}\,\Gamma^{*\,m}_{jk}\,.
\end{aligned}
\tag{1.5}
$$

We shall call the tensor defined by (1.5) the "relative" *curvature tensor*, since it depends on the derivatives $\partial \xi^l/\partial x^k$ of the field $\xi^l(x^k)$[1]. It is a peculiarity of Finsler spaces that such quantities appear, giving rise very often to serious difficulties; in general such terms have to be interpreted in the light of the special problem under consideration. We shall see that for such problems the relative curvature tensor is most useful: but, clearly, when we wish to study the nature of the Finsler space *per se* another tensor, independent of derivatives such as $\partial \xi^l/\partial x^k$ is required. Such a tensor is easily derived as follows.

Suppose that the vector field ξ^l is "stationary" at the point under consideration, i. e. that it satisfies the relation $\xi^l_{;h}(x,\,\xi) = 0$. From (2.5.1) it follows that we then have at this particular point

$$
\frac{\partial \xi^l}{\partial x^h} = - \Gamma^{*\,l}_{rh}(x,\,\xi)\,\xi^r = - \frac{\partial G^l(x,\,\xi)}{\partial \dot{x}^h}
\tag{1.6}
$$

in view of (3.1.27′)[2]. Firstly, we note that if we substitute (1.6) in (1.1) we obtain Cartan's covariant derivative $X^i_{|h}$ as a result of definition

---

[1] Rund [5], p. 91.

[2] This construction is described in more detail by Rund [5], p. 99, where with respect to each direction ξ^i at the point x^k under consideration such a — in general— non-integrable vector field is defined geometrically by means of a normal coordinate system.

(3.2.8), provided we regard the line-element ξ^i as the element of support $\dot{x}^i$, so that we may write $\xi^i = \dot{x}^i$ without danger of ambiguity. Secondly, the tensor (1.5) becomes [1]

$$K^i_{jhk} = \left(\frac{\partial \Gamma^{*i}_{jh}}{\partial x^k} - \frac{\partial \Gamma^{*i}_{jh}}{\partial \dot{x}^l} \frac{\partial G^l}{\partial \dot{x}^k} \right) - \left(\frac{\partial \Gamma^{*i}_{jk}}{\partial x^h} - \frac{\partial \Gamma^{*i}_{jk}}{\partial \dot{x}^l} \frac{\partial G^l}{\partial \dot{x}^h} \right) \\ + \Gamma^{*i}_{mk} \Gamma^{*m}_{jh} - \Gamma^{*i}_{mh} \Gamma^{*m}_{jk} \, . \tag{1.7}$$

Clearly, this curvature tensor depends only on the line-element, i. e. only on position and direction. However, it suffers from the disadvantage that formula (1.4) is no longer true if $\tilde{K}^i_{jhk}$ is replaced by K^i_{jhk}, not even when we use Cartan's derivative (3.2.8). In order to derive the commutation formula corresponding to the latter derivative, we observe that in view of (1.6) the 5th term on the right-hand side of (1.3) has to be replaced by a more appropriate term, while now the $\partial \xi^l / \partial x^h$ are no longer functions of position only. Hence in (1.3) the term in question must be replaced by the expression

$$- \frac{\partial X^i}{\partial \dot{x}^l} \frac{\partial}{\partial x^k} \left(\frac{\partial G^l}{\partial \dot{x}^h} \right) + \frac{\partial X^i}{\partial \dot{x}^m} \frac{\partial}{\partial \dot{x}^l} \left(\frac{\partial G^m}{\partial \dot{x}^h} \right) \frac{\partial G^l}{\partial \dot{x}^k} \, . \tag{1.8}$$

Using (3.1.27'), (3.3.6) and (3.3.16) we see that (1.8) may be written as

$$- \frac{\partial X^i}{\partial \dot{x}^l} \left[\frac{\partial \Gamma^{*l}_{hr}}{\partial x^k} \dot{x}^r - \left(\Gamma^{*l}_{hs} + \frac{\partial \Gamma^{*l}_{hr}}{\partial \dot{x}^s} \dot{x}^r \right) \Gamma^{*s}_{kj} \dot{x}^j \right] ,$$

or, again,

$$- \frac{\partial X^i}{\partial \dot{x}^j} \left[\frac{\partial \Gamma^{*j}_{hr}}{\partial x^k} - \frac{\partial \Gamma^{*j}_{hr}}{\partial \dot{x}^l} \frac{\partial G^l}{\partial \dot{x}^k} - \Gamma^{*j}_{mh} \Gamma^{*m}_{rk} \right] \dot{x}^r \, . \tag{1.9}$$

Thus the (symmetric) 5th term in (1.3) is replaced by (1.9), and on interchange of the indices h, k we find in view of definition (1.7) [2]

$$X^i_{|hk} - X^i_{|kh} = K^i_{jhk} X^j - \frac{\partial X^i}{\partial \dot{x}^j} K^j_{rhk} \dot{x}^r \, . \tag{1.10}$$

For future reference we note that as a result of the equivalence of the expressions (1.8) and (1.9) the following relation is satisfied by the tensor (1.7):

$$\dot{x}^i K^j_{ihk} = \frac{\partial^2 G^j}{\partial x^k \partial \dot{x}^h} - \frac{\partial^2 G^j}{\partial \dot{x}^k \partial x^h} - G^j_{hl} \frac{\partial G^l}{\partial \dot{x}^k} + G^j_{kl} \frac{\partial G^l}{\partial \dot{x}^h} \, . \tag{1.11}$$

---

[1] Similar tensors related to the geometry of paths are derived by DOUGLAS [1], KNEBELMAN [1], BOMPIANI [1]. See also ANCOCHEA [1]. The tensor (1.7) appears also in the work of DAVIES [1], p. 263, where it is denoted by R^{*i}_{jhk}.

[2] Similar relations hold for Berwald's covariant derivatives as defined in Ch. III, § 3 (BERWALD [2], p. 53) and in the general geometry of paths (KNEBELMAN [1], p. 532). It is to be noted, however, that while formula (1.4) may be used to derive the correct variation of a vector when transported by δ-parallelism about an infinitesimal closed circuit, this is not true for formula (1.10) unless special conditions concerning the element of support prevail, which we shall discuss at a later stage.

The derivation of formula (1.4) can, of course, be generalised to yield commutation formulae for tensors of arbitrary rank depending on the vector field $\xi^l(x^k)$. Suppose that $T^{i_1\cdots i_p}{}_{j_1\ldots j_q}(x,\xi)$ is a tensor of contravariant valency p and covariant valency q (p, q being arbitrary positive integers). The reader may verify that the following relation holds:

$$T^{i_1\cdots i_p}{}_{j_1\ldots j_q;hk} - T^{i_1\cdots i_p}{}_{j_1\ldots j_q;kh}$$

$$= \sum_{\alpha=1}^{p} T^{i_1\cdots i_{\alpha-1} r i_{\alpha+1}\cdots i_p}{}_{j_1\ldots j_q}\tilde{K}^{i_\alpha}_{rhk} \tag{1.12}$$

$$- \sum_{\beta=1}^{q} T^{i_1\cdots i_p}{}_{j_1\ldots j_{\beta-1} s j_{\beta+1}\ldots j_q}\tilde{K}^{s}_{j_\beta hk}.$$

For example, for a covariant tensor $T_{ij}(x,\xi)$ of rank 2 we would have

$$T_{ij;hk} - T_{ij;kh} = -T_{is}\tilde{K}^{s}_{jhk} - T_{sj}\tilde{K}^{s}_{ihk}. \tag{1.13}$$

If we write

$$g_{sj}\tilde{K}^{s}_{ihk} = \tilde{K}_{ijhk}, \tag{1.14}$$

we have in particular

$$g_{ij;hk} - g_{ij;kh} = -\tilde{K}_{jihk} - \tilde{K}_{ijhk}, \tag{1.15}$$

and hence, from (2.5.17),

$$\tilde{K}_{jihk}(x,\dot{x})\,\dot{x}^i = -\tilde{K}_{ijhk}(x,\dot{x})\,\dot{x}^i. \tag{1.15a}$$

Again, if we wish to replace the tensor $\tilde{K}^{i}_{jhk}$ by K^{i}_{jhk} in (1.12), we have to consider two additional terms of the type (1.8). By means of (1.11) these may be combined to give an expression involving the curvature tensor. In this manner we obtain the general formula

$$T^{i_1\cdots i_p}{}_{j_1\ldots j_q|hk} - T^{i_1\cdots i_p}{}_{j_1\ldots j_q|kh}$$

$$= -\frac{\partial T^{i_1\cdots i_p}{}_{j_1\ldots j_q}}{\partial \dot{x}^l} K^{l}_{rhk}\dot{x}^r$$

$$+ \sum_{\alpha=1}^{p} T^{i_1\cdots i_{\alpha-1} r i_{\alpha+1}\cdots i_p}{}_{j_1\ldots j_q} K^{i_\alpha}_{rhk} \tag{1.16}$$

$$- \sum_{\beta=1}^{q} T^{i_1\cdots i_p}{}_{j_1\ldots j_{\beta-1} s j_{\beta+1}\ldots j_q} K^{s}_{j_\beta hk}.$$

Instead of (1.13) we now have

$$T_{ij|hk} - T_{ij|kh} = -\frac{\partial T_{ij}}{\partial \dot{x}^l} K^{l}_{rhk}\,\dot{x}^r - T_{is} K^{s}_{jhk} - T_{sj} K^{s}_{ihk}. \tag{1.16a}$$

Further commutation formulae which will frequently be useful are those which involve successive differentiation with respect to direction and covariant differentiation of the type (3.2.6). The general formula reads as follows:

$$\left(\frac{\partial T^{i_1\cdots i_p}{}_{j_1\ldots j_q}}{\partial \dot{x}^h}\right)_{|k} - \frac{\partial}{\partial \dot{x}^h}\left(T^{i_1\cdots i_p}{}_{j_1\ldots j_q|k}\right)$$

$$= \frac{\partial T^{i_1\cdots i_p}{}_{j_1\ldots j_q}}{\partial \dot{x}^l}\, C^l_{hk|r}\, \dot{x}^r - \sum_{\alpha=1}^{p} T^{i_1\cdots i_{\alpha-1}\, r\, i_{\alpha+1}\cdots i_p}{}_{j_1\ldots j_q}\left(\frac{\partial \Gamma^{*\,i_\alpha}_{rk}}{\partial \dot{x}^h}\right) \qquad (1.17)$$

$$+ \sum_{\beta=1}^{q} T^{i_1\cdots i_p}{}_{j_1\ldots j_{\beta-1}\, s\, j_{\beta+1}\cdots j_q}\left(\frac{\partial \Gamma^{*\,s}_{j_\beta k}}{\partial \dot{x}^h}\right).$$

The deduction of these relations is straight-forward and may be left to the reader. We merely remark that the first term on the right-hand side arises from an expression of the type

$$\frac{\partial T^{i_1\cdots i_p}{}_{j_1\ldots j_q}}{\partial \dot{x}^l}\,(G^l_{hk} - \Gamma^{*\,l}_{hk}), \qquad (1.18)$$

to which we have applied equations (3.3.8)[1]. A particularly useful example of (1.17) is the identity

$$\left(\frac{\partial X^i}{\partial \dot{x}^h}\right)_{|k} - \frac{\partial}{\partial \dot{x}^h}\left(X^i_{|k}\right) = \frac{\partial X^i}{\partial \dot{x}^l}\, A^l_{hk|r}\, l^r - \frac{\partial \Gamma^{*\,i}_{rk}}{\partial \dot{x}^h}\, X^r. \qquad (1.18\text{a})$$

2°. The Three Curvature Tensors of Cartan

Let us consider the commutation formulae arising from the covariant derivatives of the type (3.2.7) of CARTAN: e. g.

$$DX^i = \left(F\,\frac{\partial X^i}{\partial \dot{x}^h} + A^i_{kh}X^k\right)Dl^h + X^i_{|h}\,dx^h. \qquad (1.19)$$

Clearly two distinct processes of partial differentiation are involved here, namely the process $X^i_{|h}$ [as defined by (3.2.8)] and the process

$$X^i|_h \equiv \left(F\,\frac{\partial X^i}{\partial \dot{x}^h} + A^i_{kh}X^k\right), \qquad (1.20)$$

so that we should write (1.19) in the form

$$DX^i = X^i|_h Dl^h + X^i_{|h}\,dx^h. \qquad (1.21)$$

In order to obtain a complete set of commutation formulae, we have to consider mixed derivatives involving *one or both* of the processes denoted by $|_h$ or $_{|h}$. At first sight this may appear to be somewhat laborious, but

---

[1] Similar relationships have been derived by BERWALD [2], p. 53, and KNEBEL-MAN [1], p. 532. However, the commutation formulae discussed by these authors do not, of course, involve the $\Gamma^{*\,i}_{hk}$ but only the G^i_{hk}, the latter always replacing the former wherever they occur in the analysis. Hence a term of the type (1.18) would be zero identically, and as a consequence those formulae of BERWALD and KNEBEL-MAN which correspond to (1.17) do not contain a term analogous to our first term on the right-hand side of (1.17). See § 6.

the analysis of the first part of this section will enable us to derive the necessary results without further long calculations.

Firstly, let us consider the commutation formulae corresponding to repeated application of the operation defined by (1.20). From the latter relation we find directly[1]

$$X^i|_{hk} - X^i|_{kh} = F\left\{F_{\dot{x}^k}\frac{\partial X^i}{\partial \dot{x}^h} - F_{\dot{x}^h}\frac{\partial X^i}{\partial \dot{x}^k}\right\}$$
$$+ X^r\left\{F\left(\frac{\partial A^i_{rh}}{\partial \dot{x}^k} - \frac{\partial A^i_{rk}}{\partial \dot{x}^h}\right) + A^i_{km}A^m_{rh} - A^i_{mh}A^m_{rk}\right\}. \tag{1.22}$$

But if we differentiate equation (3.2.4) with respect to direction and observe the symmetry arising from the definition (1.3.4), we see that

$$F\left(\frac{\partial A^i_{rh}}{\partial \dot{x}^k} - \frac{\partial A^i_{rk}}{\partial \dot{x}^h}\right) = F_{\dot{x}^k}A^i_{rh} - F_{\dot{x}^h}A^i_{rk}. \tag{1.23}$$

Substituting this result in (1.22) and noting (1.20), we find

$$X^i|_{hk} - X^i|_{kh} = \{F_{\dot{x}^k}X^i|_h - F_{\dot{x}^h}X^i|_k\} + S^i_{jkh}X^j, \tag{1.24}$$

where, following CARTAN[2], we have written

$$S^i_{jkh} = A^i_{kr}A^r_{jh} - A^i_{rh}A^r_{jk}. \tag{1.25}$$

This tensor is the first of Cartan's curvature tensors; we shall discuss its geometrical meaning presently[3].

Secondly, let us consider the commutation formulae involving both of the above-mentioned processes. From (3.2.8) and (1.20) we have

$$X^i|_{h|k} = F\left(\frac{\partial X^i}{\partial \dot{x}^h}\right)_{|k} + A^i_{hr|k}X^r + A^i_{hr}X^r_{|k}, \tag{1.26}$$

where we have made use of (3.2.13). Also, from (1.20) it follows that

$$X^i_{|k}|_h = F\frac{\partial}{\partial \dot{x}^h}(X^i_{|k}) + A^i_{hr}X^r_{|k} - A^r_{hk}X^i_{|r}. \tag{1.27}$$

We subtract this equation from (1.26), observing (1.18a), and obtain

$$X^i|_{h|k} - X^i_{|k}|_h = F\frac{\partial X^i}{\partial \dot{x}^l}A^l_{hk|r}l^r - \left(F\frac{\partial \Gamma^{*i}_{rk}}{\partial \dot{x}^h} - A^i_{hr|k}\right)X^r + A^r_{hk}X^i_{|r}.$$

On eliminating $\partial X^i/\partial \dot{x}^l$ by means of (1.20) we finally deduce the formula

$$X^i|_{h|k} - X^i_{|k}|_h = -P^i_{jkh}X^j + X^i|_j A^j_{hk|r}l^r + X^i_{|j}A^j_{hk}, \tag{1.28}$$

where, again following CARTAN[4], we have put

$$P^i_{jkh} = F\frac{\partial \Gamma^{*i}_{jk}}{\partial \dot{x}^h} + A^i_{jm}A^m_{hk|r}l^r - A^i_{jh|k}. \tag{1.29}$$

---

[1] Strictly speaking, we should write $X^i|_h|_k$ instead of $X^i|_{hk}$ in (1.22).

[2] CARTAN [1], p. 34, formula (XVI).

[3] A most elegant geometrical treatment of this tensor is given by VARGA [8]: we shall, however, defer an account of this treatment until the next chapter.

[4] CARTAN [1], p. 35, formula (XVII)'.

This tensor is the second of Cartan's curvature tensors. We may easily reduce it to a simpler form as follows. From (3.3.14), (3.2.13) and (3.1.15) we have

$$F \frac{\partial \Gamma_{jk}^{*i}}{\partial \dot{x}^h} = A_{kh|j}^i + A_{jh|k}^i - g^{ir} A_{jkh|r} -$$
$$- A_{km}^i A_{jh|r}^m l^r - A_{jm}^i A_{kh|r}^m l^r + A_{jk}^m A_{mh|r}^i l^r . \qquad (1.30)$$

Substitution of this relation in the definition (1.29) yields

$$P_{jkh}^i = A_{kh|j}^i - g^{im} A_{jkh|m} - A_{km}^i A_{jh|r}^m l^r + A_{jk}^m A_{mh|r}^i l^r . \qquad (1.31)$$

From this formula it is evident that the second curvature tensor is derivable directly from the covariant derivatives of the tensor A_{kh}^i.

Thirdly, the commutation formulae involving only the covariant derivatives of the type $_{|h}$ are already known in view of equations (1.10). We may, however, express the latter relations in a slightly different form by again eliminating the term $\partial X^i / \partial \dot{x}^l$ by means of (1.20). In this manner we obtain

$$X_{|hk}^i - X_{|kh}^i = (K_{jhk}^i + A_{jm}^i K_{rhk}^m l^r) X^j - X^i|_j K_{rhk}^j l^r , \qquad (1.32)$$

where we have made use of (3.1.15) and (3.2.4). The final form of the required commutation formulae is then

$$X_{|hk}^i - X_{|kh}^i = R_{jhk}^i X^j - K_{rhk}^j l^r X^i|_j , \qquad (1.33)$$

where we have put

$$R_{jhk}^i = K_{jhk}^i + C_{jm}^i K_{rhk}^m \dot{x}^r , \qquad (1.34)$$

so that

$$R_{jhk}^i \dot{x}^j = K_{jhk}^i \dot{x}^j . \qquad (1.34\,\text{a})$$

This tensor is the third curvature tensor of CARTAN: as a result of (1.11) and (1.7) we may write it explicitly in the form[1]

$$R_{jhk}^i = \left(\frac{\partial \Gamma_{jh}^{*i}}{\partial x^k} - \frac{\partial \Gamma_{jh}^{*i}}{\partial \dot{x}^l} \frac{\partial G^l}{\partial \dot{x}^k} \right)$$
$$- \left(\frac{\partial \Gamma_{jk}^{*i}}{\partial x^h} - \frac{\partial \Gamma_{jk}^{*i}}{\partial \dot{x}^l} \frac{\partial G^l}{\partial \dot{x}^h} \right) +$$
$$+ C_{jm}^i \left(\frac{\partial^2 G^m}{\partial x^k \partial \dot{x}^h} - \frac{\partial^2 G^m}{\partial \dot{x}^k \partial x^h} - G_{hl}^m \frac{\partial G^l}{\partial \dot{x}^k} + G_{kl}^m \frac{\partial G^l}{\partial \dot{x}^h} \right) +$$
$$+ \Gamma_{mk}^{*i} \Gamma_{jh}^{*m} - \Gamma_{mh}^{*i} \Gamma_{jk}^{*m} . \qquad (1.35)$$

3°. Alternative Derivation of the Curvature Tensors

For the rest of this section we shall presuppose a knowledge of the method of exterior differential forms and exterior differentiation[2]. A reader who is not familiar with these concepts may omit this treatment as the results required in the sequel

---

[1] CARTAN [1], p. 36, formula (XIX)′. See also DAVIES [1], p. 363.

[2] For the general theory of exterior differential forms the reader is referred to CARTAN [7, 8], KÄHLER [1], Ch. I; SCHOUTEN [1], Ch. II, § 12.

have already been derived in the earlier parts of the present section. Our purpose is to indicate briefly the method used by CARTAN in order to arrive at the three curvature tensors defined above. This process represents a direct generalisation of a method for the derivation of the curvature tensor of a Riemannian space which is also due to CARTAN[1].

Consider a fixed point $P(x^i)$ of F_n, a line-element $(x^i, \dot{x}^i)$ centred at P, and a displacement dx^i, which defines a neighbouring point $P_1(x^i + dx^i)$. Corresponding to a variation of the line-element from $(x^i, \dot{x}^i)$ to $(x^i + dx^i, \dot{x}^i + d\dot{x}^i)$ we have, according to (3.1.4), a differential form

$$\omega_i{}^j(d) = C_{ih}^j(x, \dot{x})\, d\dot{x}^h + \Gamma_{ih}^j(x, \dot{x})\, dx^h. \tag{1.36}$$

In accordance with the notation of CARTAN we shall, for the present, denote the covariant differential Dl^h of the unit vector l^h in the direction of the element of support by $\widetilde{\omega}^h$. In view of (3.2.2), (3.2.4) and (3.1.18) we may then write the form (1.36) as

$$\omega_i{}^j(d) = A_{ih}^j\, \widetilde{\omega}^h + \Gamma_{ih}^{*j}\, dx^h. \tag{1.37}$$

Now, in $T_n(P)$ we may, by means of the tangent vectors to the co-ordinate curves $x^i = \text{const.}$ of F_n, construct a basis consisting of n vectors $e_{(1)}, e_{(2)}, \ldots, e_{(n)}$ such that the displacement $\overline{PP_1}$ is represented in $T_n(P)$ by

$$\overline{PP_1} = dx^i\, e_{(i)}.$$

Furthermore, corresponding to the basis defined in $T_n(P)$, we may similarly construct a basis in $T_n(P_1)$ by means of a parallel displacement (3.1.3), so that the increments of the basis vectors are given by

$$de_{(i)} = \omega_i{}^k(d)\, e_{(k)}. \tag{1.38}$$

In this manner we obtain a moving frame of reference ("repère mobile"[2]); for instance, if we proceed along a given curve a frame is defined at each point of the curve.

In general, however, such increments do not give rise to integrable systems of partial differential equations. Let us consider a second displacement δx^i in $T_n(P)$, defining another point P_2 of F_n, such that

$$\overline{PP_2} = \delta x^i\, e_{(i)},$$

while the element of support at P_2 is of the form $\dot{x}^i + \delta \dot{x}^i$. The basis at P_2 is then defined by the increments

$$\delta e_{(i)} = \omega_i{}^k(\delta)\, e_{(k)}, \tag{1.38a}$$

where the form $\omega_i{}^k(\delta)$ is given by

$$\omega_i{}^j(\delta) = C_{ih}^j(x, \dot{x})\, \delta \dot{x}^h + \Gamma_{ih}^j(x, \dot{x})\, \delta x^h,$$

---

[1] CARTAN [7], Ch. VII—VIII.
[2] CARTAN [7], p. 34 and pp. 179—181.

corresponding to (1.36). Further, one may now endeavour to construct an infinitesimal "parallelogram" by transporting $\overline{PP_1}$ by a parallel displacement from P to P_2, the resulting displacement at P_2 defining a point P_3 of F_n. Similarly, $\overline{PP_2}$ may be transported by parallelism to P_1, defining a point P_3'. In general, however, P_3 and P_3' do not coincide, and the displacement $\overline{P_3'P_3}$ gives rise to the so-called *torsion* of F_n. Clearly, the components of this displacement are

$$(d\,x^k\,\omega_k{}^i(\delta) - \delta\,x^k\,\omega_k{}^i(d))\,,$$

so that the torsion is determined by the exterior product[1]

$$\Omega^i = [d\,x^k\,\omega_k{}^i]\,.$$

If we substitute in this expression from (1.37), we find that the term involving the Γ_{kh}^{*i} drops out as a result of the symmetry of these coefficients in k and h. Hence we are left with

$$\Omega^i = A_{kh}^i[d\,x^k\,\widetilde{\omega}^h]\,. \tag{1.39}$$

Further, the equations (1.38) and (1.38a) suggest that from the point of view of integrability the following expressions have to be evaluated:

$$\begin{aligned}
\delta\,d\,e_{(i)} - d\,\delta\,e_{(i)} = {}&(\delta\omega_i{}^k(d) - d\omega_i{}^k(\delta))\,e_{(k)} +\\
&+ (\omega_i{}^k(d)\,\omega_k{}^j(\delta) - \omega_i{}^k(\delta)\,\omega_k{}^j(d))\,e_{(j)}\,.
\end{aligned} \tag{1.40}$$

The coefficient of $e_{(k)}$ on the right-hand side is an exterior differential, which we shall write as[2]

$$-(\omega_i{}^k)' \equiv \delta\omega_i{}^k(d) - d\omega_i{}^k(\delta)\,, \tag{1.41}$$

while the coefficient of $e_{(j)}$ is once more an exterior product:

$$[\omega_i{}^k\omega_k{}^j] = \omega_i{}^k(d)\,\omega_k{}^j(\delta) - \omega_i{}^k(\delta)\,\omega_k{}^j(d)\,.$$

Thus (1.40) may be written as

$$\delta\,d\,e_{(i)} - d\,\delta\,e_{(i)} = ([\omega_i{}^h\omega_h{}^k] - (\omega_i{}^k)')\,e_{(k)}\,. \tag{1.42}$$

---

[1] The exterior product of two forms ω, π is defined by

$$[\omega,\,\pi] = \omega(d)\,\pi(\delta) - \omega(\delta)\,\pi(d)\,,$$

or, more generally, if $d_1, d_2, \ldots, d_p$ represent commuting symbols of differentiation, and the forms ω_i are defined by

$$\omega_i = a_{ik}\,d\,x^k\,,$$

the definition reads

$$\begin{aligned}
[\omega_1, \omega_2, \ldots, \omega_p] &= a_{1k_1}\,a_{2k_2}\cdots a_{pk_p}\,[d\,x^{k_1}, d\,x^{k_2}, \ldots, d\,x^{k_p}]\\
&= \begin{vmatrix} \omega_1(d_1), & \ldots, & \omega_1(d_p)\\ \cdots & \cdots & \cdots\\ \omega_p(d_1), & \ldots, & \omega_p(d_p) \end{vmatrix}.
\end{aligned}$$

[2] Compare CARTAN [7], p. 51, where the notation $d\omega_i{}^k$ instead of $(\omega_i{}^k)'$ is used.

This expression represents the curvature of the space (since it indicates that the parallel displacement is not, in general, integrable). Thus a representation of the curvature is given by the exterior form defined by[1]

$$\Omega_i{}^k = [\omega_i{}^h \omega_h{}^k] - (\omega_i{}^k)' . \tag{1.43}$$

Alternatively, if we write in covariant form

$$\Omega_{ij} = g_{jk}\,\Omega_i{}^k ,$$

we have from (1.43)

$$\Omega_{ij} = [\omega_i{}^h\,\omega_{hj}] - (g_{jk}\,\omega_i{}^k)' + [dg_{jk}\,\omega_i{}^k] .$$

Thus, on using (3.1.5), we find

$$\Omega_{ij} = [\omega_j{}^k\,\omega_{ik}] - (\omega_{ij})' = -\Omega_{ji} . \tag{1.43a}$$

Clearly, if the right-hand sides of (1.43) or (1.43a) are evaluated, the $\omega_j{}^k$ and their differentials being expressed by means of (1.37), we shall obtain terms involving the products $[\widetilde{\omega}{}^k\,\widetilde{\omega}{}^h]$, $[dx^k\,\widetilde{\omega}{}^h]$ and $[dx^k dx^h]$. On collecting these terms we may express (1.43) in the form

$$\Omega_i{}^j = S_{ikh}^j[\widetilde{\omega}{}^k\,\widetilde{\omega}{}^h] + P_{ikh}^j[dx^k\,\widetilde{\omega}{}^h] + R_{ikh}^j[dx^k dx^h], \tag{1.44}$$

where in the first and third terms on the right-hand side the summation is taken only over all the *combinations* (k, h). However, the factorisation in (1.44) is not unique, since the forms $\widetilde{\omega}{}^h$ are not independent as a result of the relation $l_h\widetilde{\omega}{}^h = 0$[2]. This difficulty may be overcome by arranging the terms in (1.44) such that

$$P_{ikh}^j(x, \dot{x})\,\dot{x}^h = 0 , \tag{1.45}$$

while similarly we may have

$$S_{ikh}^j(x, \dot{x})\,\dot{x}^h = 0 . \tag{1.46}$$

Under these circumstances it may be shown by means of an explicit evaluation — which is quite complicated — of the right-hand side of (1.43) that the tensors defined by (1.44) are in fact the curvature tensors (1.25), (1.31) and (1.35) as anticipated by our notation[3].

Equation (1.44) enables us to give a simple preliminary geometrical interpretation of these tensors in terms of the infinitesimal "parallelogram". Firstly, if

---

[1] For instance, in Riemannian geometry equation (1.43) gives

$$\Omega_i{}^k(d, \delta) = \tfrac{1}{2} R_{ihj}^k(dx^h\,\delta x^j - dx^j\,\delta x^h) ,$$

where R_{ihj}^k denotes the curvature tensor of the Riemannian space (CARTAN [7], p. 182). There is a discrepancy in sign between corresponding equations in the works [1] and [7] of CARTAN: this may be traced to a slight difference in notation regarding the curvature tensors.

[2] This relation results from covariant differentiation of the identity $g_{ij}(x, \dot{x})\,l^i l^j = 1$.

[3] We note that (1.45) and (1.46) are consistent with the definitions (1.25) and (1.31) in view of (3.2.4) and (3.2.14a).

the line-elements $(x^i + d\,x^i, \dot{x}^i + d\,\dot{x}^i)$, $(x^i + \delta\,x^i, \dot{x}^i + \delta\,\dot{x}^i)$ have the same centre x^i while the corresponding increments of l^i are given by $d\,l^i = \alpha^i$, $\delta l^i = \beta^i$, the increment associated with the corresponding cycle reduces to

$$\Omega_i{}^j = S^j_{ikh}\, \alpha^k\, \beta^h \,.$$

Secondly, if $\dot{x}^i + d\,\dot{x}^i$ results from $(x^i, \dot{x}^i)$ by parallel displacement from x^i to $x^i + d\,x^i$ with $d\,x^i = \alpha^i$, while $\dot{x}^i + \delta\,\dot{x}^i$ results from a mere rotation given by $\delta l^i = \beta^i$, the corresponding increment is given by

$$\Omega_i{}^j = P^j_{ikh}\, \alpha^k\, \beta^h \,.$$

Finally, if both line-elements result from $(x^i, \dot{x}^i)$ by parallel displacement from x to $x^i + d\,x^i$ and $x^i + \delta\,x^i$ respectively, with $d\,x^i = \alpha^i$ and $\delta\,x^i = \beta^i$, the corresponding increment is seen to be

$$\Omega_i{}^j = R^j_{ikh}\, \alpha^k\, \beta^h \,.$$

It is clear that these special results may also be deduced from the commutation formulae developed at the beginning of the present section.

§ 2. Identities Satisfied by the Curvature Tensors

As in Riemannian geometry, the structure of the curvature tensors is such that they satisfy a large number of identities, the most important of which we shall endeavour to derive in the present section. Apart from the identities (1.45) and (1.46), further identities which are obvious from the definitions of the tensors concerned are:

$$\tilde{K}^i_{jhk} = -\tilde{K}^i_{jkh} \; ; \qquad K^i_{jhk} = -K^i_{jkh} \; ; \tag{2.1}$$

$$S^i_{jhk} = -S^i_{jkh} ; \tag{2.2}$$

$$R^i_{jhk} = -R^i_{jkh} \,. \tag{2.3}$$

If we write

$$S_{ijkh} = g_{rj} S^r_{ikh} \,, \tag{2.4}$$

and

$$P_{ijkh} = g_{rj} P^r_{ikh} \,, \tag{2.5}$$

we have

$$S_{jihk} = -S_{ijhk} \,, \tag{2.6}$$

and

$$P_{jihk} = -P_{ijhk} \,, \tag{2.7}$$

together with

$$P_{ijkh}(x, \dot{x})\, \dot{x}^i = A_{jkh|i}(x, \dot{x})\, \dot{x}^i \,. \tag{2.8}$$

However, if we put

$$K_{ijhk} = g_{rj} K^r_{ihk} \,, \tag{2.9}$$

we see that these tensors are not, in general, skew-symmetric in i, j as in Riemannian geometry. For if we consider the special case of (1.16a) which arises when we put $T_{ij} = g_{ij}$, we have, in view of (2.9) and the fact that $g_{ij|h} = 0$ identically,

$$K_{jihk} = -K_{ijhk} - 2\,C_{ijl} K^l_{rhk}\, \dot{x}^r \,. \tag{2.10}$$

The corresponding formula for the relative curvature tensor $\tilde{K}_{jihk}$ is (1.15)[1]. In particular we have

$$K_{jihk}(x, \dot{x})\, \dot{x}^j = -K_{ijhk}(x, \dot{x})\, \dot{x}^j \tag{2.11}$$

in consequence of $(1.3.5)$.

For many purposes it is necessary to have an explicit representation of the tensor (2.9). This may easily be found as follows: if we differentiate the identities $\Gamma_{jh}^{*i} = g^{is}\Gamma_{jsh}^{*}$ with respect to x^r and $\dot{x}^l$ we find that

$$\frac{\partial \Gamma_{jh}^{*i}}{\partial x^r} - \frac{\partial \Gamma_{jh}^{*i}}{\partial \dot{x}^l}\frac{\partial G^l}{\partial \dot{x}^r} = g^{is}\left(\frac{\partial \Gamma_{jsh}^{*}}{\partial x^r} - \frac{\partial \Gamma_{jsh}^{*}}{\partial \dot{x}^l}\frac{\partial G^l}{\partial \dot{x}^r}\right)$$
$$+ \Gamma_{jsh}^{*}\left(\frac{\partial g^{is}}{\partial x^r} - \frac{\partial g^{is}}{\partial \dot{x}^l}\frac{\partial G^l}{\partial \dot{x}^r}\right).$$

But we have

$$0 = g^{is}{}_{|r} = \frac{\partial g^{is}}{\partial x^r} - \frac{\partial g^{is}}{\partial \dot{x}^l}\frac{\partial G^l}{\partial \dot{x}^r} + g^{ps}\Gamma_{pr}^{*i} + g^{ip}\Gamma_{pr}^{*s},$$

so that the above identity may be written in the form

$$\frac{\partial \Gamma_{jh}^{*i}}{\partial x^r} - \frac{\partial \Gamma_{jh}^{*i}}{\partial \dot{x}^l}\frac{\partial G^l}{\partial \dot{x}^r} + \Gamma_{mr}^{*i}\Gamma_{jh}^{*m}$$
$$= g^{is}\left(\frac{\partial \Gamma_{jsh}^{*}}{\partial x^r} - \frac{\partial \Gamma_{jsh}^{*}}{\partial \dot{x}^l}\frac{\partial G^l}{\partial \dot{x}^r} - \Gamma_{sr}^{*m}\Gamma_{jmh}^{*}\right).$$

A similar relation is obtained by interchanging h and r. On subtracting the latter from the above equation, it follows from (1.7) that the left-hand side becomes the tensor K_{jhr}^{i}: and hence, in view of (2.9), we find (after suitable interchange of suffixes)

$$K_{ijhk} = \left(\frac{\partial \Gamma_{ijh}^{*}}{\partial x^k} - \frac{\partial \Gamma_{ijh}^{*}}{\partial \dot{x}^l}\frac{\partial G^l}{\partial \dot{x}^k}\right) - \left(\frac{\partial \Gamma_{ijk}^{*}}{\partial x^h} - \frac{\partial \Gamma_{ijk}^{*}}{\partial \dot{x}^l}\frac{\partial G^l}{\partial \dot{x}^h}\right)$$
$$+ \Gamma_{jh}^{*m}\Gamma_{imk}^{*} - \Gamma_{jk}^{*m}\Gamma_{imh}^{*}. \tag{2.12}$$

The corresponding formula for the relative curvature tensor (1.5) reads:

$$\tilde{K}_{ijhk} = \left(\frac{\partial \Gamma_{ijh}^{*}}{\partial x^k} + \frac{\partial \Gamma_{ijh}^{*}}{\partial \dot{x}^l}\frac{\partial \xi^l}{\partial x^k}\right) - \left(\frac{\partial \Gamma_{ijk}^{*}}{\partial x^h} + \frac{\partial \Gamma_{ijk}^{*}}{\partial \dot{x}^l}\frac{\partial \xi^l}{\partial x^h}\right)$$
$$+ \Gamma_{jh}^{*m}\Gamma_{imk}^{*} - \Gamma_{jk}^{*m}\Gamma_{imh}^{*} - g_{jm;k}\Gamma_{ih}^{*m} + g_{jm;h}\Gamma_{ik}^{*m}. \tag{2.12a}$$

From (2.12) we may deduce directly the corresponding formula for the tensor defined by

$$R_{ijhk} = g_{jr}R_{ihk}^{r}. \tag{2.13}$$

For from (1.34) we have

$$R_{ijhk} = K_{ijhk} + C_{ijm}K_{rhk}^{m}\, \dot{x}^r, \tag{2.14}$$

---

[1] RUND [5], p. 95. Here the writer erroneously assumed that the skew-symmetry of the K_{ijhk} in i, j would follow from (1.15) in the transition from the "relative" to the well-defined curvature tensor. The correct equation is, of course, our equation (2.10).

and from (1.11) and (2.12) it then follows that[1]

$$R_{ijhk} = \left(\frac{\partial \Gamma_{ijh}^*}{\partial x^k} - \frac{\partial \Gamma_{ijh}^*}{\partial \dot{x}^l} \frac{\partial G^l}{\partial \dot{x}^k} \right) - \left(\frac{\partial \Gamma_{ijk}^*}{\partial x^h} - \frac{\partial \Gamma_{ijk}^*}{\partial \dot{x}^l} \frac{\partial G^l}{\partial \dot{x}^h} \right)$$
$$+ C_{ijm} \left(\frac{\partial^2 G^m}{\partial x^k \partial \dot{x}^h} - \frac{\partial^2 G^m}{\partial \dot{x}^k \partial x^h} - G_{hl}^m \frac{\partial G^l}{\partial \dot{x}^k} + G_{kl}^m \frac{\partial G^l}{\partial \dot{x}^h} \right) \quad (2.15)$$
$$+ \Gamma_{jh}^{*m} \Gamma_{imk}^* - \Gamma_{jk}^{*m} \Gamma_{imh}^* \,.$$

Furthermore, from (2.14) we have

$$R_{ijhk} + R_{jihk} = K_{ijhk} + K_{jihk} + 2C_{ijm} K_{rhk}^m \dot{x}^r \,,$$

and hence in consequence of (2.10) we deduce that

$$R_{ijhk} = -R_{jihk} \,. \tag{2.16}$$

Let us consider the sum $K_{jhk}^i + K_{kjh}^i + K_{hkj}^i$ (which is known to vanish identically in Riemannian geometry). On examining the expression (1.7) we notice that formally the tensor K_{jhk}^i is identical with its Riemannian counterpart except for the terms involving the directional derivatives of the Γ_{hk}^{*i}. Thus in writing down the sum under consideration, we would have a number of additional terms, namely

$$- \frac{\partial \Gamma_{jh}^{*i}}{\partial \dot{x}^l} \frac{\partial G^l}{\partial \dot{x}^k} + \frac{\partial \Gamma_{jk}^{*i}}{\partial \dot{x}^l} \frac{\partial G^l}{\partial \dot{x}^h} - \frac{\partial \Gamma_{kj}^{*i}}{\partial \dot{x}^l} \frac{\partial G^l}{\partial \dot{x}^h} + \frac{\partial \Gamma_{kh}^{*i}}{\partial \dot{x}^l} \frac{\partial G^l}{\partial \dot{x}^j}$$
$$- \frac{\partial \Gamma_{hk}^{*i}}{\partial \dot{x}^l} \frac{\partial G^l}{\partial \dot{x}^j} + \frac{\partial \Gamma_{hj}^{*i}}{\partial \dot{x}^l} \frac{\partial G^l}{\partial \dot{x}^k} \,,$$

while the sum of the remaining terms is zero as in Riemannian geometry. But in view of the symmetry of the Γ_{hk}^{*i} in h, k we see that the above expression also vanishes identically: hence we have

$$K_{jhk}^i + K_{kjh}^i + K_{hkj}^i = 0 \,. \tag{2.17}$$

An obvious consequence of this identity is

$$K_{jihk} + K_{kijh} + K_{hikj} = 0 \,, \tag{2.17a}$$

and if we apply (2.10) in turn to each of the tensors on the left-hand side, we find

$$K_{ijhk} + K_{ikjh} + K_{ihkj}$$
$$= -2\{ C_{ijl} K_{rhk}^l + C_{ikl} K_{rjh}^l + C_{ihl} K_{rkj}^l \} \dot{x}^r \,. \tag{2.18}$$

In particular, if we substitute in this result from (1.34) we have the identity[2]

$$R_{ijhk} + R_{ikjh} + R_{ihkj}$$
$$+ \{ C_{ijl} K_{rhk}^l + C_{ikl} K_{rjh}^l + C_{ihl} K_{rkj}^l \} \dot{x}^r = 0 \,. \tag{2.19}$$

---

[1] Cartan [1], p. 36, formula XIX.

[2] This identity is stated by Cartan [1], p. 37. Equation (2.17) is given by Rund [5], p. 96. Clearly (2.17) is satisfied also by the relative curvature tensor $\widetilde{K}_{jhk}^i$.

From (2.1) and (2.18) we deduce

$$K_{ihjk} - K_{ikjh} = K_{ijhk} + 2Q_{ijhk}, \qquad (2.20)$$

where we have put, for the sake of brevity,

$$Q_{ijhk} = \{C_{ijl}K^l_{rhk} + C_{ikl}K^l_{rjh} + C_{ihl}K^l_{rkj}\}\, \dot{x}^r, \qquad (2.21)$$

noting that

$$Q_{ijhk} = -Q_{ijkh} \qquad (2.22)$$

in consequence of (2.21) and (2.1). Similarly

$$K_{jkih} - K_{jhik} = K_{jikh} + 2Q_{jikh}.$$

On adding this equation to (2.20) and observing (2.10) and (2.1) we find

$$\begin{aligned}
K_{ihjk} &- K_{ikjh} + K_{jkih} - K_{jhik} \\
&= 2K_{ijhk} + 2C_{ijl}K^l_{rhk}\,\dot{x}^r + 2(Q_{ijhk} - Q_{jihk}) \\
&= 2R_{ijhk} + 2(Q_{ijhk} - Q_{jihk}),
\end{aligned} \qquad (2.23)$$

where we have used (1.34).

A similar equation is obtained by interchanging the pairs of indices $i,\, j$ and $h,\, k$. On subtracting the equation thus obtained from (2.23), we deduce that

$$\begin{aligned}
2\{(R_{ijhk} &- R_{hkij}) + (Q_{ijhk} - Q_{jihk}) - (Q_{hkij} - Q_{khij})\} \\
&= (K_{ihjk} + K_{hijk}) - (K_{ikjh} + K_{kijh}) + \\
&\quad + (K_{jkih} + K_{kjih}) - (K_{jhik} + K_{hjik}) \\
&= -2\{C_{ihl}K^l_{rjk} - C_{ikl}K^l_{rjh} + C_{jkl}K^l_{rih} - C_{jhl}K^l_{rik}\}\, \dot{x}^r,
\end{aligned} \qquad (2.24)$$

in view of (2.10). But from (2.21) it follows directly that

$$\begin{aligned}
(Q_{ijhk} - Q_{jihk}) + (Q_{khij} - Q_{hkij}) = 2\{&C_{ikl}K^l_{rjh} - \\
- C_{ihl}K^l_{rjk} &- C_{jkl}K^l_{rih} + C_{jhl}K^l_{rik}\}\, \dot{x}^r.
\end{aligned}$$

On substituting this result in (2.24) we finally have

$$\begin{aligned}
R_{ijhk} - R_{hkij} = \{&C_{ihl}K^l_{rjk} - C_{ikl}K^l_{rjh} + \\
&+ C_{jkl}K^l_{rih} - C_{jhl}K^l_{rik}\}\, \dot{x}^r.
\end{aligned} \qquad (2.25)$$

The corresponding formula for the tensor K_{ijhk} results directly from (1.34) and (2.25):

$$\begin{aligned}
K_{ijhk} - K_{hkij} = \{&C_{ihl}K^l_{rjk} - C_{ikl}K^l_{rjh} \\
&+ C_{jkl}K^l_{rih} - C_{jhl}K^l_{rik} + C_{hkl}K^l_{rij} - C_{ijl}K^l_{rhk}\}\, \dot{x}^r.
\end{aligned} \qquad (2.26)$$

In view of (2.1) and (1.3.5) we have in particular

$$(R_{ijhk} - R_{hkij})\, \dot{x}^h = (C_{jkl}K^l_{rih} - C_{ikl}K^l_{rjh})\, \dot{x}^r\, \dot{x}^h, \qquad (2.27)$$

and

$$(K_{ijhk}- K_{hkij})\,\dot{x}^h = (C_{jkl}K^l_{rih}- C_{ikl}K^l_{rjh}- C_{ijl}K^l_{rhk})\,\dot{x}^r\,\dot{x}^h\,,\qquad (2.27\,\mathrm{a})$$

together with

$$(R_{ijhk}- R_{hkij})\,\dot{x}^i\,\dot{x}^h = 0\,.\qquad\qquad (2.28)$$

It should be noted that in the above identities all terms of the type $\dot{x}^r\,K^i_{rhk}$ may be replaced by a corresponding term involving R^i_{rhk} by means of (1.34 a).

§ 3. The Bianchi Identities

Let us consider an arbitrary covariant vector field $Y_i(x^k)$ which we assume to be independent of direction. The commutation formulae (1.16) for such a field would read

$$Y_{i|jk}- Y_{i|kj}= -Y_r K^r_{ijk}\,.$$

Covariant differentiation of this relation with respect to x^h yields

$$Y_{i|jkh}- Y_{i|kjh}= - Y_{r|h}K^r_{ijk}- Y_r K^r_{ijk|h}\,.$$

Two similar equations may be obtained by cyclic interchange of the indices j, k, h. These relations are added and rearranged: this gives

$$(Y_{i|jkh}- Y_{i|jhk}) + (Y_{i|hjk}- Y_{i|hkj}) + (Y_{i|khj}- Y_{i|kjh})$$
$$= - (Y_{r|h}K^r_{ijk}+ Y_{r|k}K^r_{ihj}+ Y_{r|j}K^r_{ikh}) -$$
$$- Y_r(K^r_{ijk|h}+ K^r_{ihj|k}+ K^r_{ikh|j})\,.$$

The brackets on the left-hand side may be expressed by means of (1.16) in terms of the curvature tensors. If this is done, the first term in brackets on the right-hand side cancels with a similar term on the left, and we are left with

$$\left(\frac{\partial Y_{i|j}}{\partial \dot{x}^l}\,K^l_{skh}+ \frac{\partial Y_{i|h}}{\partial \dot{x}^l}\,K^l_{sjk}+ \frac{\partial Y_{i|k}}{\partial \dot{x}^l}\,K^l_{shj}\right)\dot{x}^s$$
$$= - Y_{i|r}(K^r_{jkh}+ K^r_{hjk}+ K^r_{khj})\qquad\qquad (3.1)$$
$$+ Y_r(K^r_{ijk|h} + K^r_{ihj|k}+ K^r_{ikh|j})\,.$$

In this equation the coefficient of $Y_{i|r}$ vanishes identically as a result of (2.17), while for the case under consideration we have

$$\frac{\partial Y_{i|j}}{\partial \dot{x}^l}= - \frac{\partial \Gamma^{*r}_{ij}}{\partial \dot{x}^l}\,Y_r\,.$$

We therefore have from (3.1) (since Y_r is essentially arbitrary)[1]

$$K^r_{ijk|h}+ K^r_{ihj|k}+ K^r_{ikh|j} +$$
$$+ \left(\frac{\partial \Gamma^{*r}_{ij}}{\partial \dot{x}^l}\, K^l_{skh}+ \frac{\partial \Gamma^{*r}_{ih}}{\partial \dot{x}^l}\, K^l_{sjk}+ \frac{\partial \Gamma^{*r}_{ik}}{\partial \dot{x}^l}\, K^l_{shj}\right) \dot{x}^s= 0 \, . \tag{3.2}$$

This equation represents our first form of the Bianchi identities. The same argument, applied to the relative curvature tensor yields similarly

$$\tilde{K}^r_{ijk;h}+ \tilde{K}^r_{ihj;k}+ \tilde{K}^r_{ikh;j}= 0 \, , \tag{3.2a}$$

where we note that the semi-colon refers as before to partial δ-differentiation.

A second form of the Bianchi identities, involving the curvature tensors R^i_{jhk} and P^i_{jhk} may be derived as follows from (3.2). Let us write (1.34) in the form

$$K^r_{ijk}= R^r_{ijk}- A^r_{im}K^m_{sjk}\, l^s \, .$$

Differentiating this equation covariantly with respect to x^h, noting the identity (3.2.12), we have

$$K^r_{ijk|h}= R^r_{ijk|h}- A^r_{im|h}K^m_{sjk}\, l^s- A^r_{im}K^m_{sjk|h}\, l^s \, .$$

Two similar equations result from a cyclic interchange of the indices j, k, h. On adding the three equations thus obtained, we find

$$K^r_{ijk|h}+ K^r_{ihj|k}+ K^r_{ikh|j}= R^r_{ijk|h}+ R^r_{ihj|k}+ R^r_{ikh|j}-$$
$$- A^r_{im|h}K^m_{sjk}\, l^s- A^r_{im|k}K^m_{shj}\, l^s- A^r_{im|j}K^m_{skh}\, l^s-$$
$$- A^r_{im}(K^m_{sjk|h}+ K^m_{shj|k}+ K^m_{skh|j})\, l^s \, .$$

The identity (3.2) is now applied to the left-hand side of this equation as well as to the last term in brackets on the right-hand side. This gives

$$0 = R^r_{ijk|h}+ R^r_{ihj|k}+ R^r_{ikh|j}+$$
$$+ F \left(\frac{\partial \Gamma^{*r}_{ij}}{\partial \dot{x}^l}\, K^l_{skh}+ \frac{\partial \Gamma^{*r}_{ih}}{\partial \dot{x}^l}\, K^l_{sjk}+ \frac{\partial \Gamma^{*r}_{ik}}{\partial \dot{x}^l}\, K^l_{shj}\right) l^s-$$
$$- A^r_{im|h}K^m_{sjk}\, l^s- A^r_{im|k}K^m_{shj}\, l^s- A^r_{im|j}K^m_{skh}\, l^s +$$
$$+ A^r_{im}\left(F \frac{\partial \Gamma^{*m}_{ij}}{\partial \dot{x}^l}\, K^l_{pkh}+ F \frac{\partial \Gamma^{*m}_{ih}}{\partial \dot{x}^l}\, K^l_{pjk}+ F \frac{\partial \Gamma^{*m}_{ik}}{\partial \dot{x}^l}\, K^l_{phj}\right) l^s\, l^p \, .$$

---

[1] Clearly we can deduce (3.2) by a similar argument even if we do not assume that the vector field Y_i is independent of direction, although the calculations would be considerably more involved. In fact, we would obtain in (3.2) an additional term:

$$\frac{\partial Y_i}{\partial \dot{x}^r}\, \{(K^r_{sjk|h}+ K^r_{shj|k}+ K^r_{skh|j}) +$$
$$+ \dot{x}^m(C^r_{ih|m}K^l_{sjk}+ C^r_{ik|m}K^l_{shj}+ C^r_{jl|m}K^l_{skh})\} \, \dot{x}^s \, ,$$

but in view of (3.3.15) it is seen that the coefficient of $\partial Y_i/\partial \dot{x}^r$ is equivalent to the left-hand side of (3.2) contracted with $\dot{x}^i$.

Collecting terms, we see that this may be written as

$$0 = R^r_{ijk\backslash h} + R^r_{ihj\backslash k} + R^r_{ikh\backslash j} +$$
$$+ l^s K^l_{skh}\left(F\frac{\partial \Gamma^{*r}_{ij}}{\partial \dot{x}^l} - A^r_{il\backslash j} + A^r_{im}F\frac{\partial \Gamma^{*m}_{sj}}{\partial \dot{x}^l}l^s\right) +$$
$$+ l^s K^l_{sjk}\left(F\frac{\partial \Gamma^{*r}_{ih}}{\partial \dot{x}^l} - A^r_{il\backslash h} + A^r_{im}F\frac{\partial \Gamma^{*m}_{sh}}{\partial \dot{x}^l}l^s\right) +$$
$$+ l^s K^l_{shj}\left(F\frac{\partial \Gamma^{*r}_{ik}}{\partial \dot{x}^l} - A^r_{il\backslash k} + A^r_{im}F\frac{\partial \Gamma^{*m}_{sk}}{\partial \dot{x}^l}l^s\right).$$

But from (1.30) and (3.2.14a) we have

$$F\frac{\partial \Gamma^{*m}_{rj}}{\partial \dot{x}^l}l^r = A^m_{jl\backslash r}l^r,$$

and thus, by definition (1.29), the coefficient of $l^s K^l_{skh}$ is simply

$$F\frac{\partial \Gamma^{*r}_{ij}}{\partial \dot{x}^l} - A^r_{il\backslash j} + A^r_{im}A^m_{jl\backslash s}l^s = P^r_{ijl}.$$

Similarly the third and fourth lines of the above equation may be simplified, and using (1.34a) we finally obtain

$$0 = R^r_{ijk\backslash h} + R^r_{ihj\backslash k} + R^r_{ikh\backslash j} +$$
$$+ l^m(R^l_{mkh}P^r_{ijl} + R^l_{mjk}P^r_{ihl} + R^l_{mhj}P^r_{ikl}). \tag{3.3}$$

This equation represents the second form of the Bianchi identities[1].

Again, one might alternatively approach these equations by means of the method of exterior forms. For on taking the exterior derivative of (1.43a) according to the usual rules

$$(\Omega_{ij})' = [(\omega_j{}^k)'\,\omega_{ik}] - [\omega_j{}^k(\omega_{ik})'],$$

and on substituting from (1.43) on the right-hand side, one obtains

$$(\Omega_{ij})' - [\omega_j{}^k\,\Omega_{ik}] + [\omega_i{}^k\,\Omega_{jk}] = 0. \tag{3.4}$$

Similarly, from (1.39) and (1.43) we find

$$(\Omega^i)' - [d\,x^k\,\Omega_k{}^i] + [\omega_h{}^i\,\Omega^h] = 0. \tag{3.5}$$

As in Riemannian geometry[2] these equations also lead to the identities (3.3) after further calculation.

§ 4. Geodesic Deviation

As a first application of the theory developed in the previous sections we shall consider the question of geodesic deviation. Problems of this nature were first discussed by LEVI-CIVITA (for the case of Riemannian spaces) and it was evident that in this manner certain aspects of the geometry of a space with non-vanishing curvature tensor could be

---

[1] CARTAN [1], p. 37, formula XXIV. The form (3.2a) was given by RUND [5], p. 97. Between the formula (3.3) and formula XXIV of CARTAN there is a discrepancy in sign for which the writer cannot give an explanation, as CARTAN did not explicitly indicate the steps taken in the course of his calculation.

[2] CARTAN [7], pp. 210—211.

described by means of some very striking theorems. A similar approach applicable to Finsler spaces will be developed in the present section.

The geometrical background to the simplest form of the equations of geodesic deviation may be sketched briefly as follows. Consider two vectors $\dot{x}^i_{(0)}$, $\dot{x}^i_{(1)}$ of the tangent space $T_n(P)$ of a point P of F_n, these vectors issuing from the origin of $T_n(P)$. We suppose that the angle between these vectors at the origin is small. Let us construct the geodesics Γ_0 and Γ_1 of F_n passing through P, tangent to $\dot{x}^i_{(0)}$ and $\dot{x}^i_{(1)}$ respectively. On each geodesic we mark off points Q_0, Q_1, such that the geodesic arc-lengths PQ_0 and PQ_1 are equal, these arc-lengths being denoted by s. The displacement $Q_0 Q_1$, regarded as an infinitesimal vector ζ^i (to a first approximation), will be a function of s, so that we may form its first and second δ-derivatives, namely $\delta\zeta^i/\delta s$, $\delta^2\zeta^i/\delta s^2$. It will be seen that in general

$$\frac{\delta^2\,\zeta^i}{\delta s^2} \neq 0 ,$$

in contradistinction to a euclidean space (or a Minkowskian space) in which these second derivatives vanish identically.

Thus these derivatives represent an invariant measure of a kind of "deviation", *namely the deviation of the geometry of F_n in the neighbourhood of the point P as compared with the geometry of the tangent space at P.* It will be seen that this invariant measure is expressed by means of the curvature tensor K^j_{ihk}. Furthermore, we shall indicate that the analysis resulting from this construction gives rise to a large number of more particular theorems of a purely geometrical nature, while the intimate relationship of the equations of geodesic deviation to the accessory problem (i. e. the theory of the second variation) of the calculus of variations will become evident in the next section.

It is, however, possible to develop the theory of a more general construction which includes the one described above as a special case, and it is to our advantage to do so here. Thus in order to derive the equations of geodesic deviation in their most general form we shall first proceed as follows.

Let F_2 be a two-dimensional subspace of the Finsler space F_n, and suppose that F_2 may be represented parametrically by the equations $x^i = x^i(u, v)$, so that u, v are the Gaussian parameters of the surface. We shall assume that the corresponding Jacobian is of rank 2, the functions $x^i(u, v)$ being at least of class C^4. The parameter lines on F_2 are the curves $u = \text{const.}$, $v = \text{const.}$, and we shall denote their tangent vectors by η^i, ξ^i respectively: thus

$$\xi^i = \frac{\partial x^i}{\partial u}, \qquad \eta^i = \frac{\partial x^i}{\partial v}, \tag{4.1}$$

so that

$$\frac{\partial \xi^i}{\partial v} = \frac{\partial^2 x^i}{\partial u \, \partial v} = \frac{\partial \eta^i}{\partial u} \, . \tag{4.2}$$

Let $X^i(x^k)$ be a vector field defined over a region of F_n containing F_2. If P is a point on F_2, and $\dot{x}^i$ a field (which we shall specialise presently) tangent to F_2 in a neighbourhood of P, we define at P in accordance with (2.4.9) the following δ-derivatives:

$$\frac{\delta X^i}{\delta u} = X^i_{;h}\xi^h \equiv \left(\frac{\partial X^i}{\partial x^h} + \Gamma^{*i}_{hk}(x, \dot{x}) \, X^k \right) \xi^h \, , \tag{4.3}$$

together with

$$\frac{\delta X^i}{\delta v} = X^i_{;h}\eta^h \equiv \left(\frac{\partial X^i}{\partial x^h} + \Gamma^{*i}_{hk}(x, \dot{x}) \, X^k \right) \eta^h \, . \tag{4.3a}$$

In particular, we have, in view of (4.1) and (4.2)

$$\frac{\delta \xi^i}{\delta v} = \frac{\partial^2 x^i}{\partial u \, \partial v} + \Gamma^{*i}_{hk}(x, \dot{x}) \, \xi^k \, \eta^h = \frac{\delta \eta^i}{\delta u} \, , \tag{4.4}$$

where we have used the symmetry of the Γ^{*i}_{hk} with respect to the suffixes h and k. Furthermore,

$$\frac{\delta^2 X^i}{\delta v \, \delta u} = (X^i_{;h}\xi^h)_{;k}\eta^k = X^i_{;hk}\xi^h \, \eta^k + X^i_{;h}\frac{\delta \xi^h}{\delta v} \, ,$$

so that as a result of (4.4)

$$\frac{\delta^2 X^i}{\delta v \, \delta u} - \frac{\delta^2 X^i}{\delta u \, \delta v} = (X^i_{;hk} - X^i_{;kh}) \, \xi^h \, \eta^k \, ,$$

or, from (1.4),

$$\frac{\delta^2 X^i}{\delta v \, \delta u} - \frac{\delta^2 X^i}{\delta u \, \delta v} = \tilde{K}^i_{jhk}(x, \dot{x}) \, X^j \xi^h \, \eta^k \, . \tag{4.5}$$

Here the "relative" curvature tensor $\tilde{K}^i_{jhk}$ refers to the (as yet) undefined field $\dot{x}^i(x^k)$.

We shall now consider two neighbouring parameter curves $v = \text{const.}$, $v + \varepsilon = \text{const.}$ of F_2. Let us denote these curves by C, C' respectively. Suppose C, C' are cut in the points A_1, B_1 respectively by the parameter curve $u = u_1 = \text{const.}$; similarly, two points A_2, B_2 are defined by the intersection of C, C' with the parameter curve $u = u_2 = \text{const.}$ Let A be an arbitrary point on the arc $A_1 A_2$ of C corresponding to a parameter-value u; to this point let there correspond a point B on the arc $B_1 B_2$ of C' whose parameter-value u^* is given by

$$u^* = u + f(u) \, , \tag{4.6}$$

where we suppose that f is at least of class C^2 and such that $(u^* - u)$ is of the order of magnitude of ε. Neglecting quantities of the order of ε^2, the difference z^i between the coordinates of the points A, B is given by

$$z^i = \frac{\partial x^i}{\partial v}\varepsilon + \frac{\partial x^i}{\partial u}f(u) = \varepsilon \, \eta^i + f(u) \, \xi^i \, . \tag{4.7}$$

We shall now endeavour to evaluate the second covariant derivatives of z^i with respect to u; these will be defined uniquely if we identify the field $\dot{x}^i$ with the field ξ^i, which is defined over F_2. From (4.4) we have

$$\frac{\delta^2 \eta^i}{\delta u^2} = \frac{\delta^2 \xi^i}{\delta u\,\delta v} = \frac{\delta^2 \xi^i}{\delta v\,\delta u} - \widetilde{K}^i_{jhk}(x,\,\xi)\,\xi^j\,\xi^h\,\eta^k\,,$$

where in the second step we have applied (4.5) with $X^i = \xi^i$, $\dot{x}^i = \xi^i$. But as a result of the skew-symmetry of the curvature tensor with respect to the indices h, k, we have $\widetilde{K}^i_{jhk}(x,\,\xi)\,\xi^h\,\xi^k = 0$, and hence we deduce from (4.7) that the last equation may be written in the form

$$\varepsilon\,\frac{\delta^2 \eta^i}{\delta u^2} = \varepsilon\,\frac{\delta^2 \xi^i}{\delta v\,\delta u} - \widetilde{K}^i_{jhk}(x,\,\xi)\,\xi^j\,\xi^h\,z^k\,. \tag{4.8}$$

Also, in virtue of the fact that ε is constant for the two curves C, C', it follows from (4.7) that

$$\frac{\delta^2 z^i}{\delta u^2} = \varepsilon\,\frac{\delta^2 \eta^i}{\delta u^2} + f(u)\,\frac{\delta^2 \xi^i}{\delta u^2} + 2f'(u)\,\frac{\delta \xi^i}{\delta u} + f''(u)\,\xi^i\,, \tag{4.9}$$

the dash denoting differentiation with respect to u. Thus from (4.8) and (4.9) we have

$$\frac{\delta^2 z^i}{\delta u^2} + \widetilde{K}^i_{jhk}(x,\,\xi)\,\xi^j\,\xi^h\,z^k - f''(u)\,\xi^i$$
$$= f(u)\,\frac{\delta^2 \xi^i}{\delta u^2} + 2f'(u)\,\frac{\delta \xi^i}{\delta u} + \varepsilon\,\frac{\delta^2 \xi^i}{\delta v\,\delta u}\,. \tag{4.10}$$

This equation represents under the most general conditions the change of the variation vector z^i as we proceed along C. By successively specialising these conditions we obtain various forms of the equations of geodesic deviation. Firstly, let the curves $v = $ const. be geodesics of F_n, while u is the arc-length along these curves. Then, by (4.1) and (2.3.20) we have $\delta \xi^i / \delta u = 0$ along C, and since the same is true along C', $\dfrac{\partial}{\partial v}\left(\dfrac{\delta \xi^i}{\delta u}\right) = 0$, so that $\dfrac{\delta^2 \xi^i}{\delta v\,\delta u} = 0$. Under these circumstances (4.10) reduces to the form

$$\frac{\delta^2 z^i}{\delta u^2} + K^i_{jhk}(x,\,\xi)\,\xi^j\,\xi^h\,z^k - f''(u)\,\xi^i = 0\,. \tag{4.11}$$

This relation represents the exact analogue of the equation of geodesic deviation of Levi-Civita[1]; and it enables us to derive the geo-

---

[1] Levi-Civita [1], p. 215. Similar equations are stated by Berwald [12], p. 303, in connection with the second variation of the length integral along a geodesic. The above geometrical generalisation of the method of Levi-Civita is equivalent to results given by Rund [7, 8]. Furthermore, it is to be noted that in (4.11) we have replaced $\widetilde{K}^i_{jhk}$ by K^i_{jhk}. For if the curves $v = $ const. are geodesics, we have

$$(\partial \xi^i / \partial x^h)\,\xi^h = (\partial \xi^i / \partial x^h)\,(\partial x^h / \partial u) = d\xi^i / du$$

along these curves, and on substituting for the latter from the equations $d\xi^i / du + 2G^i(x,\,\xi) = 0$ of the geodesics, it follows from the homogeneity of the function $G^i(x,\,\xi)$ and equation (2.5.12) that the tensor $\widetilde{K}^i_{jhk}\,\xi^j\,\xi^h$ defined by (1.5) becomes the tensor $K^i_{jhk}\,\xi^j\,\xi^h$ in accordance with its definition (1.6).

metrical meaning of the function $f(u)$ for this special case. In accordance with (1.4.1) let us put

$$y_i = g_{ij}(x, \xi)\, \xi^j, \tag{4.12}$$

so that $y_i\, \xi^i = 1$, since $ds^2 = du^2 = g_{ij}(x, \xi)\, dx^i\, dx^j$ along C. Again, since C is a geodesic of F_n, it follows from (4.12) and (2.5.8) that

$$\frac{\delta y_i}{\delta u} = 0 \tag{4.13}$$

along C. On multiplying (4.11) by y_i and noting that in virtue of (2.11)

$$K_{jihk}(x, \xi)\, \xi^j\, \xi^i = 0\,,$$

we find as a result of (4.12) and (4.13)

$$\frac{\delta}{\delta u}\left(y_i \frac{\delta z^i}{\delta u}\right) - f''(u) = 0\,,$$

or

$$\frac{d}{du}\left(y_i \frac{\delta z^i}{\delta u} - f'(u)\right) = 0\,. \tag{4.14}$$

This result represents a first integral of equations (4.11).

But on differentiating (4.7) multiplied by y_i with respect to u, we observe that as a result of (4.12) and (4.13) we have

$$\varepsilon \frac{d}{du}(y_i\, \eta^i) = \frac{\delta}{\delta u}(y_i\, z^i - f(u)) = y_i \frac{\delta z^i}{\delta u} - f'(u)\,,$$

so that equation (4.14) becomes

$$\frac{d}{du}(y_i\, \eta^i) = C_1\,, \tag{4.15}$$

where C_1 is a constant of integration. Using this result it is easily shown[1] that $C_1 = 0$ under the geometrical conditions specified above, and thus it follows from (4.14) that $f'(u)$ is the projection of the vector $\delta z^i/\delta u$ onto the tangent vector ξ^i of the geodesic C at the point P.

If we choose the function $f(u)$ such that $f''(u) = 0$, equations (4.11) obviously reduce to the form[2]

$$\frac{\delta^2 z^i}{\delta u^2} + K^i_{jhk}(x, \xi)\, \xi^j\, \xi^h\, z^k = 0\,, \tag{4.16}$$

---

[1] RUND [7], p. 6. A direct proof of this statement may also be established on the basis of the fact that the tangent vectors to the geodesics $v = $ const. form a unit vector field. We shall omit this discussion since no further use will be made of this simplification.

[2] In view of (1.34) it is clear that equation (4.16) may also be written in terms of Cartan's curvature tensor in the alternative form

$$\frac{\delta^2 z^i}{\delta u^2} + R^i_{jhk}(x, \xi)\, \xi^j\, \xi^h\, z^k = 0\,.$$

This particular case of (4.11) is obtained by CARTAN [1], p. 40 (equation XXV). Similar equations may also be derived in the general geometry of paths (Ch. III, § 3): in fact they may be obtained almost directly from the equations of variation of the differential equations of the paths. This process is discussed by BERWALD [10, IV] and KOSAMBI [1—4]. It should be noted that these equations of variation are intimately connected with the inverse problem in the calculus of variations as is evident in particular from the work of DAVIS [1, 2]. See also DAVIES [1], SU [1].

which now admits the first integral

$$y_i \frac{\delta z^i}{\delta u} = f'(u) = \text{const.} \,, \qquad (4.17)$$

so that if we use (4.13) once more we have

$$y_i z^i = f'(u)\, u + C_2 \,, \qquad (4.17\,\text{a})$$

where C_2 is another constant. If, in particular, we further choose $f(u)$ such that $f'(u) = 0$, this relation expresses the fact that we can establish a correspondence between the geodesics C and C' such that the projection of z^i onto the unit tangent vector ξ^i of C remains constant. A *normal* correspondence is one for which z^i is normal with respect to ξ^i, in which case (4.17a) reduces to

$$y_i z^i = 0 \,, \qquad (4.17\,\text{b})$$

in accordance with the definition of normality.

Equation (4.16) is of great significance from several points of view. Firstly, as in classical differential geometry, it leads to a discussion of problems "in the large", such as the theorem of BONNET concerning diameters of two-dimensional convex surfaces[1]. Secondly, equation (4.16) is intimately connected with the theory of the second variation of the length integral, and hence with the accessory problem in the calculus of variations as will be seen in the next section. In order to exhibit the latter relationship more clearly it is necessary to derive an alternative form of these equations. Let X^i be a vector field defined along the geodesic C such that the direction of X^i coincides with that of z^i, while X^i is normalised so as to satisfy the relation

$$g_{ij}(x, \xi)\, X^i X^j = 1 \qquad (4.18)$$

along C. On differentiating this equation with respect to u and noting that $\delta g_{ij}(x, \xi)/\delta u = 0$ along a geodesic by (2.5.8a), we have

$$g_{ij}(x, \xi)\, X^i \frac{\delta X^j}{\delta u} = 0. \qquad (4.19)$$

Also, by definition of X^i, there exists a scalar function $z(u)$ such that

$$z^i = z X^i \qquad (4.20)$$

along C, so that we have

$$\left.\begin{aligned}
\frac{\delta z^i}{\delta u} &= z' X^i + z \frac{\delta X^i}{\delta u} \,; \\[2mm]
\frac{\delta^2 z^i}{\delta u^2} &= z'' X^i + 2 z' \frac{\delta X^i}{\delta u} + z \frac{\delta^2 X^i}{\delta u^2} \,.
\end{aligned}\right\} \qquad (4.21)$$

If we substitute this result in (4.16), multiplying the resulting equation by $g_{ij}(x, \xi)\, X^j$, and noting (4.18), (4.19) and (4.20), we obtain the follow-

____

[1] BLASCHKE [4], p. 218.

ing differential equation of the second order for the scalar function $z(u)$:

$$z'' + z\left[g_{ij}(x, \xi)\, X^j\, \frac{\delta^2 X^i}{\delta u^2} + K_{jihk}\, \xi^j\, \xi^h\, X^i\, X^k\right] = 0 , \qquad (4.22)$$

or, alternatively, as is easily verified by differentiation of (4.19),

$$z'' + z\left[K_{jihk}(x, \xi)\, \xi^j\, \xi^h\, X^i\, X^k - g_{ij}(x, \xi)\, \frac{\delta X^i}{\delta u} \cdot \frac{\delta X^j}{\delta u}\right] = 0 . \qquad (4.22\,\text{a})$$

Except for the last term on the left-hand side, this relation has the same form as the equation of JACOBI in the classical theory of surfaces[1], provided we define the scalar Riemannian curvature for two directions (ξ, X) in a manner analogous to that of Riemannian geometry[2] by means of the formula

$$R(x, \xi, X) = \frac{K_{jihk}(x, \xi)\, \xi^j\, \xi^h\, X^i\, X^k}{(g_{jh}(x, \xi)\, g_{ik}(x, \xi) - g_{ji}(x, \xi)\, g_{hk}(x, \xi))\, \xi^j\, \xi^h\, X^i\, X^k} . \qquad (4.23)$$

For assuming that we are dealing with a normal variation (4.17b) [i. e. $g_{ij}(x, \xi)\, \xi^j\, z^i = z\, g_{ij}(x, \xi)\, \xi^j\, X^i = 0$], and since ξ^i is a unit vector field while the X^i satisfy (4.18), the denominator in (4.23) is simply unity, and equation (4.22a) becomes[3]

$$z'' + z\left[R(x, \xi, X) - g_{ij}(x, \xi)\, \frac{\delta X^i}{\delta u} \cdot \frac{\delta X^j}{\delta u}\right] = 0 . \qquad (4.24)$$

Thus the tensor equations (4.16) have been reduced to a *single* invariant equation of similar significance. For if there exists a solution $z = z(u)$ of (4.24) which has a zero for $u = u_1$ say, then, in view of (4.20), $z^i = 0$ for $u = u_1$.

As a result of assumption C (Ch. I, § 1) the last term on the left-hand side of (4.24) vanishes if and only if $\delta X^i/\delta u = 0$, which is not generally true, since X^i depends on the choice of not only the geodesic C but also of C'. However, for the case $n = 2$ we shall see that $\delta X^i/\delta u = 0$[4]. This is easily proved as follows: Firstly, we note that in virtue of equations (4.17b), (4.13), (4.20) and the first of (4.21) we have

$$y_i\, \frac{\delta X^i}{\delta u} = 0 .$$

---

[1] BLASCHKE [4], p. 216.

[2] EISENHART [1], p. 81.

[3] Alternative forms of equation (4.24) are given by RUND [11], pp. 188—189.

[4] This is a special case of a more general theorem (RUND [11], p. 185) concerning the geometry of the two-dimensional subspace F_2 containing the geodesics C, C'. In fact, if $X^\alpha (\alpha, \beta, \ldots = 1, 2)$ are the components of X^i in F_2 (endowed with a coordinate system u^α and with an induced Finsler metric $g_{\alpha\beta}$ as defined in Chapter V), then $\delta X^\alpha/\delta u = 0$, the covariant differentiation referring to the induced $\Gamma^{*\alpha}_{\beta\gamma}$, irrespective of whether $\delta X^i/\delta u = 0$ or not.

Secondly, since by hypothesis the vectors X^i and ξ^i are non-parallel, we may, for the case $n = 2$, express the vector $\delta X^i/\delta u$ as a linear combination of X^i and ξ^i:

$$\frac{\delta X^i}{\delta u} = \lambda X^i + \mu \, \xi^i \,.$$

On multiplying this equation by y_i, summing over i, we see that as a result of (4.12) and (4.17b) the component μ vanishes. Thus $\delta X^i/\delta u = \lambda X^i$, and on substituting this result in (4.19) and observing condition C once more, we find that the component $\lambda = 0$. This proves our assertion, as a consequence of which equation (4.24) for the case of a two-dimensional Finsler space reduces to the form

$$z'' + z \, R(x, \xi) = 0 \,. \tag{4.24a}$$

Thus the classical equation of JACOBI for a two-dimensional Riemannian space[1] may be generalised directly to the case of a two-dimensional Finsler space and applied in a similar manner to the derivation of theorems such as the one mentioned above.

For instance, it is not difficult to show that (4.24a) leads to generalisations of the formulae of BERTRAND and PUISEUX of classical differential geometry[2]. Consider a fixed point O of a two-dimensional Finsler space F_2 with coordinates $x^i_{(0)}$ $(i = 1, 2)$, and a fixed direction $\dot{x}^i_{(0)}$ through O. Denote by Γ_0 the geodesic of F_2 tangent to $\dot{x}^i_{(0)}$ at O and construct the family of geodesics through O whose (directed) tangent vectors are inclined at angles φ, $0 \leqq \varphi \leqq \Phi$, to $\dot{x}^i_{(0)}$. Here φ is the angle of LANDSBERG as defined by equation (1.7.12). By marking off equal distances $s = r = $ const. (measured from O) along the geodesics of this family, we obtain a "sector" of radius r and angle Φ at the vertex. Within this sector the tangent vectors to the geodesics through O form a vector field, the members of which define in a very natural manner the elements of support, with respect to which we measure the length σ of the arc $s = r$. The latter is, in fact, an orthogonal trajectory to the family of geodesics. Also, the area Ω of the sector may be defined by means of (1.8.13), using the same elements of support. It may then be shown that[3]

$$R(x_{(0)}, \dot{x}_{(0)}) = 6 \lim_{\substack{r \to 0 \\ \Phi \to 0}} \left(\frac{r\, \Phi - \sigma}{r^3\, \Phi} \right), \tag{4.25}$$

together with

$$R(x_{(0)}, \dot{x}_{(0)}) = 12 \lim_{\substack{r \to 0 \\ \Phi \to 0}} \left(\frac{r^2\, \Phi - 2\, \Omega}{r^4\, \Phi} \right). \tag{4.25a}$$

---

[1] BLASCHKE [4], p. 216, eqn. (67), where this relation is deduced from the accessory problem with respect to the length integral on a surface. Applications of (4.24a) are discussed by MAYER [1], DUSCHEK and MAYER [2] for the case of Riemannian spaces.

[2] See for instance, BLASCHKE [4], p. 153.

[3] BERWALD [10, I], p. 54 et seq. For the case of an F_2 of constant curvature (see § 7 of the present chapter) the same formulae are discussed by MOÓR [2], p. 13 et seq. The proof is quite similar to that of the classical formulae.

A further striking interpretation of R may be obtained as follows. Let C_0 be a curve of F_2 through the line-element $(x_{(0)}, \dot{x}_{(0)})$ at O. We may construct a family of curves C equidistant from C_0 together with the orthogonal trajectories to this family. Consider one of these trajectories which intersects C_0 at a point P, such that the arc-length OP measured along C_0 is s_0. The arc-length measured along this trajectory is denoted by S, measured from P. The four points O, P, Q, R corresponding to the values $(0, 0)$, $(s_0, 0)$, (s_0, S), $(0, S)$ of s_0 and S respectively, define a curvilinear "quadrilateral" on F_2. Again, the tangents to the family of curves C equidistant from C_0 define a field of elements of support, with respect to which all measurements are made. In particular, the area [as given by (1.8.13)] of the quadrilateral will be denoted by Ω. It may then be shown that[1]

$$R(x_{(0)}, \dot{x}_{(0)}) = \lim_{\substack{s_0 \to 0 \\ S \to 0}} \frac{s_0^2 - \sigma^2}{\Omega^2}, \qquad (4.26)$$

where σ is the arc-length of the segment RQ of the quadrilateral. Furthermore, if C_0 is a geodesic, and if ϱ^{-1} is the curvature at R of the curve C of the family containing the segment RQ the above relation reduces to the form[2]

$$R(x_{(0)}, \dot{x}_{(0)}) = \lim_{S \to 0} \frac{1}{\varrho S}. \qquad (4.26a)$$

A further detailed discussion of the equation (4.24 a) of JACOBI is given by FUNK [2]. Let φ_1 and φ_2 be two distinct particular integrals of (4.24 a), and put

$$\psi = \frac{\varphi_1}{\varphi_2}. \qquad (4.27)$$

It is then found in the usual manner that the Schwarzian derivative[3] is given by the curvature R:

$$\frac{\psi'''}{\psi'} - \frac{3}{2}\left(\frac{\psi''}{\psi'}\right)^2 = 2R. \qquad (4.28)$$

This result may be used to derive further geometrical relationships involving R similar to those described above[4].

These results naturally raise the question as to the existence of a Gauss-Bonnet theorem for Finsler spaces. It was shown, however, by BUSEMANN[5] that there exists no universal angular measure (i. e. a measure depending solely on the local Minkowskian metric) for two-dimensional Finsler spaces such that a Gauss-Bonnet theorem always holds. A number of formulae have been suggested which would reduce to the classical Gauss-Bonnet formula if the space were locally euclidean, but none of these can claim to possess the intrinsic geometrical or topological significance which one would normally expect of a Gauss-Bonnet theorem[6].

In conclusion, we remark that the well-known generalisation of the Gauss-Bonnet formula to $2p$-dimensional Riemannian spaces due to ALLENDOERFER and WEIL may be adapted to a special class of Finsler spaces, namely those for which the curvature tensor S_{ijhk} (§ 1) vanishes identically[7]. It is remarkable that this type of Finsler space is characterised by the fact that its angular metric is of constant unit curvature (see Ch. V, § 8).

---

[1] BERWALD [10, I], p. 47 et seq. See also BERWALD [1].

[2] Here ϱ is defined with respect to the angle (1.7.12): see Ch. V, § 1.

[3] See KAMKE [1], p. 120.

[4] FUNK [2], p. 187 et seq.

[5] BUSEMANN [12]; [10], p. 408.

[6] NAZIM [1, 2]; BUSEMANN [8]; RUND [3]; also BLISS [2].

[7] This result is due to LICHNEROWICZ [6, 7].

§ 5. The First and Second Variations of the Length Integral

The formulae of the previous sections will enable us to evaluate without difficulty the first and second variations of the fundamental integral (1.1.7), thus establishing a new link with the classical calculus of variations. We adhere to the geometrical construction of § 4; i. e. we consider once more a two-dimensional subspace F_2 on which a local coordinate system is defined by the curves $u = $ const. and $v = $ const. We shall not, however, assume that the curves $v = $ const. are geodesics, but we shall regard the variable u as representing the arc-length s along those curves. The notation will remain the same. Clearly the difference between the length integrals (between fixed values $u = u_1$ and $u = u_2$) taken along the neighbouring curves $v = 0$ and $v = \varepsilon$ is given (to a first approximation relative to ε) by the so-called "first variation":

$$\delta I = \varepsilon \int_{u_1}^{u_2} \left[\frac{\partial F(x, \xi)}{\partial v} \right]_{v = 0} du . \tag{5.1}$$

By (1.3.2) we have

$$F^2(x, \xi) = g_{ij}(x, \xi)\, \xi^i\, \xi^j ,$$

$(\xi^i = \dfrac{\partial x^i}{\partial u}$ being a unit vector since $u = $ arc-length), so that

$$F \frac{\partial F}{\partial v} = g_{ij}(x, \xi)\, \xi^i \frac{\partial \xi^j}{\partial v} + \frac{1}{2} \frac{\partial g_{ij}}{\partial x^k} \xi^i \xi^j \eta^k ,$$

where we have made use of (1.3.5). If we apply (4.2), (2.2.9) and (2.2.10) to this result we find that

$$F \frac{\partial F}{\partial v} = g_{ij}(x, \xi)\, \xi^j \left[\frac{\partial \eta^i}{\partial u} + \gamma_h{}^i{}_k(x, \xi)\, \eta^h \xi^k \right] . \tag{5.2}$$

But the following identity results from (1.3.5) and (2.3.14):

$$g_{ij}(x, \xi)\, \xi^j P^i_{hk}(x, \xi) = g_{ij}(x, \xi)\, \xi^j\, \gamma_h{}^i{}_k(x, \xi) ,$$

so that in view of (5.2) and the definition (2.3.17)

$$F \frac{\partial F}{\partial v} = \frac{\partial F}{\partial v} = g_{ij}(x, \xi)\, \xi^j \frac{\delta \eta^i}{\delta u} , \tag{5.2a}$$

and hence (5.1) reads

$$\delta I = \varepsilon \int_{u_1}^{u_2} g_{ij}(x, \xi)\, \xi^j \frac{\delta \eta^i}{\delta u}\, du . \tag{5.3}$$

Interesting alternative forms of the first variation may be derived instantly. Firstly, recalling the definition (1.6.4) of the Minkowskian cosine, we see that (5.3) is equivalent to

$$\delta I = \varepsilon \int_{u_1}^{u_2} \left| \frac{\delta \eta^i}{\delta u} \right| \cos \left(\xi, \frac{\delta \eta}{\delta u} \right) du . \tag{5.3a}$$

Secondly, we observe that as a result of (2.5.8) we have along C

$$g_{ij}(x, \xi)\, \xi^j\, \frac{\delta \eta^i}{\delta u} = \frac{\delta}{\delta u}\, (g_{ij}(x, \xi)\, \xi^i\, \eta^j) - g_{ij}(x, \xi)\, \eta^i\, \frac{\delta \xi^j}{\delta u}\,,$$

so that (5.3) becomes

$$\delta I = \varepsilon \left\{ [g_{ij}(x, \xi)\, \xi^i\, \eta^j]_{u_1}^{u_2} - \int_{u_1}^{u_2} g_{ij}(x, \xi)\, \eta^i\, \frac{\delta \xi^j}{\delta u}\, du \right\}. \tag{5.3b}$$

From (5.3), (5.3a), (5.3b) the following theorems result[1]:

1. The first variation vanishes if the variation vector η^i is transported by parallel displacement along the curve C.

2. The first variation vanishes if the covariant derivative of the variation vector η^i is normal (i. e. transversal) with respect to the tangent vector of the curve C.

3. The first variation is positive or negative according as the angle between the covariant derivative of η^i and the tangent vector ξ^i to C is acute or obtuse.

4. The first variation vanishes if C is a geodesic (i. e. $\delta \xi^i/\delta u = 0$) and if either the variation vector vanishes at the end-points or is normal with respect to the tangent vector ξ^i of C at the end-points.

The second variation of the length-integral is defined to be

$$\delta^2 I = \frac{\varepsilon^2}{2} \int_{u_1}^{u_2} \left[\frac{\partial^2 F(x, \xi)}{\partial v^2} \right]_{v = 0} du\,, \tag{5.4}$$

which may easily be evaluated with the aid of (5.2a). On differentiating the latter equation and taking heed of (2.5.8) once more, we obtain

$$\left(\frac{\partial F}{\partial v} \right)^2 + F\, \frac{\partial^2 F}{\partial v^2} = g_{ij}(x, \xi)\, \frac{\delta \xi^j}{\delta v}\, \frac{\delta \eta^i}{\delta u} + g_{ij}(x, \xi)\, \xi^j\, \frac{\delta^2 \eta^i}{\delta v\, \delta u}\,.$$

Using (4.4) and (5.2a) we see that (5.4) may therefore be written in the form

$$\delta^2 I = \frac{\varepsilon^2}{2} \int_{u_1}^{u_2} \left\{ [g_{ij}(x, \xi) - g_{ih}(x, \xi)\, g_{jk}(x, \xi)\, \xi^h\, \xi^k]\, \frac{\delta \eta^i}{\delta u}\, \frac{\delta \eta^j}{\delta u} + \right.$$
$$\left. + g_{ij}(x, \xi)\, \xi^j\, \frac{\delta^2 \eta^i}{\delta v\, \delta u} \right\} du\,. \tag{5.5}$$

From this relation we deduce that the second variation vanishes if the variation vector is transported by parallel displacement along C.

Again, alternative forms of (5.5) may be derived. From (1.14) and (4.5) we have

$$g_{ij}(x, \xi)\, \xi^j\, \frac{\delta^2 \eta^i}{\delta v\, \delta u} = g_{ij}(x, \xi)\, \xi^j\, \frac{\delta^2 \eta^i}{\delta u\, \delta v} + \check{K}_{ijhk}(x, \xi)\, \eta^i\, \xi^j\, \xi^h\, \eta^k\,, \tag{5.6}$$

---

[1] These theorems are generalisations of corresponding results in Riemannian geometry due to SYNGE [2]. For the case of a Finsler space the first variation is discussed by STOKES [1], FREEMAN [1], RUND [11], p. 192.

where the "relative" curvature tensor $\tilde{K}_{ijhk}$ refers to the field defined by the ξ^i. Furthermore, along C we have

$$g_{ij}(x,\xi)\,\xi^j\,\frac{\delta^2\eta^i}{\delta u\,\delta v} = \frac{\delta}{\delta u}\left(g_{ij}(x,\xi)\,\xi^j\,\frac{\delta\eta^i}{\delta v}\right) - g_{ij}(x,\xi)\,\frac{\delta\xi^j}{\delta u}\,\frac{\delta\eta^i}{\delta v}\,. \qquad (5.7)$$

When (5.6) and (5.7) are substituted in (5.5) we see that the second variation becomes

$$\delta^2 I = \frac{\varepsilon^2}{2}\int_{u_1}^{u_2}\left\{[g_{ij}(x,\xi) - g_{ih}(x,\xi)\,g_{jk}(x,\xi)\,\xi^h\,\xi^k]\,\frac{\delta\eta^i}{\delta u}\,\frac{\delta\eta^j}{\delta u} + \right.$$
$$+ \tilde{K}_{ijhk}(x,\xi)\,\eta^i\,\eta^k\,\xi^j\,\xi^h - g_{ij}(x,\xi)\,\frac{\delta\xi^j}{\delta u}\,\frac{\delta\eta^i}{\delta v}\Big\}\,du + \qquad (5.8)$$
$$+ \frac{\varepsilon^2}{2}\left[g_{ij}(x,\xi)\,\xi^j\,\frac{\delta\eta^i}{\delta v}\right]_{u_1}^{u_2}\,.$$

An important special case of (5.8) arises when the curve C is a geodesic. For in this case $\delta\xi^j/\delta u = 0$, and we have[1]

$$\delta^2 I = \frac{\varepsilon^2}{2}\left[g_{ij}(x,\xi)\,\xi^j\,\frac{\delta\eta^i}{\delta v}\right]_{u_1}^{u_2} +$$
$$+ \frac{\varepsilon^2}{2}\int_{u_1}^{u_2} du\left\{[g_{ij}(x,\xi) - g_{ih}(x,\xi)\,g_{jk}(x,\xi)\,\xi^h\,\xi^k]\,\frac{\delta\eta^i}{\delta u}\cdot\frac{\delta\eta^j}{\delta u} + \right. \qquad (5.8\text{a})$$
$$+ K_{ijhk}(x,\xi)\,\eta^i\,\eta^k\,\xi^j\,\xi^h\Big\}\,.$$

This equation is equivalent to an expression for the second variation given by DAVIES[2].

If we specialise still further and assume that the variation is a normal one, i. e. that

$$g_{ij}(x,\xi)\,\xi^j\,\eta^i = 0 \qquad (5.9)$$

[this, of course, implies a restriction on the (u,v) coordinate system on F_2], we obtain as a result of (2.5.8) by differentiation of (5.9)

$$g_{ij}(x,\xi)\,\frac{\delta\eta^i}{\delta u}\,\xi^j = -g_{ij}(x,\xi)\,\eta^i\,\frac{\delta\xi^j}{\delta u} = 0\,, \qquad (5.10)$$

---

[1] Again, for this special case we may replace the relative curvature tensor $\tilde{K}_{ijhk}$ by the tensor K_{ijhk}. This is easily seen if we write the last term on the right-hand side of (5.6) in the form $g_{ij}(x,\xi)\,\xi^j\,\tilde{K}^i_{rhk}\,\eta^r\,\xi^h\,\eta^k$. Then from (1.5) and (1.7) we have

$$(\tilde{K}^i_{rhk} - K^i_{rhk})\,\xi^h = \frac{\partial\Gamma^{*i}_{rh}}{\partial\dot{x}^l}\left(\frac{\partial\xi^l}{\partial x^k} + \frac{\partial G^l}{\partial\dot{x}^k}\right)\xi^h - \frac{\partial\Gamma^{*i}_{rk}}{\partial\dot{x}^l}\left(\frac{d\xi^l}{du} + 2G^l\right)$$

as a result of the homogeneity of the G^l. The last term vanishes since C is a geodesic; and since

$$g_{ij}(x,\xi)\,\xi^j\,\xi^h(\partial\Gamma^{*i}_{rh}/\partial\dot{x}^l) = 0$$

by (3.3.15) and (3.2.14a), the result follows directly. Furthermore, from (1.34a) we see that we may replace $K_{ijhk}\,\xi^j\,\xi^h$ by $R_{ijhk}\,\xi^j\,\xi^h$.

[2] DAVIES [3], p. 246. In this paper an elegant geometrical interpretation of the first term on the right-hand side is derived. Similar calculations are given by AUSLANDER [1, 2] and FREEMAN [1]. The second variation may also be studied by means of the Lie derivative (Ch. V, § 5), as was shown by LAPTEW [1].

since C is a geodesic. Furthermore, differentiating (5.9) covariantly with respect to v, noting (2.5.8), it follows similarly that if the variation η^i vanishes at given end-points $u = u_1$ and $u = u_2$ of C, we have

$$\left[g_{ij}(x, \xi)\, \xi^j\, \frac{\delta \eta^i}{\delta v} \right]_{u_1}^{u_2} = 0 \, .$$

Under these circumstances it is seen that in virtue of the latter equation and (5.10) the second variation (5.8a) reduces to

$$\delta^2 I = \frac{\varepsilon^2}{2} \int\limits_{u_1}^{u_2} \left\{ g_{ij}(x, \xi)\, \frac{\delta \eta^i}{\delta u}\, \frac{\delta \eta^j}{\delta u} + K_{ijhk}(x, \xi)\, \eta^i\, \eta^k\, \xi^j\, \xi^h \right\} du \, . \quad (5.11)$$

Now let us define a vector X^i parallel to η^i such that

$$\eta^i = \eta\, X^i \, , \qquad g_{ij}(x, \xi)\, X^i X^j = 1 \, . \quad (5.12)$$

It then follows from (5.9) and (5.12) that

$$\{ g_{jh}(x, \xi)\, g_{ik}(x, \xi) - g_{ji}(x, \xi)\, g_{hk}(x, \xi) \}\, \xi^j\, \xi^h\, X^i X^k = 1 \, ,$$

and hence, from (4.23) and (2.11),

$$K_{ijhk}(x, \xi)\, \eta^i\, \eta^k\, \xi^j\, \xi^h = - \eta^2 K_{jihk}(x, \xi)\, \xi^j\, \xi^h\, X^i X^k$$
$$= - \eta^2 R(x, \xi, X) \, . \quad (5.13)$$

Also[1], differentiating both equations (5.12) with respect to u, noting (2.5.8a), we find

$$\frac{\delta \eta^i}{\delta u} = \eta' X^i + \eta\, \frac{\delta X^i}{\delta u} \, , \qquad g_{ij}(x, \xi)\, X^i\, \frac{\delta X^j}{\delta u} = 0 \, ,$$

so that

$$g_{ij}(x, \xi)\, \frac{\delta \eta^i}{\delta u}\, \frac{\delta \eta^j}{\delta u} = \eta'^2 + \eta^2 g_{ij}(x, \xi)\, \frac{\delta X^i}{\delta u}\, \frac{\delta X^j}{\delta u} \, . \quad (5.14)$$

On substituting (5.13) and (5.14) in (5.11) we obtain

$$\delta^2 I = \frac{\varepsilon^2}{2} \int\limits_{u_1}^{u_2} \left\{ \eta'^2 + \eta^2 \left(g_{ij}(x, \xi)\, \frac{\delta X^i}{\delta u}\, \frac{\delta X^j}{\delta u} - R(x, \xi, X) \right) \right\} du \, . \quad (5.15)$$

This expression represents a generalisation of the form of the second variation in Riemannian geometry due to SYNGE[2].

Also, for any *definite* F_2 containing the geodesic C, the integrand of (5.15) (apart from the functions η and η') is a given function of the arc-

---

[1] From (5.13), (5.11) and condition C of Ch. I, § 1, we may immediately deduce the following theorem: A curve C being a geodesic, the second variation of its length is positive for all variations with fixed end-points (and its length is therefore a relative minimum with respect to variations of that type), if the function R corresponding to every element (ξ, X) containing the direction of C is zero or negative. For $n = 2$, see BOLZA [1]; for n-dimensional Riemannian spaces, SYNGE [2], p. 260. Further theorems due to SYNGE [2] concerning the curvature of the two-dimensional subspace containing C may be generalised similarly.

[2] SYNGE [2], p. 261; RUND [11], p. 198.

length u, and we may thus write down the Euler equations of the accessory problem[1], i. e. the so-called equations of JACOBI:

$$\eta'' + \eta \left\{ R(x, \xi, X) - g_{ij}(x, \xi) \frac{\delta X^i}{\delta u} \frac{\delta X^j}{\delta u} \right\} = 0 . \tag{5.16}$$

But it is seen that this relation is formally identical with the scalar form (4.24) of the equation of geodesic deviation; for if we put $f = 0$ in (4.7) the functions $z(u)$ and $\eta(u)$ differ only by the constant factor ε, while the vectors X^i have the same meaning in both equations. *This result establishes the relation between the theory of geodesic deviation and the theory of the accessory problems in the calculus of variations.* It should be noted, however, that the equation (5.16) is in a sense more general than (4.24), for (4.24) refers to a subspace F_2 containing a family $v = $ const. of geodesics of F_n, while in (5.15) the subspace F_2 need only contain one geodesic, namely C. For the case $n = 2$ no such distinction need be made, and since $\delta X^i/\delta u = 0$ in this case, equation (5.16) reduces to

$$\eta'' + \eta R(x, \xi) = 0 . \tag{5.16a}$$

Conjugate points on C are defined by consecutive zeros of any solution of Jacobi's equation (5.16); by means of Sturm's theorem[2] we deduce the following result[3]: If the function R corresponding to every element (ξ, X) containing the direction of a geodesic C is less than a positive number A, the distance between a pair of conjugate points on C cannot be less then $\pi/\sqrt{A}$; while if R is always greater than a positive number B, the distance between a pair of conjugate points for variations in an F_2 generated by parallel propagation of a unit normal vector to C is less than $\pi/\sqrt{B}$.[4]

§ 6. The Curvature Tensors Arising from Berwald's Connection

It is clear that the connection coefficients $G_{hk}^i(x, \dot{x})$ (as defined in Chapter III, § 3) which were used by BERWALD for the definition of covariant derivatives will also lead to a set of curvature tensors. Again,

---

[1] CARATHÉODORY [1], p. 260 et seq. The accessory problem is concerned with the extreme values, in the sense of the calculus of variations, of the integral (5.15).

[2] BOLZA [1], p. 62. See also MORSE [1], Ch. IV.

[3] This result is a direct generalisation of a well-known theorem of Riemannian geometry: SYNGE [2], p. 264; [3]; SCHOENBERG [1]; MYERS [3]. A detailed derivation and discussion of related theorems is given by AUSLANDER [3], this work being based on "global" methods.

[4] For a more detailed discussion of conjugate points the reader is referred to treatises on the calculus of variations, especially to MORSE [1]. A very complete description of conjugate loci in n-dimensional Finsler spaces is given by WHITEHEAD [4], while HOUSEHOLDER [1] investigates the dependence of the position of focal points on the curvature, the latter being based on the definitions of CARTAN. See also RINOW [1].

we could derive the corresponding commutation formulae in order to define these tensors. However, such a process would be tedious, and since there exists a very close relationship between these tensors and the curvature tensors defined above, we may circumvent such calculations. In fact, we may start from the special form (4.16) of the equations of geodesic deviation, which BERWALD[1] writes in the form

$$\frac{\delta^2 z^j}{\delta u^2} + H_k^j(x, \dot{x})\, z^k = 0 \,, \tag{6.1}$$

(where we have replaced ξ^i by $\dot{x}^i$). The tensor $H_k^j(x, \dot{x})$ is called the "deviation tensor", being defined by

$$H_k^j(x, \dot{x}) = K_{ihk}^j(x, \dot{x})\, \dot{x}^i \dot{x}^h = R_{ihk}^j(x, \dot{x})\, \dot{x}^i \dot{x}^h \,, \tag{6.2}$$

the second identity resulting from (1.34). Equation (1.11) permits us to write down immediately an explicit expression for this tensor:

$$H_k^i(x, \dot{x}) = 2\,\frac{\partial G^i}{\partial x^k} - \frac{\partial^2 G^i}{\partial x^h \partial \dot{x}^k}\,\dot{x}^h + 2 G_{kl}^i G^l - \frac{\partial G^i}{\partial \dot{x}^l}\,\frac{\partial G^l}{\partial \dot{x}^k} \,, \tag{6.3}$$

where we have made use of the fact that the function $G^l(x, \dot{x})$ is positively homogeneous of degree 2 in the $\dot{x}^i$.

Two further tensors may be defined similarly. Firstly, we write

$$H_{jk}^i(x, \dot{x}) = \frac{1}{3}\left(\frac{\partial H_k^i(x, \dot{x})}{\partial \dot{x}^j} - \frac{\partial H_j^i(x, \dot{x})}{\partial \dot{x}^k}\right). \tag{6.4}$$

We differentiate (6.3) with respect to $\dot{x}^j$, and in the resulting equation we interchange the indices j and k. After some simplification we find that we may express the definition (6.4) in the form

$$H_{jk}^i = \frac{\partial^2 G^i}{\partial x^k \partial \dot{x}^j} - \frac{\partial^2 G^i}{\partial x^j \partial \dot{x}^k} + G_{kr}^i\,\frac{\partial G^r}{\partial \dot{x}^j} - G_{rj}^i\,\frac{\partial G^r}{\partial \dot{x}^k}\,. \tag{6.4a}$$

From (1.11) and (1.34a) it then follows that

$$H_{jk}^i(x, \dot{x}) = K_{hjk}^i(x, \dot{x})\,\dot{x}^h = R_{hjk}^i(x, \dot{x})\,\dot{x}^h \,. \tag{6.5}$$

Secondly, we write[2]

$$H_{hjk}^i = \frac{\partial H_{jk}^i}{\partial \dot{x}^h} = \frac{1}{3}\left(\frac{\partial^2 H_k^i}{\partial \dot{x}^h \partial \dot{x}^j} - \frac{\partial^2 H_j^i}{\partial \dot{x}^h \partial \dot{x}^k}\right), \tag{6.6}$$

and from (6.4a) we find that

$$H_{hjk}^i = \frac{\partial G_{hj}^i}{\partial x^k} - \frac{\partial G_{hk}^i}{\partial x^j} + G_{hj}^r G_{rk}^i - G_{hk}^r G_{rj}^i + \\ + G_{rhk}^i\,\frac{\partial G^r}{\partial \dot{x}^j} - G_{rhj}^i\,\frac{\partial G^r}{\partial \dot{x}^k}\,, \tag{6.7}$$

---

[1] BERWALD [10, IV], p. 758. We have permitted ourselves a slight change in notation, since BERWALD uses the symbol K_{jkh}^i instead of H_{jkh}^i, but if we were to retain the former notation this would cause confusion with the curvature tensors introduced in § 1.

[2] BERWALD calls the tensors (6.4) and (6.6) the „Grundtensor der Krümmung" and „Krümmungstensor" respectively.

where we have put

$$G^i_{hjk} = \frac{\partial G^i_{hj}}{\partial \dot{x}^k} = \frac{\partial G^i_{kj}}{\partial \dot{x}^h}, \qquad (6.8)$$

noting that

$$\frac{\partial G^i_{hj}}{\partial \dot{x}^k}\, \dot{x}^k = G^i_{hjk}\, \dot{x}^k = 0 . \qquad (6.8a)$$

Unfortunately, the tensor (6.7) does not bear a simple direct relationship to the tensors defined previously. We note that we have from (6.5) and (6.6)

$$H^i_{hjk} = K^i_{hjk} + \dot{x}^r \frac{\partial K^i_{rjk}}{\partial \dot{x}^h} . \qquad (6.9)$$

By means of a straight-forward calculation one may deduce

$$\begin{aligned}
H_{ijkh} = K_{ijkh} &+ A_{imk|r} A^m_{jh|s}\, l^r\, l^s - \\
&- A_{imh|r} A^m_{jk|s}\, l^r\, l^s + A_{ijk|r|h}\, l^r - A_{ijh|r|k}\, l^r,
\end{aligned} \qquad (6.9a)$$

or, equivalently,

$$H_{ijkh} = R_{ijkh} - 2A^m_{ij} R_{smkh}\, l^s + \dot{x}^s \frac{\partial R_{sjkh}}{\partial \dot{x}^i}, \qquad (6.9b)$$

in the notation of Chapter III. The inverse formula reads

$$\begin{aligned}
R_{ijkh} = \frac{1}{2}\, (H_{ijkh} - H_{jikh}) - \\
- \frac{1}{4}\, g^{mp} \left\{ g_{im(k)}\, g_{jp(h)} - g_{im(h)}\, g_{jp(k)} \right\},
\end{aligned} \qquad (6.9c)$$

where (k) denotes covariant differentiation with respect to x^k in the sense of BERWALD (Ch. III, § 3)[1]. In particular,

$$H_{ijkl}(x, \dot{x})\, \dot{x}^i = K_{ijkl}(x, \dot{x})\, \dot{x}^i . \qquad (6.9d)$$

The commutation formulae[2] involving the tensor H^i_{jhk} are very similar to those developed for K^i_{ihk} in § 1. The reader may easily verify that the following identities hold:

$$T_{(h)\,(k)} - T_{(k)\,(h)} = - \frac{\partial T}{\partial \dot{x}^i}\, H^i_{hk} , \qquad (6.10)$$

or

$$T_{ij(h)\,(k)} - T_{ij(k)\,(h)} = - \frac{\partial T_{ij}}{\partial \dot{x}^r}\, H^r_{hk} - T_{rj} H^r_{ihk} - T_{ir} H^r_{jhk} . \qquad (6.10a)$$

---

[1] The definitions (6.2), (6.4) and (6.7) are given in BERWALD [2]. The relations (6.5), (6.9a), (6.9b) and (6.9c) between these curvature tensors and those of CARTAN are derived (in a slightly different form) in BERWALD [9]. It should be noted that the definitions (6.2), (6.4) and (6.7) may be applied equally well to the theory of the general geometry of paths, as is done by BERWALD in [10, IV]. More generally, VARGA [12] derives formulae for the difference between the curvature tensors resulting from two distinct metric connections defined over the same space of line-elements.

[2] Commutation formulae involving both CARTAN's and BERWALD's derivatives are given by DAVIES [1], § 1. Further identities involving the tensor (6.7) are derived by ISPAS [1, 2]. Some of these identities are generalisations of the well-known identities of VEBLEN.

Corresponding to the formulae (1.17) we have

$$\left(\frac{\partial T}{\partial \dot{x}^k}\right)_{(h)} - \frac{\partial T_{(h)}}{\partial \dot{x}^k} = 0 \, , \tag{6.11}$$

or

$$\left(\frac{\partial T^i_j}{\partial \dot{x}^k}\right)_{(h)} - \frac{\partial T^i_{j(h)}}{\partial \dot{x}^k} = T^i_r G^r_{jkh} - T^r_j G^i_{rkh} \, , \tag{6.11a}$$

or

$$\left(\frac{\partial T^i_{jl}}{\partial \dot{x}^k}\right)_{(h)} - \frac{\partial T^i_{jl(h)}}{\partial \dot{x}^k} = T^i_{rl} G^r_{jkh} + T^i_{jr} G^r_{lkh} - T^r_{jl} G^i_{rkh} \, . \tag{6.11b}$$

In the sequel we shall have occasion to use certain identities satisfied by the tensor H^i_{jhk}: in fact, these are the alternative forms of the Bianchi identities. They may be derived as follows. Firstly, we may, by means of (6.6), verify that

$$H^i_{hjk} + H^i_{khj} + H^i_{jkh} = 0 \tag{6.12}$$

identically. Secondly, let $T(x, \dot{x})$ be an arbitrary continuously differentiable function of position and direction. We apply (6.10) to this function, and on differentiating the result covariantly with respect to x^l and noting (6.11) we find

$$T_{(i)(k)(l)} - \dot{T}_{(k)(i)(l)} = -\frac{\partial T_{(l)}}{\partial \dot{x}^r} H^r_{ik} - \frac{\partial T}{\partial \dot{x}^r} H^r_{ik(l)} \, .$$

By means of (6.6) and (3.3.9) we may write this equation in the form

$$T_{(i)(k)(l)} - T_{(k)(i)(l)} = -\frac{\partial T_{(l)}}{\partial \dot{x}^r} H^r_{sik} \dot{x}^s - \frac{\partial T}{\partial \dot{x}^r} H^r_{sik(l)} \dot{x}^s \, .$$

Also, from (6.10a) and (6.6) we have

$$T_{(k)(i)(l)} - T_{(k)(l)(i)} = -\frac{\partial T_{(k)}}{\partial \dot{x}^r} H^r_{sil} \dot{x}^s - T_{(r)} H^r_{kil} \, .$$

Adding these identities, we obtain

$$T_{(i)(k)(l)} - T_{(k)(l)(i)} = -\frac{\partial T_{(l)}}{\partial \dot{x}^r} H^r_{sik} \dot{x}^s - \frac{\partial T_{(k)}}{\partial \dot{x}^r} H^r_{sil} \dot{x}^s -$$
$$- \frac{\partial T}{\partial \dot{x}^r} H^r_{sik(l)} \dot{x}^s - T_{(r)} H^r_{kil} \, .$$

Two similar equations result from a cyclic permutation of the indices i, k, l. These are added to the first equation, and after simplification of the left-hand side we are left with

$$0 = \frac{\partial T}{\partial \dot{x}^r} \left(H^r_{sik(l)} + H^r_{sli(k)} + H^r_{skl(i)}\right) \dot{x}^s +$$
$$+ \dot{x}^s \left(\frac{\partial T_{(l)}}{\partial \dot{x}^r} H^r_{sik} + \frac{\partial T_{(k)}}{\partial \dot{x}^r} H^r_{sli} + \frac{\partial T_{(i)}}{\partial \dot{x}^r} H^r_{skl} +\right.$$
$$+ \frac{\partial T_{(k)}}{\partial \dot{x}^r} H^r_{sil} + \frac{\partial T_{(i)}}{\partial \dot{x}^r} H^r_{slk} + \left.\frac{\partial T_{(l)}}{\partial \dot{x}^r} H^r_{ski}\right) +$$
$$+ T_{(r)} \left(H^r_{kil} + H^r_{ilk} + H^r_{lki}\right) \, .$$

However, from (6.6) it is evident that the tensor H^r_{sjk} is skew-symmetric in j and k. Hence the second coefficient of $\dot{x}^s$ in the above relation vanishes, while the same is true for the coefficient of $T_{(r)}$ in virtue of (6.12). Also, since $\partial T/\partial \dot{x}^r$ is essentially arbitrary, it follows that we have

$$(H^r_{s\,i\,k\,(l)} + H^r_{s\,l\,i\,(k)} + H^r_{s\,k\,l\,(i)})\,\dot{x}^s = 0\,, \tag{6.13}$$

or, by (6.6) and (3.3.9),

$$H^r_{i\,k\,(l)} + H^r_{l\,i\,(k)} + H^r_{k\,l\,(i)} = 0\,. \tag{6.13a}$$

Furthermore, in view of (6.2) and (6.5) we may write

$$H^j_{k\,i}\,\dot{x}^k = -H^j_{i\,k}\,\dot{x}^k = H^j_i\,, \tag{6.14}$$

and thus, on multiplying (6.13) by $\dot{x}^i$, we find

$$H^r_{k\,(l)} - H^r_{l\,(k)} + H^r_{k\,l\,(i)}\,\dot{x}^i = 0\,. \tag{6.13b}$$

Finally, we deduce from (6.11b) and (6.6) that

$$\frac{\partial H^r_{i\,k\,(l)}}{\partial \dot{x}^j} = H^r_{j\,i\,k\,(l)} - H^r_{m\,k}\,G^m_{i\,j\,l} - H^r_{i\,m}\,G^m_{k\,j\,l} + H^m_{i\,k}\,G^r_{m\,j\,l}\,.$$

Again two similar equations are obtained by cyclic permutation of the indices i, k, l. On adding the three relations thus obtained, we find after simplification

$$\frac{\partial H^r_{i\,k\,(l)}}{\partial \dot{x}^j} + \frac{\partial H^r_{l\,i\,(k)}}{\partial \dot{x}^j} + \frac{\partial H^r_{k\,l\,(i)}}{\partial \dot{x}^j}$$

$$= (H^r_{j\,i\,k\,(l)} + H^r_{j\,l\,i\,(k)} + H^r_{j\,k\,l\,(i)}) + H^m_{i\,k}\,G^r_{m\,j\,l} + H^m_{l\,i}\,G^r_{m\,j\,k} + H^m_{k\,l}\,G^r_{m\,j\,i}\,.$$

Hence, if we differentiate (6.13a) with respect to $\dot{x}^j$, we obtain the identity:

$$\begin{aligned} H^r_{j\,i\,k\,(l)} + H^r_{j\,l\,i\,(k)} + H^r_{j\,k\,l\,(i)} + \\ + H^m_{i\,k}\,G^r_{m\,j\,l} + H^m_{l\,i}\,G^r_{m\,j\,k} + H^m_{k\,l}\,G^r_{m\,j\,i} = 0\,. \end{aligned} \tag{6.13c}$$

Equations (6.13) to (6.13c) are alternative forms of the *Bianchi identities*[1] *for the tensors* H^i_j, $H^i_{k\,j}$, $H^i_{l\,k\,j}$.

The following direct consequences of (6.13c) will be found useful in the sequel. On multiplying (6.13c) by $\dot{x}^l$ and noting (6.8a) and (6.14) we find

$$(H^r_{j\,i\,k\,(l)} + H^r_{j\,l\,i\,(k)} + H^r_{j\,k\,l\,(i)})\,\dot{x}^l = H^m_k\,G^r_{m\,j\,i} - H^m_i\,G^r_{m\,j\,k}\,. \tag{6.15}$$

Contraction of r and j finally yields

$$(H^r_{r\,i\,k\,(l)} + H^r_{r\,l\,i\,(k)} + H^r_{r\,k\,l\,(i)})\,\dot{x}^l = H^m_k\,G^r_{r\,m\,i} - H^m_i\,G^r_{r\,m\,k}\,. \tag{6.15a}$$

---

[1] BERWALD [2], p. 54.

Also, the following contractions will be used: We write

$$H_i = H^h_{ih} , \qquad (6.16a)$$

together with

$$H_{ij} = H^h_{ijh} = \frac{\partial H_j}{\partial \dot{x}^i} , \qquad (6.16b)$$

the second relation being a direct consequence of definition (6.6). Since H_i is homogeneous of the first degree in its directional arguments, we also have

$$H_{ij} \dot{x}^i = H_j . \qquad (6.16c)$$

Furthermore, a new scalar $H(x, \dot{x})$ may be defined by putting

$$H = \frac{1}{n-1} H^i_i . \qquad (6.17)$$

This scalar will appear in several important formulae. For instance, we note that in virtue of (6.2) and the skew-symmetry of the K^i_{ihk} in h and k we have

$$H^i_j(x, \dot{x}) \dot{x}^j = 0 . \qquad (6.18)$$

Differentiating this identity with respect to $\dot{x}^k$ we find

$$\frac{\partial H^i_j}{\partial \dot{x}^k} \dot{x}^j + H^i_k = 0 , \qquad (6.19)$$

so that by (6.17) contraction of i and k yields

$$\frac{\partial H^r_j}{\partial \dot{x}^r} \dot{x}^j + (n-1) H = 0 . \qquad (6.19a)$$

This relation allows us to write (6.16c) in a more useful form. Firstly, differentiation of (6.19a) with respect to $\dot{x}^i$ gives

$$\frac{\partial^2 H^r_j}{\partial \dot{x}^i \partial \dot{x}^r} \dot{x}^j + \frac{\partial H^r_i}{\partial \dot{x}^r} + (n-1) \frac{\partial H}{\partial \dot{x}^i} = 0 . \qquad (6.19b)$$

Secondly, from (6.16a), (6.16b) and (6.4) we obtain

$$\begin{aligned}
H_{ij} \dot{x}^j &= \frac{1}{3} \left(\frac{\partial^2 H^r_r}{\partial \dot{x}^i \partial \dot{x}^j} - \frac{\partial^2 H^r_j}{\partial \dot{x}^i \partial \dot{x}^r} \right) \dot{x}^j \\
&= \frac{1}{3} (n-1) \frac{\partial H}{\partial \dot{x}^i} - \frac{1}{3} \frac{\partial^2 H^r_j}{\partial \dot{x}^i \partial \dot{x}^r} \dot{x}^j ,
\end{aligned} \qquad (6.20)$$

the second relation resulting from (6.17) and the fact that H is homogeneous of the second degree. By means of (6.19b) we see that this equation reduces to the form

$$H_{ij} \dot{x}^j = \frac{2}{3} (n-1) \frac{\partial H}{\partial \dot{x}^i} + \frac{1}{3} \frac{\partial H^r_i}{\partial \dot{x}^r} .$$

But on contracting (6.4) we also have, using (6.16a) and (6.17),

$$\frac{1}{3} \frac{\partial H^r_i}{\partial \dot{x}^r} = \frac{1}{3} (n-1) \frac{\partial H}{\partial \dot{x}^i} - H^r_{ir} = \frac{1}{3} (n-1) \frac{\partial H}{\partial \dot{x}^i} - H_i , \qquad (6.21)$$

so that finally

$$H_{ij}\,\dot{x}^j = (n-1)\,\frac{\partial H}{\partial \dot{x}^i} - H_i\,. \tag{6.22}$$

Also, from (6.20) and (6.19a) we may deduce that

$$H_{ij}\,\dot{x}^i\,\dot{x}^j = \frac{2}{3}\,(n-1)\,H - \frac{1}{3}\,\frac{\partial H_j^r}{\partial \dot{x}^r}\,\dot{x}^j = (n-1)\,H\,. \tag{6.23}$$

A comparison with (6.16c) shows that

$$H_i\,\dot{x}^i = (n-1)\,H\,. \tag{6.24}$$

In conclusion we remark that if we contract the indices i, k in (6.7) and interchange the indices h and j in the resulting equation, we obtain by subtraction

$$H_{hjk}^k - H_{jhk}^k = \frac{\partial G_{jk}^k}{\partial x^h} - \frac{\partial G_{hk}^k}{\partial x^j} +$$
$$+ \; G_{jk}^r G_{rh}^k - G_{hk}^r G_{rj}^k + \frac{\partial G^r}{\partial \dot{x}^j}\,G_{hrk}^k - \frac{\partial G^r}{\partial \dot{x}^h}\,G_{jrk}^k\,.$$

Applying (6.7) once more, we deduce by means of (6.16b):

$$H_{jh} - H_{hj} = H_{khj}^k\,. \tag{6.25}$$

§ 7. Spaces of Constant Curvature

Let us return to the definition (4.23) of the Riemannian curvature $R(x, \dot{x}, X)$ of the space defined at a point x^i with respect to a 2-direction $(\dot{x}, X)$ at that point:

$$R(x, \dot{x}, X) = \frac{K_{ijhk}(x, \dot{x})\,\dot{x}^i\,\dot{x}^h\,X^j\,X^k}{[g_{ih}(x, \dot{x})\,g_{jk}(x, \dot{x}) - g_{ij}(x, \dot{x})\,g_{hk}(x, \dot{x})]\,\dot{x}^i\,\dot{x}^h\,X^j\,X^k}\,. \tag{7.1}$$

In Riemannian geometry this scalar represents the Gaussian curvature of the two-dimensional geodesic subspace defined by the element $(\dot{x}, X)$. The same is true for the present more general case if we adopt a definition for the Gaussian curvature of a two-dimensional subspace which coincides with the one given by FINSLER[1].

Following SYNGE[2] we shall call a point of F_n at which $R(x, \dot{x}, X)$ is independent of the choice of X^i an *isotropic* point of F_n. We shall denote the corresponding value of $R(x, \dot{x}, X)$ by $R(x, \dot{x})$. Also, since the

---

[1] FINSLER [1], p. 105. This result is due to VARGA [10], p. 120. The osculating Riemannian metric is used in the course of a fairly simple proof which is based on the scalar form of the equations of geodesic deviation for $n = 2$. See Ch. V, § 6.

[2] SYNGE and SCHILD [1], p. 111. The corresponding definition in Riemannian geometry demands that $R(x, \dot{x}, X)$ be independent of both $\dot{x}^i$ and X^i, but it is clear that we have to adapt the definition with respect to Finsler spaces as stated above. BERWALD [10, IV], p. 774, calls Finsler spaces in which every point is isotropic "spaces of scalar curvature".

curvature tensor as well as the metric tensors appearing in (7.1) are homogeneous of degree zero, we may replace the $\dot{x}^i$ by the unit vectors l^i in the direction of the $\dot{x}^i$. Clearly, then, if the point x^i is isotropic, it follows from (7.1) that we must have at that point (for any vector X^i)

$$\{R(x, \dot{x}) (g_{jk}(x, \dot{x}) - l_j l_k) - K_{ijhk}(x, \dot{x}) l^i l^h\} X^j X^k = 0 ,$$

and, since by hypothesis the coefficients of $X^j X^k$ are independent of the X^i, which are arbitrary, it follows that

$$2R(x, \dot{x}) (g_{jk}(x, \dot{x}) - l_j l_k) - [K_{ijhk}(x, \dot{x}) + K_{ikhj}(x, \dot{x})] l^i l^h = 0 . \quad (7.2)$$

But from (2.27a) and (1.3.5) we have

$$[K_{ijhk}(x, \dot{x}) - K_{hkij}(x, \dot{x})] l^i l^h = C_{kjl} K^l_{rih} \dot{x}^r l^i l^h .$$

The right-hand side vanishes since the K^l_{rih} are skew-symmetric in i and h, and hence, by interchange of dummy-suffixes we find

$$K_{ijhk}(x, \dot{x}) l^i l^h = K_{ikhj}(x, \dot{x}) l^i l^h .$$

Condition (7.2) thus reduces to

$$R(x, \dot{x}) (g_{jk}(x, \dot{x}) - l_j l_k) = K_{ijhk}(x, \dot{x}) l^i l^h , \quad (7.3)$$

or

$$R(x, \dot{x}) (\delta^j_k - l_k l^j) = K^j_{ihk}(x, \dot{x}) l^i l^h . \quad (7.3a)$$

From this equation we may easily deduce the corresponding value of $R(x, \dot{x})$. Again, if we define as in Riemannian geometry the tensor K_{ih} by contraction of the curvature tensor:

$$K_{ih}(x, \dot{x}) = K^j_{ihj}(x, \dot{x}) , \quad (7.4)$$

it follows from (7.3a) when we contract k and j that

$$R(x, \dot{x}) = \frac{1}{(n-1)} K_{ih}(x, \dot{x}) l^i l^h = \frac{K_{ih}(x, \dot{x}) \dot{x}^i \dot{x}^h}{(n-1) F^2(x, \dot{x})} . \quad (7.5)$$

Equation (7.3a), with R given by (7.5), therefore represents a necessary condition that the space F_n be isotropic at the point x^i. Conversely, it is seen by substitution of (7.3a) and (7.5) in (7.1) that the condition is also sufficient[1].

We may write (7.3a) in a more useful form as follows. From (6.2), (6.4) and (6.5) we have

$$K^i_{hjk} \dot{x}^h = \frac{1}{3} \left\{ \frac{\partial}{\partial \dot{x}^j} (K^i_{hlk} \dot{x}^h \dot{x}^l) - \frac{\partial}{\partial \dot{x}^k} (K^i_{hlj} \dot{x}^h \dot{x}^l) \right\} .$$

---

[1] It is easily verified that every two-dimensional Finsler space is isotropic (although in a rather trivial sense). This result follows immediately from the definition (7.1), for in two dimensions the arbitrary vector X^i may be expressed in the form $X^i = \lambda \xi^i + \mu \eta^i$ (§ 5), and on substituting this in (7.1) we find that the right-hand side is independent of λ and μ. We shall therefore assume $n > 2$ for the remainder of this section, unless the contrary is stated.

We substitute (7.3a) in this equation, carrying out the differentiations as indicated, noting at the same time that

$$\frac{\partial l_i}{\partial \dot{x}^j} = \frac{1}{F}\left(g_{ij} - l_i\, l_j\right), \qquad \frac{\partial l^i}{\partial \dot{x}^j} = \frac{1}{F}\left(\delta^i_j - l^i\, l_j\right).$$

The resulting equation reads

$$K^h_{\,i\,k\,j}\, l^i = \frac{F}{3}\left\{\frac{\partial R}{\partial \dot{x}^k}\left(\delta^h_j - l^h\, l_j\right) - \frac{\partial R}{\partial \dot{x}^j}\left(\delta^h_k - l^h\, l_k\right)\right\} + R\left(l_k\, \delta^h_j - l_j\, \delta^h_k\right).$$

$$(7.6)$$

From this form of the condition (7.3a) we may now deduce the following formulae which will be found useful in the sequel. Since $R(x, \dot{x})$ is homogeneous of degree zero in $\dot{x}^i$ [equation (7.5)] we have

$$K^h_{r\,s\,k}\, l^r\, l^s = R\left(\delta^h_k - l_k\, l^h\right),\qquad (7.7)$$

together with

$$K^h_{r\,s\,h}\, l^r\, l^s = (n-1)\, R,\qquad (7.7\,\text{a})$$

so that in view of (3.2.12)

$$(K^h_{r\,s\,k}\, l^r\, l^s)_{|j} = R_{|j}\left(\delta^h_k - l_k\, l^h\right),\qquad (7.7\,\text{b})$$

together with

$$(K^h_{r\,s\,h}\, l^r\, l^s)_{|k} = (n-1)\, R_{|k}.\qquad (7.7\,\text{c})$$

In particular, it follows from (7.7b) that

$$(K^h_{r\,s\,k}\, l^r\, l^s)_{|h} = R_{|k} - R_{|h}\, l^h\, l_k.\qquad (7.7\,\text{d})$$

We are now in a position to state the following generalisation of Schur's theorem of Riemannian geometry[1]: *If a Finsler space F_n (with $n > 2$) is isotropic at each point of a region, and if the scalar $R(x, \dot{x})$ defined by equation (7.5) is independent of its directional arguments $\dot{x}^i$, then the Riemannian curvature is constant throughout that region.*

In order to prove this theorem we observe that as a result of (2.8) we have

$$P^r_{i\,j\,l}(x, \dot{x})\, l^i = A^r_{j\,l|i}\, l^i,$$

so that if we multiply the Bianchi identities (3.3) by l^i, noting (1.34a) and (3.2.12), we obtain

$$(K^i_{r\,k\,h|l} + K^i_{r\,h\,l|k} + K^i_{r\,l\,k|h})\, l^r + $$
$$+ (A^i_{k\,m|s}K^m_{r\,h\,l} + A^i_{h\,m|s}K^m_{r\,l\,k} + A^i_{l\,m|s}K^m_{r\,k\,h})\, l^r\, l^s = 0.$$

$$(7.8)$$

In this equation we contract j and h, while multiplying by l^k. Using (2.1) and (3.2.4a) we find after suitable interchange of indices

---

[1] In essence this theorem is due to BERWALD [10, IV], p. 778. It was stated by BERWALD in terms of his curvature tensors (§ 6) and the scalars resulting from the latter; however, in virtue of the relations (6.2) to (6.6) Berwald's theorem is equivalent to the one stated above.

$$(K^h_{r\,s\,h|k} - K^h_{r\,s\,k|h} - K^h_{r\,k\,h|s})\, l^r\, l^s +$$
$$+ (A^h_{k\,m|p} K^m_{r\,s\,h} - A_{m|p} K^m_{r\,s\,k})\, l^p\, l^r\, l^s = 0 \;. \tag{7.9}$$

We now substitute from equations (7.6) to (7.7d) in this relation, and, after some simplification, we see that it reduces to

$$(n-2) \left\{ R_{|k} - \frac{1}{3} F \left(\frac{\partial R}{\partial \dot x^k} \right)_{|j} l^j - R_{|j}\, l^j\, l_k \right\} + R (A^h_{h\,k|j}\, l^j - A_{k|j}\, l^j) = 0 \;,$$

or, finally,

$$F \left(\frac{\partial R}{\partial \dot x^k} \right)_{|j} l^j = 3\,(R_{|k} - R_{|j}\, l^j\, l_k) \;. \tag{7.10}$$

Now, in accordance with the hypothesis of the theorem, let us suppose that R is a function of position only, i. e.

$$R_{|k} = \frac{\partial R}{\partial x^k} \;, \qquad \frac{\partial R_{|k}}{\partial \dot x^j} = 0 \;.$$

Equation (7.10) becomes

$$R_{|k} = R_{|j}\, l^j\, l_k \;, \tag{7.10a}$$

and on differentiating this relation with respect to $\dot x^h$, we thus obtain

$$R_{|j} \left\{ (\delta^j_h - l^j\, l_h)\, l_k + l^j\, (g_{h\,k} - l_h\, l_k) \right\} = 0 \;,$$

or, using (7.10a),

$$-\,(R_{|j}\, l^j)\, (l_h\, l_k - g_{h\,k}) = 0 \;.$$

Observing (7.10a) once more[1], we note that $R_{|k} = 0$, which proves the theorem enunciated above.

The following fact[2] is also worthy of notice: *if the curvature tensor of a Finsler space F_n (with $n > 2$) satisfies the relation*

$$K^h_{i\,k\,j}\, l^i = R\,(l_k\, \delta^h_j - l_j\, \delta^h_k) \;, \tag{7.11}$$

the scalar $R(x, \dot x)$ is constant. For if we multiply this condition by l^k we obtain equation (7.3a), so that (7.11) is a sufficient condition that the space be isotropic. Furthermore, comparing (7.11) with (7.6) it follows that under these conditions

$$\frac{\partial R}{\partial \dot x^k}\,(\delta^h_j - l^h\, l_j) = \frac{\partial R}{\partial \dot x^j}\,(\delta^h_k - l^h\, l_k) \;,$$

or

$$(n-2)\,\frac{\partial R}{\partial \dot x^k} = 0 \;.$$

Thus the hypotheses of the generalisation of Schur's theorem are satisfied and our statement follows immediately from the application of the latter.

---

[1] Note that the coefficient of $R_{|j}\, l^j$ is $F\,\partial^2 F / \partial \dot x^h\, \partial \dot x^k$, which cannot vanish for all values h, k as a result of condition C (Ch. I, § 1).

[2] BERWALD, loc. cit.

Incidentally, it is of interest to observe that if $R(x, \dot{x})$ is independent of direction, differentiation of (7.6) with respect to $\dot{x}^l$ yields, after some simplification,

$$\frac{\partial}{\partial \dot{x}^l}(K^h_{ikj}\,\dot{x}^i) = R(\delta^h_j\,g_{lk} - \delta^h_k\,g_{jl})\,,$$

and thus from (6.5) and (6.6) we have

$$H_{likj} = R(g_{ij}\,g_{lk} - g_{ik}\,g_{jl})\,. \tag{7.6a}$$

This equation is analogous to the relation which characterises an isotropic point in Riemannian geometry[1], provided we regard the tensor defined by equation (6.7) (and not the tensors K^h_{ikj} or R^h_{ikj}) as the natural generalisation of the curvature tensor of Riemannian geometry. Again, putting

$$H_{ik} = H^r_{ikr}\,,$$

it follows from (7.6a) that under these circumstances

$$H_{ik} = (n-1)\,R\,g_{ik}\,. \tag{7.6b}$$

From the generalisation of Schur's theorem it follows that Finsler spaces F_n (with $n > 2$) of constant Riemannian curvature satisfy equations (7.6a) or (7.6b).

Isotropic Finsler spaces may also be characterised by the properties of vectors which have been transported by parallelism about an infinitesimal circuit. Since these properties are somewhat complicated, we refer the reader to a detailed description by VARGA[2].

More specialised cases arise when certain functions involving the curvature tensors K^i_{jhk} vanish. Firstly we note that in view of (1.34) *the tensor R^i_{jhk} vanishes when K^i_{jhk} vanishes. The converse is also true*, for in consequence of (1.34a) we may write (1.34) in the form

$$K^i_{jhk}(x, \dot{x}) = R^i_{jhk}(x, \dot{x}) - C^i_{jm}R^m_{rhk}(x, \dot{x})\,\dot{x}^r\,.$$

Secondly, it follows from (6.2) and (6.5) that Berwald's tensors H^i_k, H^i_{jk} vanish with K^i_{jhk}, but from (6.9b) we see that in general H^i_{hjk} vanishes only if K^i_{jhk} vanishes for all directional arguments $\dot{x}^i$. The converse is, however, not necessarily true[3].

Let us consider a vector field $\xi^i(x^k)$ in F_n: if this field is to be stationary, i. e. if it is to satisfy the partial differential equations

$$\xi^i_{|k} = \frac{\partial \xi^i}{\partial x^k} + \Gamma^{*i}_{hk}(x, \xi)\,\xi^h = 0\,, \tag{7.12}$$

---

[1] SYNGE and SCHILD [1], p. 112. For the two-dimensional case in Finsler geometry, see MOÓR [2], p. 7.

[2] VARGA [9], pp. 154—155.

[3] BERWALD [9], § 2.

the following integrability conditions must hold:

$$\frac{\partial}{\partial x^m}\left(\Gamma^{*\,i}_{h\,k}(x,\,\xi)\,\xi^h\right) - \frac{\partial}{\partial x^k}\left(\Gamma^{*\,i}_{h\,m}(x,\,\xi)\,\xi^h\right) = 0\,.$$

Observing (7.12) and the definition (1.7) it is easily seen that this condition reduces to the form[1]

$$\xi^j K^i_{j\,h\,k}(x,\,\xi) = 0\,. \tag{7.13}$$

This relation is therefore a sufficient condition for the existence of a vector field $\xi^i(x^k)$ which is stationary with respect to itself [in the sense of equation (7.12)[2]]. *Furthermore, the "parallel displacement" of a vector X^i with respect to the field ξ^k as defined by the equations*

$$\frac{\partial X^i}{\partial x^k} = -\Gamma^{*\,i}_{h\,k}(x,\,\xi)\,X^h \quad \text{or} \quad dX^i = -\Gamma^{*\,i}_{h\,k}(x,\,\xi)\,X^h\,dx^k \tag{7.14}$$

is integrable if

$$K^i_{j\,h\,k}(x,\,\xi) = 0\,. \tag{7.15}$$

For again the integrability conditions of (7.14) may be reduced to the form

$$X^j K^i_{j\,h\,k}(x,\,\xi) = 0\,.$$

Assuming that (7.15) is satisfied for a stationary vector field $\xi^i(x^k)$ one may show *that there exist coordinate systems in which the $\Gamma^{*\,i}_{h\,k}(x,\,\xi)$ vanish.* This may be done as follows: at a given point P of F_n we select n arbitrary but linearly independent vectors $X^i_{(\alpha)}(\alpha = 1,\,\ldots,\,n)$. Let these vectors undergo a parallel displacement relative to the field ξ^k as defined by (7.14). In view of our assumptions there are defined n vectors $X^i_{(\alpha)}$ at each point in a neighbourhood of P and by continuity these vectors will remain linearly independent in this neighbourhood[3]. A special coordinate system (x^α) may now be defined by putting

$$\frac{\partial x^i}{\partial x^\alpha} = X^i_{(\alpha)} \tag{7.16}$$

---

[1] This result could, of course, have been foreseen in the light of § 1, 3°, and in particular as a result of equation (1.44). Nevertheless, the following more detailed discussion may serve to give the reader a deeper insight into the structure of Finsler spaces.

[2] The same result is derived also by VARGA [7], p. 375. In fact, it is shown by VARGA that if we construct the tensor $\bar{g}_{ij}(x) = g_{ij}(x,\,\xi)$ and regard this tensor as the metric tensor of a non-euclidean space of curvature k, then the relations

$$R^j_{i\,h\,k} = k\,(\delta^j_k\,g_{i\,h} - \delta^j_h\,g_{i\,k})$$

must be satisfied. These equations together with (7.13) represent the necessary and sufficient conditions that the Finsler space represents a generalised non-euclidean space (in the above sense).

[3] The $X^i_{(\alpha)}$ might be chosen to satisfy

$$g_{ij}(x,\,\xi)\,X^i_{(\alpha)}X^j_{(\beta)} = \delta^\alpha_\beta$$

at P.

at each point in this region. We note that the relevant integrability conditions, i. e.

$$\frac{\partial X^i_{(\alpha)}}{\partial x^\beta} = \frac{\partial X^i_{(\beta)}}{\partial x^\alpha}, \qquad (\alpha, \beta, \ldots = 1, \ldots, n),$$

are satisfied, for

$$\frac{\partial X^i_{(\alpha)}}{\partial x^\beta} = \frac{\partial X^i_{(\alpha)}}{\partial x^k}\frac{\partial x^k}{\partial x^\beta} = - \Gamma^{*\,i}_{h\,k}(x, \xi)\, X^h_{(\alpha)} X^k_{(\beta)} \tag{7.17}$$

in virtue of (7.14) and (7.16), so that the left-hand side is symmetric in α and β. But the general transformation law (2.4.6a) for the $\Gamma^{*\,i}_{h\,k}$ reads

$$\Gamma^{*\,\gamma}_{\alpha\,\beta}(x, \xi) = \frac{\partial x^\gamma}{\partial x^i}\left(\frac{\partial^2 x^i}{\partial x^\alpha\,\partial x^\beta} + \Gamma^{*\,i}_{h\,k}(x, \xi)\frac{\partial x^h}{\partial x^\alpha}\frac{\partial x^k}{\partial x^\beta}\right),$$

and as a result of (7.16) and (7.17) it follows that $\Gamma^{*\,\gamma}_{\alpha\,\beta}(x, \xi) = 0$. In fact, one may go further. For, it follows from (3.1.22) that $\Gamma^\gamma_{\alpha\,\beta} = 0$, and hence from (3.1.7) that

$$\frac{\partial g_{\alpha\beta}(x, \xi)}{\partial x^\gamma} = 0.$$

Taking into account (1.3.2) we see that this relation leads to the equations

$$\frac{\partial F(x, \xi)}{\partial x^\gamma} = 0, \tag{7.18}$$

in the special coordinate system (7.16). If this equation were true in any *one* coordinate system for *all* directions ξ, the space would be Minkowskian. But even if condition (7.15) were to hold for all directions ξ^i at each point of F_n this would not necessarily lead to this particular state of affairs. For let us suppose that there exists another vector field $\eta^i(x^k)$ stationary with respect to itself, while $K^i_{j\,h\,k}(x, \eta) = 0$. As before, we may show that there exist coordinate systems (x^a) $(a = 1, \ldots, n)$ in which $\partial F(x, \eta)/\partial x^a = 0$, but such a coordinate system would not necessarily coincide with the (x^α) system. Even if the same initial set of vectors $X^i_{(a)}$ is chosen at P, the fields $X^i_{(\alpha)}$, $X^i_{(a)}$ will, in general, differ at points other than P, since their construction [by equation (7.14)] involves $\Gamma^{*\,i}_{h\,k}(x, \xi)$ and $\Gamma^{*\,i}_{h\,k}(x, \eta)$ respectively. They will coincide, however, if the $\Gamma^{*\,i}_{h\,k}$ are independent of direction, a condition which is seen to be satisfied as a result of (3.3.14) if $C^i_{h\,k|r} = 0$. We therefore have the theorem[1]: *a Finsler space is Minkowskian if the tensors* $K^i_{j\,h\,k}$ *(or* $R^i_{j\,h\,k}$*) and* $C^i_{h\,k|r}$ *(or* $A^i_{h\,k|r}$*) vanish identically.* Clearly a similar theorem holds for Berwald's curvature tensor: The Finsler space is Minkowskian if the tensors $H^i_{j\,h\,k}$ and $G^i_{j\,h\,k}$ vanish identically[2].

---

[1] Cartan [1], p. 39, where this result is stated without proof. The construction of the coordinate system (7.16) leading to the equations (7.18) is given by Rund [5], p. 101, but it was erroneously assumed that (7.15) is a sufficient condition for the space to be Minkowskian. See also Varga [1], p. 161, Vagner [13], p. 126.

[2] A glance at the equations (3.3.8a) and (3.3.14) shows that the condition $A^i_{h\,j|k} = 0$ implies $G^i_{h\,j\,k} = 0$. The converse is also true, as may be shown by a short calculation for the expression of the $G^i_{h\,j\,k}$ in terms of the $A^i_{j\,k}$ and their various derivatives (Berwald [9], equation (11)).

For further details concerning Finsler spaces of constant curvature the reader is referred to VARGA [7] and [9]. These spaces may be characterised in different ways by the properties of vectors which have been transported by parallelism about infinitesimal circuits. Such properties are discussed also by BERWALD [1] and [10, IV].

Furthermore, an interesting result due to BLASCHKE [2] is relevant to the topics treated in the present section. The "bisector" (called "Symmetrale") of two points P, Q of a two-dimensional Finsler space F_2 is the locus of those points M whose distances from P and Q (measured along the geodesics joining M, P and M, Q respectively) are equal. If this locus is to be a geodesic of F_2, the metric function of F_2 is the metric function of a Riemannian space of constant curvature.

VARGA [14] and [15] describes Finsler spaces which possess a distant parallelism of the line elements. Finally, it is shown by WALKER [1] that a completely symmetric Finsler space [1] is a Riemannian space of constant curvature.

§ 8. The Projective Curvature Tensors

In this section we shall introduce the so-called "projective" transformations or changes, which will lead us to various projective curvature tensors. Strictly speaking, this topic forms an integral part of the general geometry of paths (Chapter III, § 3), for it is in connection with the latter that these concepts were originally defined: nevertheless, we shall see that the projective curvature tensors play an important role in the theory of special classes of Finsler spaces. Thus, while we shall derive the analysis of the first two parts of this section against the background of the general geometry of paths, the reader may easily translate this theory into the terminology of Finsler geometry, regarding the paths as geodesics.

1°. The generalised Weyl Tensor

Let us suppose that in a general space of paths the differential equations of the paths with respect to some special parameter s are of the form

$$\frac{d^2 x^i}{ds^2} + 2G^i\left(x, \frac{dx}{ds}\right) \equiv \frac{dx'^i}{ds} + 2G^i(x, x') = 0 , \quad \left(x'^i = \frac{dx^i}{ds}\right). \quad (8.1)$$

By means of equations (6.3), (6.4) and (6.6) we may define curvature tensors H^i_k, H^i_{jk}, H^i_{hjk} with respect to the functions G^i, i. e. these tensors may be constructed even if we do not presuppose the existence of a metric. For this reason we shall adhere to our original notation for these tensors, for we shall eventually apply our results to metric spaces.

Clearly, if we subject equation (8.1) to an arbitrary transformation of its parameter s: $t = t(s)$, where $dt/ds > 0$, the special form of (8.1)

---

[1] A space is said to have spherical symmetry about a point P if there exist coordinates x^i, all zero at P, such that the space admits all orthogonal transformations of the x^i, i. e. if the structure (metric, paths) is transformed into itself by every orthogonal transformation. The term "completely symmetric" implies spherical symmetry about every point.

cannot in general be preserved. In fact, denoting differentiation with respect to t as usual by a dot, we find that the differential equations of the paths assume the form

$$\frac{\ddot{x}^i + 2G^i(x,\dot{x})}{\dot{x}^i} = \frac{\ddot{x}^k + 2G^k(x,\dot{x})}{\dot{x}^k} \qquad (i, k = 1, \ldots, n) . \qquad (8.2)$$

It is obvious that these equations remain unchanged if we replace the functions $G^i(x,\dot{x})$ by new functions $\overline{G}^i(x,\dot{x})$, the latter being defined by

$$\overline{G}^i(x,\dot{x}) = G^i(x,\dot{x}) - P(x,\dot{x})\,\dot{x}^i , \qquad (8.3)$$

where $P(x,\dot{x})$ is an arbitrary scalar function positively homogeneous of the first degree in the $\dot{x}^i$. Clearly this relation represents the most general modification of the function G^i which will leave equations (8.2) unchanged. Using the nomenclature of the restricted geometry of paths, we shall call (8.3) a *projective change* of the functions G^i. Alternatively, equations (8.3) may be regarded as representing a mapping[1] of two path-spaces, characterised by the functions G^i and $\overline{G}^i$ respectively, onto each other [corresponding line-elements being denoted by the same coordinates $(x^i, \dot{x}^i)$]. The special parameters s and $\bar{s}$ of equations (8.1) are related by the equation[2]

$$\bar{s} = A + B \int e^{-2\int P(x,\,dx/ds)\,ds}\,ds , \qquad (8.3a)$$

where A and B are arbitrary constants.

In analogy to (6.3) one may define curvature tensors $\overline{H}_k^j$ with respect to the new functions $\overline{G}_{hk}^j$:

$$\overline{H}_k^j = 2\,\frac{\partial \overline{G}^j}{\partial x^k} - \frac{\partial^2 \overline{G}^j}{\partial x^h\,\partial \dot{x}^k}\,\dot{x}^h + 2\overline{G}_{kl}^j\,\overline{G}^l - \frac{\partial \overline{G}^j}{\partial \dot{x}^l}\,\frac{\partial \overline{G}^l}{\partial \dot{x}^k} . \qquad (8.4)$$

From (8.3) we find

$$\overline{G}_{kl}^j = G_{kl}^j - \frac{\partial P}{\partial \dot{x}^k}\,\delta_l^j - \frac{\partial P}{\partial \dot{x}^l}\,\delta_k^j - \frac{\partial^2 P}{\partial \dot{x}^k\,\partial \dot{x}^l}\,\dot{x}^j . \qquad (8.5)$$

On substituting these and similar expressions in (8.4), we may simplify the resulting relation, a great many of these simplifications being possible as a result of the homogeneity properties of the G^i (second degree) and of

---

[2] This is easily verified as follows. For suppose that we have found a parameter s such that the paths (defined by the G^i) are represented by (8.1). If we introduce a parameter transformation $\bar{s} = \bar{s}(s)$, together with the projective change (8.3), we find

$$\left(\frac{d^2x^i}{d\bar{s}^2} + 2\overline{G}^i\left(x, \frac{dx}{d\bar{s}}\right)\right)\left(\frac{d\bar{s}}{ds}\right)^2 = -\frac{dx^i}{d\bar{s}}\left(2P\left(x, \frac{dx}{ds}\right)\frac{d\bar{s}}{ds} + \frac{d^2\bar{s}}{ds^2}\right) .$$

This reduces to the form (8.1) if the right-hand side is equated to zero. Thus a differential equation for $\bar{s}$ as a function of s is obtained. Clearly (8.3a) represents its most general solution.

the function P (first degree). It will be found that we are left with

$$\overline{H}^j_k = H^j_k - \dot{x}^j \left(2 \frac{\partial P}{\partial x^k} - \frac{\partial^2 P}{\partial \dot{x}^k \partial x^h} \dot{x}^h + 2 \frac{\partial^2 P}{\partial \dot{x}^k \partial \dot{x}^l} G^l - \frac{\partial G^l}{\partial \dot{x}^k} \frac{\partial P}{\partial \dot{x}^l} + P \frac{\partial P}{\partial \dot{x}^k} \right) +$$
$$+ \delta^j_k \left(\frac{\partial P}{\partial x^h} \dot{x}^h - 2 \frac{\partial P}{\partial \dot{x}^l} G^l + P^2 \right). \tag{8.6}$$

By compounding the covariant derivatives $\left(\dfrac{\partial P}{\partial \dot{x}^k} \right)_{(h)} \dot{x}^h$ and $P_{(k)}$, it is

seen that the above equation may be written in tensor form:

$$\overline{H}^j_k = H^j_k + \dot{x}^j \left[\left(\frac{\partial P}{\partial \dot{x}^k} \right)_{(h)} \dot{x}^h - 2 P_{(k)} - P \frac{\partial P}{\partial \dot{x}^k} \right] +$$
$$+ \delta^j_k (P_{(h)} \dot{x}^h + P^2). \tag{8.7}$$

The variation of the tensor H^j_{hk} under the projective change (8.3) may be calculated similarly. If we differentiate (8.6) with respect to $\dot{x}^h$, and interchange the indices h and k, we find on subtraction, by means of (6.4), after some simplification

$$\overline{H}^j_{hk} = H^j_{hk} - \delta^j_h \left(\frac{\partial P}{\partial x^k} - \frac{\partial P}{\partial \dot{x}^l} \frac{\partial G^l}{\partial \dot{x}^k} + P \frac{\partial P}{\partial \dot{x}^k} \right) +$$
$$+ \delta^j_k \left(\frac{\partial P}{\partial x^h} - \frac{\partial P}{\partial \dot{x}^l} \frac{\partial G^l}{\partial \dot{x}^h} + P \frac{\partial P}{\partial \dot{x}^h} \right) - \tag{8.7a}$$
$$- \dot{x}^j \left(\frac{\partial^2 P}{\partial x^k \partial \dot{x}^h} - \frac{\partial^2 P}{\partial \dot{x}^k \partial x^h} + \frac{\partial^2 P}{\partial \dot{x}^k \partial \dot{x}^l} \frac{\partial G^l}{\partial \dot{x}^h} - \frac{\partial^2 P}{\partial \dot{x}^h \partial \dot{x}^l} \frac{\partial G^l}{\partial \dot{x}^k} \right).$$

Again, this relation may be written in tensor form as follows:

$$\overline{H}^j_{hk} = H^j_{hk} - \delta^j_h \left(P_{(k)} + P \frac{\partial P}{\partial \dot{x}^k} \right) + \delta^j_k \left(P_{(h)} + P \frac{\partial P}{\partial \dot{x}^h} \right) -$$
$$- \dot{x}^j \left[\left(\frac{\partial P}{\partial \dot{x}^h} \right)_{(k)} - \left(\frac{\partial P}{\partial \dot{x}^k} \right)_{(h)} \right]. \tag{8.7b}$$

Contracting in (8.6) after division by $n - 1$, we find in view of (6.17)

$$\overline{H} = H - \left(2 \frac{\partial P}{\partial \dot{x}^l} G^l - \frac{\partial P}{\partial x^j} \dot{x}^j - P^2 \right),$$

or

$$\overline{H} = H + P_{(h)} \dot{x}^h + P^2. \tag{8.8}$$

Combining (8.7) and (8.8) we therefore have

$$\overline{H}^j_k - \overline{H} \delta^j_k = H^j_k - H \delta^j_k + \dot{x}^j \left[\left(\frac{\partial P}{\partial \dot{x}^k} \right)_{(h)} \dot{x}^h - 2 P_{(k)} - P \frac{\partial P}{\partial \dot{x}^k} \right]. \tag{8.9}$$

We differentiate this equation with respect to $\dot{x}^r$, and then contract r and j. Noting that the expression in square brackets is homogeneous of degree one, we observe that this operation will simply reproduce this expression $(n + 1)$ times, so that

$$\frac{\partial \overline{H}^j_k}{\partial \dot{x}^j} - \frac{\partial \overline{H}}{\partial \dot{x}^k} = \frac{\partial H^j_k}{\partial \dot{x}^j} - \frac{\partial H}{\partial \dot{x}^k} + (n + 1) \left[\left(\frac{\partial P}{\partial \dot{x}^k} \right)_{(h)} \dot{x}^h - 2 P_{(k)} - P \frac{\partial P}{\partial \dot{x}^k} \right].$$

Between this equation and (8.9) the expression in square brackets may be eliminated. We thus obtain

$$\frac{1}{n+1}\left(\frac{\partial \overline{H}{}^i_k}{\partial \dot{x}^i} - \frac{\partial \overline{H}}{\partial \dot{x}^k}\right)\dot{x}^j - (\overline{H}{}^j_k - \overline{H}\,\delta^j_k)$$
$$= \frac{1}{n+1}\left(\frac{\partial H^i_k}{\partial \dot{x}^i} - \frac{\partial H}{\partial \dot{x}^k}\right)\dot{x}^j - (H^j_k - H\,\delta^j_k)\,.$$

Hence the tensor defined by[1]

$$W^j_k = H^j_k - H\,\delta^j_k - \frac{1}{n+1}\left(\frac{\partial H^i_k}{\partial \dot{x}^i} - \frac{\partial H}{\partial \dot{x}^k}\right)\dot{x}^j \tag{8.10}$$

is invariant under the projective change (8.3), and may therefore be regarded as the *projective deviation tensor* [being the counterpart of the tensor H^j_k of the equations (6.1) of geodesic deviation].

If we contract j and k in (8.10), taking into account the fact that the functions H, H^j_k are homogeneous of the second degree, we obtain by means of (6.17) and (6.19a)

$$W^j_j = (n-1)\,H - nH + \frac{n-1}{n+1}\,H + \frac{2}{n+1}\,H = 0\,. \tag{8.11}$$

Similarly it follows from (8.10), (6.18) and (6.19a) that

$$W^j_k\,\dot{x}^k = 0\,, \tag{8.12}$$

from which we deduce by differentiation with respect to $\dot{x}^h$ that

$$\frac{\partial W^j_k}{\partial \dot{x}^h}\,\dot{x}^k = -W^j_h\,. \tag{8.12a}$$

If we differentiate (8.10) with respect to $\dot{x}^h$, afterwards contracting j and h, we find after some reduction that

$$\frac{\partial W^i_k}{\partial \dot{x}^i} = 0\,. \tag{8.12b}$$

From the tensor defined by (8.10) we may now derive further projective curvature tensors by applying the method introduced in § 6 in the course of our treatment of Berwald's curvature tensor. Firstly, in analogy with (6.4) we write

$$W^j_{hk} = \frac{1}{3}\left(\frac{\partial W^j_k}{\partial \dot{x}^h} - \frac{\partial W^j_h}{\partial \dot{x}^k}\right), \tag{8.13}$$

and secondly, following (6.6), we put

$$W^j_{ihk} = \frac{\partial W^j_{hk}}{\partial \dot{x}^i} = \frac{1}{3}\left(\frac{\partial^2 W^j_k}{\partial \dot{x}^i\,\partial \dot{x}^h} - \frac{\partial^2 W^j_h}{\partial \dot{x}^i\,\partial \dot{x}^k}\right). \tag{8.14}$$

It is essential that these tensors be expressed analytically as functions of the tensor H^i_j and its associated tensors. In order to find the required expression for (8.13) we substitute (8.10) on the right-hand side and carry

---

[1] W^j_k and the associated tensors were defined in this manner by Berwald [10, IV].

out the differentiations as indicated. Applying (6.4) and (6.6) to the result of these operations we find after a little simplification

$$W^j_{hk} = H^j_{hk} + \frac{\dot{x}^j}{n+1} H^i_{ikh} +$$
$$+ \frac{1}{3(n+1)} \left\{ \delta^j_h \left((n+2) \frac{\partial H}{\partial \dot{x}^k} - \frac{\partial H^i_k}{\partial \dot{x}^i} \right) - \delta^j_k \left((n+2) \frac{\partial H}{\partial \dot{x}^h} - \frac{\partial H^i_h}{\partial \dot{x}^i} \right) \right\} .$$

By means of (6.21) the directional derivatives of the H^i_k may be eliminated: in this manner the above expression is reduced to

$$W^j_{hk} = H^j_{hk} + \frac{\dot{x}^j}{n+1} H^i_{ikh} + \frac{\delta^j_h}{n+1} \left(\frac{\partial H}{\partial \dot{x}^k} + H_k \right) - \frac{\delta^j_k}{n+1} \left(\frac{\partial H}{\partial \dot{x}^h} + H_h \right) .$$

But from (6.24) and (6.16b) it is clear that

$$(n-1) \frac{\partial H}{\partial \dot{x}^k} = H_k + \dot{x}^i H_{ki} . \tag{8.15}$$

By eliminating the directional derivatives of H between the last two equations, we obtain the desired expression:

$$W^j_{hk} = H^j_{hk} + \frac{\dot{x}^j}{n+1} H^i_{ikh} + \frac{\delta^j_h}{n^2-1} (nH_k + \dot{x}^i H_{ki}) -$$
$$- \frac{\delta^j_k}{n^2-1} (nH_h + \dot{x}^i H_{hi}) . \tag{8.16}$$

By means of (6.25) an alternative form is given by

$$W^j_{hk} = H^j_{hk} + \frac{\dot{x}^j}{n+1} (H_{hk} - H_{kh}) +$$
$$+ \frac{\delta^j_h}{n^2-1} (nH_k + \dot{x}^r H_{kr}) - \frac{\delta^j_k}{n^2-1} (nH_h + \dot{x}^r H_{hr}) . \tag{8.16a}$$

The tensor (8.14) is found by differentiation of this expression with respect to $\dot{x}^i$. After substituting from (6.6) and (6.16b) in the resulting expression, we find

$$W^j_{ihk} = H^j_{ihk} + \frac{\delta^j_i}{n+1} (H_{hk} - H_{kh}) + \tag{8.17}$$
$$+ \frac{\dot{x}^j}{n+1} \left(\frac{\partial H_{hk}}{\partial \dot{x}^i} - \frac{\partial H_{kh}}{\partial \dot{x}^i} \right) + \frac{\delta^j_h}{n^2-1} \left(nH_{ik} + H_{ki} + \dot{x}^r \frac{\partial H_{kr}}{\partial \dot{x}^i} \right) -$$
$$- \frac{\delta^j_k}{n^2-1} \left(nH_{ih} + H_{hi} + \dot{x}^r \frac{\partial H_{hr}}{\partial \dot{x}^i} \right) .$$

This represents the generalisation of a tensor introduced by WEYL in the restricted geometry of paths.

We remark that the projective curvature tensors satisfy the following simple identities. Noting that W^i_j is homogeneous of the second degree in its directional arguments, we deduce from (8.13) and (8.12a) that

$$W^j_{hk} \dot{x}^h = \frac{1}{3} \left(2W^j_k - \frac{\partial W^j_k}{\partial \dot{x}^k} \dot{x}^h \right) = W^j_k , \tag{8.18}$$

so that in view of (8.14)

$$W^j_{ihk} \dot{x}^i = W^j_{hk} , \qquad W^j_{ihk} \dot{x}^i \dot{x}^h = W^j_k . \tag{8.18a}$$

Further, differentiation of (8.12a) yields

$$\frac{\partial^2 W_h^j}{\partial \dot{x}^i \, \partial \dot{x}^k} \, \dot{x}^h = - \frac{\partial W_k^j}{\partial \dot{x}^i} - \frac{\partial W_i^j}{\partial \dot{x}^k} \, ,$$

and hence we have from (8.14) and (8.13)

$$W_{ihk}^j \, \dot{x}^h = \frac{2}{3} \frac{\partial W_k^j}{\partial \dot{x}^i} + \frac{1}{3} \frac{\partial W_i^j}{\partial \dot{x}^k} = \frac{\partial W_k^j}{\partial \dot{x}^i} + W_{ki}^j \, . \tag{8.19}$$

A glance at the equations (8.13), (8.14), (8.18) and (8.18a) shows that if any one of the three tensors W_j^i, W_{jk}^i, W_{jkh}^i vanishes identically, the other two are also automatically zero. Also, as a result of (8.11) and (8.12b) it follows from (8.13) and (8.14) that

$$W_{hr}^r = 0 \, , \qquad W_{ihr}^r = 0 \, . \tag{8.20}$$

2°. The Projective Connection

Of fundamental importance to the projective theory of paths is a set of connection coefficients which is invariant under a projective change (8.3). These coefficients are easily derived as follows. Noting once more that the functions $P(x, \dot{x})$ in (8.3) are homogeneous of the first degree in their directional arguments, we see that contraction of the indices j and k in equation (8.5) gives

$$\overline{G}_{rl}^r = G_{rl}^r - (n+1) \frac{\partial P}{\partial \dot{x}^l} \, . \tag{8.21}$$

Differentiating this relation with respect to $\dot{x}^k$, we find

$$\overline{G}_{rlk}^r = G_{rlk}^r - (n+1) \frac{\partial^2 P}{\partial \dot{x}^l \, \partial \dot{x}^k} \, . \tag{8.21a}$$

By means of these equations we may eliminate $\partial P / \partial \dot{x}^l$ and $\partial^2 P / \partial \dot{x}^l \, \partial \dot{x}^k$ from (8.5), thus obtaining

$$\overline{G}_{kl}^j = G_{kl}^j - \frac{1}{n+1} \, [\delta_l^j (G_{rk}^r - \overline{G}_{rk}^r) + \delta_k^j (G_{rl}^r - \overline{G}_{rl}^r)] -$$
$$- \frac{\dot{x}^j}{n+1} \, (G_{rlk}^r - \overline{G}_{rlk}^r) \, .$$

Hence the quantities defined by

$$\Pi_{kl}^i = G_{kl}^j - \frac{1}{n+1} \, [\delta_l^j \, G_{rk}^r + \delta_k^j \, G_{rl}^r + \dot{x}^j \, G_{rlk}^r] \tag{8.22}$$

are invariant under the projective change (8.3) and are thus independent of the parametrisation of the paths[1].

These quantities are known as the *projective connection coefficients,* for they define a projective covariant derivative in the same manner as the G_{hk}^i give rise to a covariant derivative. However, as in the restricted

---

[1] DOUGLAS [1], p. 156.

theory of paths, the projective covariant derivatives of a tensor are not, in general, tensors[1]. Nevertheless, it is not difficult to show that every differential invariant which remains unchanged under the projective change (8.3) is expressible in terms of the Π^j_{kl} and their derivatives[2].

Differentiating (8.22) with respect to $\dot{x}^h$, we obtain

$$B^j_{klh} \equiv \frac{\partial \Pi^j_{kl}}{\partial \dot{x}^h} =$$
$$= G^j_{klh} - \frac{1}{n+1}\,[\delta^j_l G^r_{rkh} + \delta^j_k G^r_{rlh} + \delta^j_h G^r_{rlk}] - \frac{\dot{x}^j}{n+1}\, G^r_{rlkh}. \tag{8.23}$$

Clearly this expression represents *a tensor which is invariant under the projective change* (8.3).

In the restricted geometry of paths, which is characterized by the existence of a parametrisation for which $G^j_{klh} = 0$, the tensor (8.23) vanishes identically. The converse is also true, as we shall now show. Differentiating (8.5) with respect to $\dot{x}^h$, we find

$$\overline{G}^j_{klh} = G^j_{klh} - \delta^j_l \frac{\partial^2 P}{\partial \dot{x}^h \partial \dot{x}^k} - \delta^j_k \frac{\partial^2 P}{\partial \dot{x}^h \partial \dot{x}^l} - \delta^j_h \frac{\partial^2 P}{\partial \dot{x}^k \partial \dot{x}^l} - \dot{x}^j \frac{\partial^3 P}{\partial \dot{x}^h \partial \dot{x}^k \partial \dot{x}^l}. \tag{8.24}$$

We now choose a special projective change (8.3) for which the function P is to satisfy the relation

$$P = \frac{1}{n+1}\, G^r_{rs}\, \dot{x}^s, \quad \text{or} \quad \frac{\partial^2 P}{\partial \dot{x}^k \partial \dot{x}^h} = \frac{1}{n+1}\, G^r_{rkh}. \tag{8.25}$$

If we substitute (8.25) in the right-hand side of (8.24), the latter becomes identical with the right-hand side of (8.23), i. e. under the special projective change defined by (8.25) we have

$$B^j_{klh} = \overline{G}^j_{klh}.$$

Hence if $B^j_{klh} = 0$, there exists a parametrisation of the paths for which the tensor G^j_{klh} vanishes.

We may summarise these results by the

Theorem: A necessary and sufficient condition that the general space of paths be reducible by a projective change (8.3) *to a space of paths of the restricted type is that the tensor B^j_{klh} vanishes identically.*

In passing we note that corresponding to each coordinate system (x^i) there exists a so-called *projective* parameter $t_{(x)}$, determined up to a non-degenerate linear transformation, such that the $\overline{G}^j_{kl}$ determined by

---

[1] This is true even in the restricted geometry of paths (cf. EISENHART [2], p. 101), for the transformation law of the Π^j_{kl} as defined by (8.22) is the same as in the restricted theory. This is immediately obvious if we note that the definition of the projective connection coefficients in the restricted theory is formally the same as (8.22), except that it lacks the last term on the right-hand side, this term being a tensor.

[2] DOUGLAS [1], § 7.

this parameter are equal to the Π^i_{kl} defined by (8.22). For on substituting from (8.25) in (8.5) we find

$$\overline{G}^i_{kl}= G^i_{kl}-\frac{1}{n+1}\,(\delta^i_l\,G^r_{rk}+\,\delta^i_k\,G^r_{rl})-\frac{\dot{x}^j}{n+1}\,G^r_{rlk}\,.$$

Comparison with (8.22) shows that $\overline{G}^i_{kl}=\Pi^i_{kl}$ for this special parametrisation[1].

3°. Projectively Flat Spaces; Spaces with Rectilinear Geodesics

A space endowed with a system of general paths is called *projectively flat* if there exist coordinate systems in which the paths can be represented by $(n-1)$ linear equations. If the space is also a metric space (the paths being the geodesics) it is called a space with *rectilinear geodesics* if it is projectively flat in the above sense.

The fundamental significance of the tensors W^i_{klh} and B^i_{klh} is illustrated by the following

Theorem[2]: *A necessary and sufficient condition that a general n-dimensional path space be projectively flat is that for $n > 2$ the equations*

$$B^i_{klh}= 0 , \qquad W^i_{klh}= 0 , \tag{8.26}$$

be satisfied, while for $n = 2$ the second of these equations must be replaced by

$$2\,(H_{i\,(j)}- H_{j\,(i)}) + (H_{i\,h\,(j)}- H_{j\,h\,(i)})\,\dot{x}^h = 0 . \tag{8.27}$$

Proof: In the present section we shall only deal with the case $n > 2$, for 2-dimensional Finsler spaces are treated in detail in Chapter VI, § 6. We have seen that the first condition (8.26) implies that the general space of paths is of the restricted type. Under these circumstances the tensor W^i_{klh} reduces to the well-known Weyl tensor. The theorem then follows from the classical theorem[3] according to which the necessary and sufficient condition that a space with an affine connection be projectively flat is that the Weyl tensor vanishes.

The significance of the generalised Weyl tensor is illustrated even more strikingly by the

Theorem[4]: *In order that a general path space of n dimensions may be mapped by means of a projective change onto a general path space of zero curvature ($H^i_{ihk}= 0$) it is necessary and sufficient that for $n > 2$*

$$W^i_j = 0 , \tag{8.28}$$

while for $n = 2$ equations (8.27) must be satisfied.

---

[1] Douglas [1], §§ 5—6.

[2] For $n > 2$ this theorem is given by Douglas [1], p. 162. The case $n = 2$ is treated by Berwald [10, III], §§ 9—10. See also Funk [2, 3, 4] and Wirtinger [1].

[3] For the proof of this theorem of the restricted theory we refer the reader to Eisenhart [2], p. 96.

[4] Berwald [10, IV], p. 767. See also Ch. VI, § 6.

Proof: Let us put

$$Q_k = P_{(k)} + P \frac{\partial P}{\partial \dot{x}^k} . \tag{8.29}$$

Then by (6.11) we have

$$\frac{\partial Q_k}{\partial \dot{x}^h} - \frac{\partial Q_h}{\partial \dot{x}^k} = \left(\frac{\partial P}{\partial \dot{x}^h}\right)_{(k)} - \left(\frac{\partial P}{\partial \dot{x}^k}\right)_{(h)} . \tag{8.29a}$$

Substituting these two relations in (8.7b), we see that the latter may be written in the form

$$\overline{H}^j_{hk} = H^j_{hk} + \delta^j_k Q_h - \delta^j_h Q_k - \dot{x}^j \left(\frac{\partial Q_k}{\partial \dot{x}^h} - \frac{\partial Q_h}{\partial \dot{x}^k}\right) . \tag{8.30}$$

Thus, in order that it be possible to find a projective change (8.3) for which $\overline{H}^j_{hk}$ vanishes, there must exist firstly a vector Q_h satisfying the relation

$$H^j_{hk} = \delta^j_h Q_k - \delta^j_k Q_h + \dot{x}^j \left(\frac{\partial Q_k}{\partial \dot{x}^h} - \frac{\partial Q_h}{\partial \dot{x}^k}\right) , \tag{8.31}$$

and secondly a scalar function $P(x, \dot{x})$, homogeneous of the first degree in its directional arguments, which satisfies equation (8.29). Now the integrability conditions of (8.29) read

$$Q_{k(h)} - Q_{h(k)} = P_{(k)(h)} - P_{(h)(k)} + P_{(h)} \frac{\partial P}{\partial \dot{x}^k} - P_{(k)} \frac{\partial P}{\partial \dot{x}^h} +$$

$$+ P\left[\left(\frac{\partial P}{\partial \dot{x}^k}\right)_{(h)} - \left(\frac{\partial P}{\partial \dot{x}^h}\right)_{(k)}\right] .$$

By (6.10) and (8.29a) these equations are equivalent to

$$Q_{k(h)} - Q_{h(k)} = -\frac{\partial P}{\partial \dot{x}^l} H^l_{kh} + P_{(h)} \frac{\partial P}{\partial \dot{x}^k} - P_{(k)} \frac{\partial P}{\partial \dot{x}^h} +$$

$$+ P\left(\frac{\partial Q_h}{\partial \dot{x}^k} - \frac{\partial Q_k}{\partial \dot{x}^h}\right) =$$

$$= \frac{\partial P}{\partial \dot{x}^j}\left[H^j_{hk} + \delta^j_k Q_h - \delta^j_h Q_k +\right.$$

$$\left. + \dot{x}^j \left(\frac{\partial Q_h}{\partial \dot{x}^k} - \frac{\partial Q_k}{\partial \dot{x}^h}\right)\right] ,$$

the second equality resulting from an application of (8.29). But if we substitute (8.31) on the right-hand side, we find that the required integrability condition reduces to

$$Q_{k(h)} - Q_{h(k)} = 0 . \tag{8.32}$$

If, in equation (8.31), we contract the indices j and k, noting that the Q_k are positively homogeneous of the first degree in their directional arguments as a result of (8.29), we see that

$$H^r_{hr} = H_h = -n Q_h + \frac{\partial Q_r}{\partial \dot{x}^h} \dot{x}^r , \tag{8.33}$$

and hence by (6.16b)

$$H_{kh} = \frac{\partial H_h}{\partial \dot{x}^k} = -n\frac{\partial Q_h}{\partial \dot{x}^k} + \frac{\partial Q_k}{\partial \dot{x}^h} + \frac{\partial^2 Q_r}{\partial \dot{x}^h\,\partial \dot{x}^k}\,\dot{x}^r, \qquad (8.33\,\mathrm{a})$$

or

$$H_{kh}\,\dot{x}^h = -n\frac{\partial Q_r}{\partial \dot{x}^k}\,\dot{x}^r + Q_k\,. \qquad (8.33\,\mathrm{b})$$

Eliminating the directional derivatives of the Q_r between equations (8.33) and (8.33b), we obtain

$$Q_k = -\frac{1}{n^2-1}\,(nH_k + H_{kr}\,\dot{x}^r)\,.$$

From this relation it follows that the integrability conditions (8.32) may be written in the form

$$(nH_{k(h)} + H_{kr(h)}\,\dot{x}^r) - (nH_{h(k)} + H_{hr(k)}\,\dot{x}^r) = 0\,. \qquad (8.34)$$

But for $n = 2$ this condition reduces to equation (8.27): thus we have proved the theorem for this case. We now have to show that for $n > 2$ the equations (8.34) result from the conditions (8.28).

From equation (8.16) we obtain two relations by contracting the indices j and k, and j and h. This gives

$$W_{hj}^i = H_{hj}^i + \frac{\dot{x}^j}{n+1}H_{rjh}^r - \frac{1}{n+1}(nH_h + \dot{x}^r H_{hr})\,, \qquad (8.35)$$

together with

$$W_{jk}^i = H_{jk}^i + \frac{\dot{x}^j}{n+1}H_{rkj}^r + \frac{1}{n+1}(nH_k + \dot{x}^r H_{kr})\,. \qquad (8.35\,\mathrm{a})$$

From (8.13) we deduce that the W_{hk}^i are skew-symmetric in h and k. Thus it follows from (8.20) that the left-hand sides of both (8.35) and (8.35a) vanish identically. Hence if we differentiate these relations covariantly with respect to x^k and x^h respectively, and add the two identities thus obtained, we find

$$0 = H_{hj(k)}^j + H_{jk(h)}^j + \frac{\dot{x}^j}{n+1}(H_{rjh(k)}^r + H_{rkj(h)}^r) +$$
$$+ \frac{1}{n+1}[n(H_{k(h)} - H_{h(k)}) + \dot{x}^r(H_{kr(h)} - H_{hr(k)})]\,. \qquad (8.36)$$

Now contraction of (6.13a) yields

$$H_{hj(k)}^j + H_{jk(h)}^j = -H_{kh(j)}^j = H_{hk(j)}^j\,, \qquad (8.37)$$

while from (6.15a) we may deduce

$$\dot{x}^j(H_{rkj(h)}^r + H_{rjh(k)}^r) = H_{rkh(j)}^r\,\dot{x}^j - (H_h^m G_{rmk}^r - H_k^m G_{rmh}^r)\,. \qquad (8.38)$$

On substituting the identities (8.37), (8.38) in (8.36), we obtain

$$0 = H_{hk(j)}^j + \frac{\dot{x}^j}{n+1}H_{rkh(j)}^r - \frac{1}{n+1}(H_h^m G_{rmk}^r - H_k^m G_{rmh}^r) +$$
$$+ \frac{1}{n+1}[n(H_{k(h)} - H_{h(k)}) + \dot{x}^r(H_{kr(h)} - H_{hr(k)})]\,. \qquad (8.39)$$

Covariant differentiation of (8.16) with respect to x^s gives, after contraction of j and s,

$$W^j_{hk(j)} = H^j_{hk(j)} + \frac{\dot{x}^j}{n+1} H^r_{rkh(j)} +$$
$$+ \frac{1}{n^2-1} \left[n\left(H_{k(h)} - H_{h(k)}\right) + \dot{x}^r \left(H_{kr(h)} - H_{hr(k)}\right) \right] .$$

We now subtract (8.39) from this relation, thus obtaining

$$W^j_{hk(j)} = \frac{1}{n+1} \left(H^m_h G^r_{rmk} - H^m_k G^r_{rmh}\right) -$$
$$- \frac{n-2}{n^2-1} \left[n\left(H_{k(h)} - H_{h(k)}\right) + \dot{x}^r \left(H_{kr(h)} - H_{hr(k)}\right) \right] . \tag{8.40}$$

If we multiply equation (8.10) by G^r_{rmh}, noting (6.8a), we see that

$$W^m_k G^r_{rmh} = H^m_k G^r_{rmh} - H G^r_{rkh} ,$$

which, when substituted in (8.40), finally shows that (8.40) is equivalent to the identity

$$W^j_{hk(j)} = \frac{1}{n+1} \left(W^m_h G^r_{rmk} - W^m_k G^r_{rmh}\right) -$$
$$- \frac{n-2}{n^2-1} \left[n\left(H_{k(h)} - H_{h(k)}\right) + \dot{x}^r \left(H_{kr(h)} - H_{hr(k)}\right) \right] . \tag{8.41}$$

This proves Berwald's theorem: for if W^k_j vanishes, then so does W^j_{hk} and its covariant derivatives, and (8.41) reduces to the required integrability condition (8.34).

Clearly this theorem is directly applicable to the theory of Finsler spaces provided we interpret the paths as geodesics, so that if the differential equations of the paths are written in the form (8.1), the special parameter s represents the arc-length.

A further important property of the generalised Weyl tensor is exhibited by the following

Theorem: The generalised Weyl tensor vanishes identically in an isotropic Finsler space.

Proof: From (6.2) and (7.3a) we see that the condition for isotropy reads

$$H^i_j = F^2 R \left(\delta^i_j - l^i l_j\right) ,$$

where, by means of (6.5), (7.5) and (6.23) the scalar R may be expressed in the form $R = H/F^2$. Thus the above condition may be written in the form

$$H^i_j = H \left(\delta^i_j - l^i l_j\right) . \tag{8.42}$$

On differentiating this equation with respect to $\dot{x}^k$ and contracting i and k afterwards, we find after a little simplification

$$\frac{\partial H^r_j}{\partial \dot{x}^r} - \frac{\partial H}{\partial \dot{x}^j} = -(n+1) \frac{H}{F} l_j . \tag{8.43}$$

Substitution of (8.42) and (8.43) in the definition (8.10) of W_j^i immediately yields the desired result, namely $W_j^i = 0$.

As a corollary to the last two theorems we therefore have the

Theorem[1]: An n-dimensional isotropic Finsler space (with $n > 2$) may be transformed by means of a projective change into a general space of paths of zero curvature.

Finally we remark that Finsler spaces of isotropic curvature form a special subclass of the class of Finsler spaces whose projective (Weyl) curvature tensor vanishes identically. This subclass is characterised by the

Theorem[2]: Amongst the spaces of zero projective curvature the Finsler spaces of isotropic curvature are characterised for $n > 2$ by the relations

$$F \frac{\partial}{\partial \dot{x}^j} (K_{ihk}^j \, l^i \, l^h) - \frac{1}{n-1} F \frac{\partial}{\partial \dot{x}^k} (K_{ih} \, l^i \, l^h) + (K_{ih} \, l^i \, l^h) \, l_k = 0 \, . \qquad (8.44)$$

Proof: Equations (7.3a) represent the condition for isotropy.

Differentiating these equations with respect to $\dot{x}^l$, and then contracting j and l, we find

$$F \frac{\partial}{\partial \dot{x}^j} (K_{ihk}^j \, l^i \, l^h) = F \frac{\partial R}{\partial \dot{x}^k} - (n-1) \, R \, l_k \, .$$

Also, differentiation of (7.5) with respect to $\dot{x}^k$ gives

$$\frac{\partial R}{\partial \dot{x}^k} = \frac{1}{n-1} \frac{\partial}{\partial \dot{x}^k} (K_{ih} \, l^i \, l^h) \, . \qquad (8.45)$$

Eliminating $\partial R / \partial \dot{x}^k$ between these relations we have

$$F \frac{\partial}{\partial \dot{x}^j} (K_{ihk}^j \, l^i \, l^h) - \frac{F}{n-1} \frac{\partial}{\partial \dot{x}^k} (K_{ih} \, l^i \, l^h) = - (n-1) \, R \, l_k \, .$$

Replacing the term on the right-hand side by means of (7.5) we obtain the required relation (8.44).

In order to prove the converse we first note that in view of (6.2) and (6.17) we may write

$$H = \frac{1}{n-1} K_{rsi}^i \, \dot{x}^r \, \dot{x}^s \, . \qquad (8.46)$$

Now, assuming that $W_j^i = 0$, we find by substitution of (6.2) and (8.46) in (8.10) that

$$F^2 K_{rsk}^j \, l^r \, l^s = F^2 \frac{\delta_k^j}{n-1} \, K_{rsi}^i \, l^r \, l^s +$$

$$+ \frac{\dot{x}^j}{n+1} \left[\frac{\partial}{\partial \dot{x}^i} (K_{rsk}^i \, \dot{x}^r \, \dot{x}^s) - \frac{1}{n-1} \frac{\partial}{\partial \dot{x}^k} (K_{rsi}^i \, \dot{x}^r \, \dot{x}^s) \right]$$

$$= F^2 \frac{\delta_k^j}{n-1} \, K_{rsi}^i \, l^r \, l^s + \frac{\dot{x}^j}{n+1} \frac{\partial}{\partial \dot{x}^i} \left[F^2 \left(K_{rsk}^i - \frac{\delta_k^i}{n-1} \, K_{rsh}^h \right) l^r \, l^s \right] .$$

---

[1] BERWALD, loc. cit. For $n = 2$ completely different conditions prevail; see Ch. VI, § 6.

[2] See Footnote 1.

Carrying out the differentiations indicated, and remarking that due to (2.11) we have

$$l_i K^i_{rsk}\, l^r = 0\ ,$$

we find

$$K^j_{rsk}\, l^r\, l^s = \frac{\delta^j_k}{n-1}\, K^i_{rsi}\, l^r\, l^s - \frac{2\, l_k\, l^j}{n^2-1}\, K^h_{rsh}\, l^r\, l^s +$$
$$+ \frac{F\, l^j}{n+1}\left[\frac{\partial}{\partial \dot{x}^i}\,(K^i_{rsk}\, l^r\, l^s) - \frac{1}{n-1}\,\frac{\partial}{\partial \dot{x}^k}\,(K^h_{rsh}\, l^r\, l^s) \right].$$

Furthermore, assuming also that equation (8.44) holds, we see that the above relation reduces to the form

$$K^j_{rsk}\, l^r\, l^s = \frac{1}{n-1}\, K^i_{rsi}\, l^r\, l^s (\delta^j_k - l^j\, l_k)\ .$$

From (7.4), (7.5) and (7.3a) it now follows that under these circumstances the space is of isotropic curvature.

In conclusion we append the following table[1]. It may serve to summarise the more important results concerning special types of Finsler spaces $(n \geq 2)$:

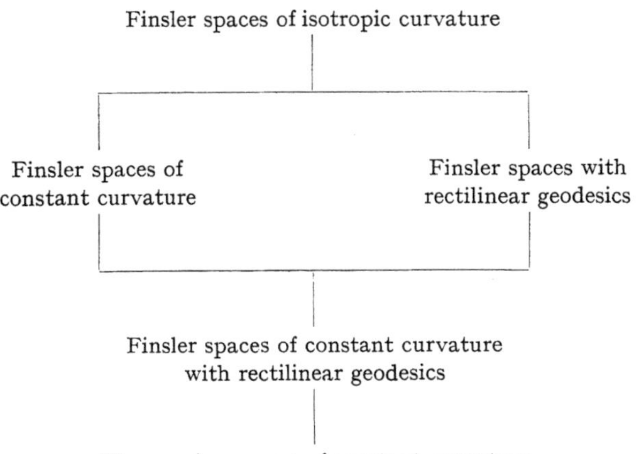

---

[1] BERWALD [10, IV], p. 756. See also BERWALD [14], FUNK [2, 3, 4].

Chapter V

The Theory of Subspaces

In the present chapter we shall discuss the differential geometry of manifolds immersed in a Finsler space F_n. Here we encounter problems of a more specialised type, and while we have so far succeeded in treating the two fundamentally different aspects of Finsler spaces (i. e. the theory of the euclidean connection on the one hand, and the "locally Minkowskian" theory on the other) from a more or less unified point of view, it will become evident in the sequel that the two theories diverge significantly.

It should also be pointed out that the theory of subspaces as based on the euclidean connection has been developed by different authors along two distinct lines of thought. The first direction, due mainly to HOMBU and DAVIES, is based on a generalisation of the D-symbolism of VAN DER WAERDEN and BORTOLOTTI, enjoying a high degree of generality. We shall devote §§ 4—5 of the present chapter to this theory. The second approach, due to BERWALD and closely connected with the work of FINSLER and CARTAN, is less formal in character and restricted to two-dimensional subspaces of an F_3, but is noteworthy because of its geometrical appeal. This theory is described in § 6.

Since the theory of hypersurfaces and subspaces in general depends to a large extent on the study of the behaviour of curves contained in them, we shall begin this chapter with a brief consideration of the general theory of curves.

§ 1. The Theory of Curves

Let C be an analytic curve of F_n defined by the equations

$$x^i = x^i(s) , \tag{1.1}$$

where we shall suppose that the parameter s represents the arc-length measured along C from an arbitrary fixed point of C. Writing, as before, $x'^i = dx^i/ds$ for the tangent vectors of C, it follows from (1.3.9) that x'^i is a unit vector, i. e.

$$g_{ij}(x, x') \, x'^i \, x'^j = 1 . \tag{1.2}$$

Applying the process of δ-differentiation with respect to s to this equation, and noting (2.5.8), we find

$$g_{ij}(x, x') \, x'^i \, \frac{\delta x'^j}{\delta s} = 0 , \tag{1.3}$$

so *that the vector $\delta x'^i/\delta s$ is normal with respect to the tangent vector x'^i of C* [1]. The unit vector in the direction of $\delta x'^i/\delta s$ is called the *principal normal*.

In order to define the curvature of C at a given point $P(x^i)$ of C, let us consider a point $Q(x^i + \Delta x^i)$ of C in the neighbourhood of P. The arc-length of the segment PQ is denoted by Δs. The tangent vector $x'^i + \Delta x'^i$ of C at Q may be obtained directly from (1.1) by differentiation. But we may also construct a second vector $x'^i + \Delta^* x'^i$ at Q, namely by transportation of the tangent vector x'^i at P to the point Q by δ-parallelism. In view of (2.3.18) and (2.3.19) the difference between these two vectors is given by

$$\Delta x'^i - \Delta^* x'^i = \Delta x'^i + \gamma_h{}^i{}_k(x, x')\, x'^h \Delta x^k . \tag{1.4}$$

Also, it is easily seen that both vectors $x'^i + \Delta x'^i$, $x'^i + \Delta^* x'^i$ are unit vectors, for as regards the first we only have to observe that (1.2) is satisfied everywhere along C, while the variation of the square of the length of the second vector is represented by

$$d(F^2(x, x')) = \delta g_{ij}(x, x')\, x'^i x'^j + 2 g_{ij}(x, x')\, \delta x'^i x'^j,$$

where the first term on the right-hand side vanishes as a consequence of (2.5.8), and the second term vanishes by definition of δ-parallelism. Thus, according to definition (1.7.10), the angle between these vectors at Q is given by

$$\Delta \varphi = [g_{ij}(x, x' + \Delta^* x')\, (\Delta x'^i - \Delta^* x'^i)\, (\Delta x'^j - \Delta^* x'^j)]^{1/2} , \tag{1.5}$$

or, if the alternative definition (1.7.2) of angle is used,

$$\Delta \theta = [g_{ij}(x, \Delta x' - \Delta^* x')\, (\Delta x'^i - \Delta^* x'^i)\, (\Delta x'^j - \Delta^* x'^j)]^{1/2} . \tag{1.5a}$$

It is evident from the geometrical construction that these angles represent some measure of the curvature at P of the curve C. In euclidean differential geometry the two angles would coincide, and the curvature of C would be defined by the derivative of this angle with respect to arc-length. This analogy now compels us to introduce two distinct radii of curvature ϱ and r at P; these are defined by

$$\frac{1}{\varrho} = \lim_{\Delta s \to 0} \frac{\Delta \varphi}{\Delta s} , \tag{1.6}$$

and

$$\frac{1}{r} = \lim_{\Delta s \to 0} \frac{\Delta \theta}{\Delta s} . \tag{1.6a}$$

But clearly, when the limit is taken, it follows from (1.4), (2.3.17) and (2.3.19) that

$$\lim_{\Delta s \to 0} \frac{\Delta x'^i - \Delta^* x'^i}{\Delta s} = \frac{\delta x'^i}{\delta s} , \tag{1.7}$$

---

[1] See Ch. I, § 6.

so that from (1.5), (1.5a), (1.6), (1.6a) we deduce the following expressions for the two types of curvature:

$$\frac{1}{\varrho} = \left[g_{ij}(x, x') \frac{\delta x'^i}{\delta s} \frac{\delta x'^j}{\delta s} \right]^{1/2}, \tag{1.8}$$

and

$$\frac{1}{r} = \left[g_{ij}\left(x, \frac{\delta x'}{\delta s}\right) \frac{\delta x'^i}{\delta s} \frac{\delta x'^j}{\delta s} \right]^{1/2}. \tag{1.8a}$$

Thus ϱ^{-1} and r^{-1} represent the length of the vector $\delta x'^i/\delta s$ according as this length is measured by means of the osculating indicatrix (corresponding to x'^i) or by means of the original indicatrix ($F = 1$) in the Minkowskian tangent space $T_n(P)$ at P[1]. Also, as a result of equations (2.3.20) we note that *both curvatures ϱ^{-1} and r^{-1} vanish identically if C is a geodesic of the space F_n.* The converse is also true, as is evident from condition C (Ch. I, § 1).

A further simple geometrical interpretation of ϱ and r is afforded by the following consideration. Let Γ be a geodesic of F_n tangent to C at P, and let us mark off points H and M, whose arc-lengths from P are Δs, on Γ and C respectively. Denoting the coordinates of H and M by $\bar{x}^i(s + \Delta s)$ and $x^i(s + \Delta s)$ we have

$$x^i(s + \Delta s) = x^i(s) + x'^i(s)\, \Delta s + \tfrac{1}{2} x''^i(s)\, (\Delta s)^2 + \cdots,$$

with a similar expansion for $\bar{x}^i$. Since, by construction, $\bar{x}^i(s) = x^i(s)$, and $\bar{x}'^i(s) = x'^i(s)$, it follows that

$$x^i(s + \Delta s) - \bar{x}^i(s + \Delta s) = \frac{(\Delta s)^2}{2} \left(x''^i + \gamma_h{}^i{}_k(x, x')\, x'^h x'^k \right) + \cdots$$
$$= \frac{(\Delta s)^2}{2} \frac{\delta x'^i}{\delta s} + \cdots,$$

where we have made use of the fact that Γ is a geodesic and also of equation (2.3.19). If we proceed to the limit and denote the length of the displacement HM by $|HM|$, the above equation yields[2]

$$\lim_{\Delta s \to 0} \frac{|HM|}{(\Delta s)^2} = \frac{1}{2\varrho} \quad \text{or} \quad \frac{1}{2r} \tag{1.9}$$

according as $|HM|$ is measured by means of the osculating indicatrix or by the indicatrix $F(x, x') = 1$. Equation (1.9) represents the generalisation of a well-known and useful theorem of classical differential geometry.

---

[1] CARTAN [1], p. 19, defines ϱ by this property. The alternative definition (1.6a) of r is given by RUND [3], Ch. VI; [4], § 6. Clearly (1.8) represents a natural definition within the framework of the theories which are based on the element of support, while (1.8a) is dictated by the local Minkowskian metric.

[2] RUND [10, II], § 5, where the same construction is introduced under slightly more general conditions which are relevant to the notion of the "relative" curvature (LIPKA [1]).

The covariant differential (3.2.7) of CARTAN may be used to derive the *Frenet formulae* for any analytic curve C of F_n[1]. The element of support is defined by the unit tangent vector to C, and, since for such a covariant differential we have $Dg_{ij}(x, x') = 0$, the analysis leading up to the Frenet formulae is formally precisely the same as for Riemannian geometry[2]. It would therefore be superfluous if we were to repeat this derivation here: and we shall content ourselves by merely stating the final result. Write

$$\xi^i_{(1)} = x'^i , \qquad \xi^i_{(k)} = \frac{D\,\xi^i_{(k-1)}}{Ds} , \qquad\qquad (k = 2, \ldots, n) ,$$

and define an orthonormal basis at P with respect to x'^i by means of the n vectors

$$\eta^i_{(1)} = \xi^i_{(1)} , \qquad \eta^i_{(p)} = \frac{1}{\sqrt{D_{p-1}D_p}} \begin{vmatrix} (1,1) & \cdots & (1,p-1) & \xi^i_{(1)} \\ \cdot & \cdots & \cdot & \cdot \\ \cdot & \cdots & \cdot & \cdot \\ (p,1) & \cdots & (p,p-1) & \xi^i_{(p)} \end{vmatrix} ,$$

$$(p, q = 2, \ldots, n) ,$$

where

$$D_0 = 1 , \qquad D_p = \begin{vmatrix} (1,1) & \cdots & (1,p) \\ \cdot & \cdots & \cdot \\ \cdot & \cdots & \cdot \\ (p,1) & \cdots & (p,p) \end{vmatrix} , \qquad (p, q) = g_{ij}(x, x')\, \xi^i_{(p)}\, \xi^j_{(q)} .$$

Then the Frenet formulae read as follows:

$$\frac{D\,\eta^i_{(p)}}{Ds} = -\frac{1}{\varrho_{(p-1)}}\, \eta^i_{(p-1)} + \frac{1}{\varrho_{(p)}}\, \eta^i_{(p+1)} , \qquad (p = 2, \ldots, n-1) ,$$

where

$$\frac{1}{\varrho_{(p)}} = \frac{\sqrt{D_{p-1}D_{p+1}}}{D_p} , \qquad \left(\frac{1}{\varrho_0} = \frac{1}{\varrho_n} = 0\right) .$$

An entirely different approach to the problem of defining a suitable measure of curvature (of a curve) is due to FINSLER, who arrived at the same invariant without previous reference to either of the notions of angle or parallelism. In fact, let P, Q be two points on the curve C, and denote by s the arc-length of the arc PQ of C. We may construct a unique geodesic through the points P, Q (provided that the latter are not too distant from each other). Let σ represent the length of the geodesic

---

[1] TAYLOR [1]. For the δ-process the corresponding generalisation of the Frenet formulae is involved and of little practical use due to the fact that δg_{ij} does not, in general, vanish identically. From the point of view of Minkowskian geometry this is quite natural, since the notion of orthogonality is not symmetric.

[2] Cf. for instance, SYNGE and SCHILD [1], pp. 73—76; SCHOUTEN [1], p. 229; EISENHART [1], p. 106.

arc PQ. The curvature k of C at P is then defined by[1]

$$k = \sqrt{\lim_{s \to 0} 24 \left(\frac{s - \sigma}{s^3} \right)}. \tag{1.10}$$

By means of a fairly long calculation involving the Weierstrass $\mathscr{E}$-function it may be shown that (1.10) is equivalent to[2]

$$k^2 = F^{-3}(x, \dot x)\, F_{\dot x^i \dot x^j}(x, \dot x)\, (\ddot x^i - \ddot \xi^i)\, (\ddot x^j - \ddot \xi^j)\,, \tag{1.11}$$

where $\ddot \xi^i$ refers to a family of geodesics containing the geodesic segment PQ, the dot denoting differentiation with respect to an arbitrary parameter t. If we revert to the arc-length as parameter, we find on applying the equations (2.2.8′) of the geodesics that

$$^i - \ddot \xi\, \ddot x^i = -\dot s^2\, \frac{\delta x'^i}{\delta s} = -F^2(x, \dot x)\, \frac{\delta x'^i}{\delta s}\,,$$

and (1.11) becomes

$$k^2 = F(x, \dot x)\, F_{\dot x^i \dot x^j}(x, \dot x)\, \frac{\delta x'^i}{\delta s}\, \frac{\delta x'^j}{\delta s}\,. \tag{1.12}$$

On substituting this result in (1.1.19c) and noting (1.3.1) we have

$$k^2 = g_{ij}(x, \dot x)\, \frac{\delta x'^i}{\delta s}\, \frac{\delta x'^j}{\delta s} - g_{ih}(x, x')\, g_{jk}(x, x')\, x'^h\, x'^k\, \frac{\delta x'^i}{\delta s}\, \frac{\delta x'^j}{\delta s}\,, \tag{1.13}$$

and hence, in view of (1.3) and (1.8) we have $k^2 = \varrho^{-2}$: *thus the definition* (1.8) *coincides with that of* FINSLER (1.10)[3].

The expression (1.11) assumes a particularly simple form for the case of a two-dimensional Finsler space. Let us write $x^1 = x$, $x^2 = y$, $\dot \xi^1 = \dot \xi$, $\dot \xi^2 = \dot \eta$. Substituting in (1.11) we obtain

$$F^3(x, \dot x)\, k^2 = F_{\dot x \dot x}(\ddot x - \ddot \xi)^2 + 2 F_{\dot x \dot y}(\ddot x - \ddot \xi)\, (\ddot y - \ddot \eta) + F_{\dot y \dot y}(\ddot y - \ddot \eta)^2\,.$$

But from (1.3.15) we have, for $n = 2$,

$$F_1 = \frac{F_{\dot x \dot x}}{\dot y^2} = -\frac{F_{\dot x \dot y}}{\dot x\, \dot y} = \frac{F_{\dot y \dot y}}{\dot x^2}\,, \tag{1.14}$$

and hence

$$F^3(x, \dot x)\, k^2 = F_1(x, \dot x)\, [\dot x\, (\ddot y - \ddot \eta) - \dot y\, (\ddot x - \ddot \xi)]^2\,. \tag{1.15}$$

---

[1] FINSLER [1], Ch. VII, p. 59; FINSLER [2], p. 5. MENGER [1] also introduced a definition of curvature for general metric spaces. It was shown by HAANTJES [1] that Menger's definition coincides in Minkowskian and even in general Finsler spaces with definition (1.10). BUSEMANN [4], pp. 172—174, adopts a somewhat different definition of curvature. There exists a relationship between the latter and (1.10) involving the curvature of Busemann's isoperimetrix. See [4], p. 181; also BUSEMANN [8], pp. 285—286 for two-dimensional spaces.

[2] FINSLER [1], p. 64; [2], p. 5. We remark that the expression (1.11) depends on the differentiability properties of C, in contradistinction to (1.10).

[3] This is not surprising as FINSLER [1], p. 65, gives an interpretation of k in terms of an angle analogous to (1.6). In Ch. I, § 7 we had observed that the angle defined by FINSLER is identical with the one used by CARTAN.

But since $\ddot{\xi}$, $\ddot{\eta}$ refer to a geodesic, they satisfy the Weierstrass form of the Euler-Lagrange equations[1]

$$F_{x\dot{y}} - F_{y\dot{x}} + F_1(\dot{\xi}\,\ddot{\eta} - \ddot{\xi}\,\dot{\eta}) = 0\,, \tag{1.16}$$

and, since by construction, $\dot{\xi} = \dot{x}$, $\dot{\eta} = \dot{y}$, we may eliminate $\ddot{\xi}$ and $\ddot{\eta}$ from (1.15), thus obtaining

$$k = \frac{1}{\varrho} = \frac{1}{\sqrt{F^3 F_1}}\,[F_{x\dot{y}} - F_{y\dot{x}} + F_1(\dot{x}\,\ddot{y} - \ddot{x}\,\dot{y})]\,. \tag{1.17}$$

This invariant appeared for the first time in the work of LANDSBERG[2] and UNDERHILL, and was called "extremal curvature" by the former. It is easily shown that if F is of the special form

$$F^2(u, v, \dot{u}, \dot{v}) = E(u, v)\,\dot{u}^2 + 2F(u, v)\,\dot{u}\,\dot{v} + G(u, v)\,\dot{v}^2$$

(as in the classical theory of surfaces), then k is identical with the well-known expression for the geodesic curvature.

§ 2. The Projection Factors

An m-dimensional subspace F_m of F_n may be represented parametrically by the equations[3]

$$x^i = x^i(u^\alpha)\,, \tag{2.1}$$

where we suppose that the variables u^α form a coordinate system of F_m. Furthermore, throughout this discussion we shall assume that the functions (2.1) are of class C^4, and in particular, introducing the notation

$$B^i_\alpha = \frac{\partial x^i}{\partial u^\alpha}\,, \tag{2.2}$$

we shall also assume that the matrix of these projection factors, namely

$$\|B^i_\alpha\|\,, \tag{2.3}$$

is of rank m.

Also, along any coordinate curve of parameter u^α in F_m, the vector whose n components are B^i_α is tangential to the curve. Corresponding to the m independent variables u^α there are m such linearly independent vector fields tangent to F_m in terms of which any vector tangent to F_m is linearly expressible[4]. In particular, if dx^i is a small displacement tangent to F_m, it follows from (2.1) and (2.2) that

$$dx^i = B^i\,du^\alpha\,,$$

---

[1] BOLZA [1], p. 203.

[2] LANDSBERG [2], p. 329; UNDERHILL [1, 2]. Further discussion of this invariant is given by BERWALD [3, 5], [10, I], pp. 38—42, and MOÓR [1].

[3] Throughout this chapter Latin indices run from 1 to n, Greek indices from 1 to m, except for μ, ν, σ, τ which run from $m + 1$ to n.

[4] Strictly speaking, this statement is the *definition* of the notion of "tangency" to F_m.

where du^α denotes the same displacement in terms of the coordinates of F_m. Thus, if we denote the components of a vector X^i tangent to F_m by X^α in terms of the u^α-system, we have

$$X^i = B^i_\alpha X^\alpha .\tag{2.4}$$

If we apply this equation to the tangent vector $\dot{x}^i$ (or $\dot{u}^\alpha$) to a curve C contained in F_m, it follows that

$$\frac{\partial \dot{x}^i}{\partial \dot{u}^\alpha} = B^i_\alpha .\tag{2.4a}$$

Clearly, the Finsler metric of the space F_n induces a Finsler metric on the space F_m. For at a given point P on F_m the m vectors B^i_α span an m-dimensional linear subspace $T_m(P)$ of the tangent space $T_n(P)$ of F_n at P, which should be regarded as the tangent space to F_m at P. If $F(x, \dot{x})$ represents the metric function of F_n for a direction $\dot{x}^i$ tangent to F_m at P, it follows from (2.4) that the corresponding metric function for F_m is given by

$$\overline{F}(u^\alpha, \dot{u}^\alpha) = F(x^i(u^\alpha), B^i_\alpha \dot{u}^\alpha) .\tag{2.5}$$

The metric tensor $g_{\alpha\beta}(u, \dot{u})$ of F_m is defined (as for F_n) by

$$g_{\alpha\beta}(u, \dot{u}) = \frac{1}{2} \cdot \frac{\partial^2 \overline{F}^2(u, \dot{u})}{\partial \dot{u}^\alpha \, \partial \dot{u}^\beta} ,\tag{2.5a}$$

and on differentiating (2.5) successively with respect to $\dot{u}^\alpha$, $\dot{u}^\beta$ and noting (1.3.1) and (2.4a), we find

$$g_{\alpha\beta}(u, \dot{u}) = g_{ij}(x, \dot{x}) B^{ij}_{\alpha\beta} ,\tag{2.6}$$

where we have used the notation

$$B^{ij\cdots k}_{\alpha\beta\cdots\gamma} = B^i_\alpha B^j_\beta \ldots B^k_\gamma .\tag{2.7}$$

In view of (2.4) we have

$$g_{\alpha\beta}(u, \dot{u}) X^\alpha X^\beta = g_{ij}(x, \dot{x}) X^i X^j ,\tag{2.8}$$

so that as a consequence of condition C of Ch. I, § 1, *the matrix* $\|g_{\alpha\beta}(u, \dot{u})\|$ *of* (2.6) *has rank m for all directions* $\dot{u}^\alpha$.

If, in analogy with (1.4.1) and (1.5.6) we define a covariant vector v_α corresponding to $\dot{u}^\alpha$ of T_m by writing

$$v_\alpha = g_{\alpha\beta}(u, \dot{u}) \dot{u}^\beta = \overline{F} \frac{\partial \overline{F}}{\partial \dot{u}^\alpha} ,\tag{2.9}$$

it follows from (2.4) and (2.6) that

$$v_\alpha = g_{ij}(x, \dot{x}) B^i_\alpha \dot{x}^j = y_i B^i_\alpha .\tag{2.10}$$

Here we note that the B^i_α are functions of position only; thus differentiating (2.10) with respect to v_β we obtain

$$\delta^\beta_\alpha = B^i_\alpha B^\beta_i ,\tag{2.11}$$

where we have written

$$B^\beta_i = \frac{\partial y_i}{\partial v_\beta} .\tag{2.11a}$$

Despite equation (2.11) this set of projection parameters is not, in general, independent of direction. Differentiating (2.11) once more with respect to v_γ, we find

$$0 = B_\alpha^i \frac{\partial B_i^\beta}{\partial v_\gamma} . \tag{2.11b}$$

The covariant vectors (2.10) are the elements of an m-dimensional linear subspace $T'_m(P)$ of the dual tangent space $T'_n(P)$ of F_n at P. The distance function in $T'_m(P)$ is given by

$$\overline{H}(u, v) = H\left(x^i(u),\, y_i(u, v)\right), \tag{2.12}$$

and on differentiating with respect to v_α, v_β we have in virtue of (2.11a)

$$\frac{1}{2} \frac{\partial^2 \overline{H}^2(u, v)}{\partial v_\alpha \partial v_\beta} = \frac{1}{2} \frac{\partial^2 H^2(x, y)}{\partial y_i \partial y_j} B_{ij}^{\alpha\beta} + H \frac{\partial H}{\partial y_i} \frac{\partial B_i^\alpha}{\partial v_\beta} .$$

But from equations (1.5.7), (2.4) and (2.11b) it follows that the second term on the right-hand side vanishes identically. Hence we have by (1.5.4)

$$g^{\alpha\beta}(u, v) \equiv \frac{1}{2} \frac{\partial^2 \overline{H}^2(u, v)}{\partial v_\alpha \partial v_\beta} = g^{ij}(x, y) B_{ij}^{\alpha\beta} . \tag{2.13}$$

Naturally, the functions $\overline{F}(u, \dot{u})$, $\overline{H}(u, v)$ have the same homogeneity properties as the original functions $F(x, \dot{x})$ and $H(x, y)$. In particular, $\dot{u}^\beta = g^{\alpha\beta}(u, v)\, v_\alpha$ is the equation inverse to (2.9). Thus from the two identities: $\partial y_i/\partial \dot{x}^j = g_{ij}$, $\partial \dot{u}^\beta/\partial v^\alpha = g^{\alpha\beta}$, we deduce from (2.11a) and (2.4a) the following explicit expression for the B_i^α:

$$B_i^\alpha = g^{\alpha\beta}(u, v)\, g_{ij}(x, \dot{x})\, B_\beta^j . \tag{2.14}$$

A covariant vector Y_i is said to be *normal* to F_m at P if in $T_n(P)$ it is normal to $T_m(P)$, i. e. if it satisfies the equations

$$Y_i B_\alpha^i = 0 . \tag{2.15}$$

These are m equations for the determination of n functions Y_i. Since the rank of the matrix (2.3) is assumed to be m, it follows that there exist $(n - m)$ linearly independent vectors $\underset{\mu}{N_i}$ $(\mu = m + 1, \ldots, n)$ normal to F_m, and these may be chosen in a multiply infinite number of ways:

$$\underset{\mu}{N_i} B_\alpha^i = 0 . \tag{2.15a}$$

With respect to a given direction $\dot{x}^i$ in $T_n(P)$ we may thus choose a set of normals satisfying the relations[1]

$$\underset{\mu}{N^i} = g^{ij}(x, \dot{x})\, \underset{\mu}{N_j} , \qquad g_{ij}(x, \dot{x})\, \underset{\mu}{N^i}\, \underset{\nu}{N^j} = \delta_\nu^\mu . \tag{2.16}$$

---

[1] Equation (2.15) is quite independent of the metric, but as soon as we require a system of mutually perpendicular *contravariant* normals, the metric tensor and hence an element of support enter the picture. In § 7, in which we shall deal with the "locally Minkowskian" theory, it will be seen that serious difficulties arise as a result of this state of affairs.

As a consequence of (2.14), (2.15) and (2.16) we note that

$$B_i^\alpha(x, \dot{x})\, \underset{\mu}{N^i}(x, \dot{x}) = 0 \,. \tag{2.17}$$

We shall often have occasion to use the expression defined by

$$\varphi_j^i = B_\alpha^i B_j^\alpha - \delta_j^i \,.$$

From (2.11) it follows that $\varphi_j^i B_\beta^j = 0$, so that φ_j^i is of the form $\sum_\mu \underset{\mu}{\lambda^i}\, \underset{\mu}{N_j}$, where the factors $\underset{\mu}{\lambda^i}$ are given by $\varphi_j^i \underset{\nu}{N^j} = \underset{\nu}{\lambda^i}$ in consequence of (2.16). But from the definition of φ_j^i and equation (2.17) we have $\varphi_j^i \underset{\nu}{N^j} = -\underset{\nu}{N^i}$. This determines the $\underset{\nu}{\lambda^i} = -\underset{\nu}{N^i}$, and hence

$$B_\alpha^i B_j^\alpha(x, \dot{x}) = \delta_j^i - N_j^i(x, \dot{x}) \,, \tag{2.18}$$

where we have written

$$N_j^i(x, \dot{x}) = \sum_{\mu=m+1}^{n} \underset{\mu}{N^i}(x, \dot{x})\, \underset{\mu}{N_j}(x, \dot{x}) \,. \tag{2.19}$$

An immediate useful consequence of this and the preceding formulae is the relation

$$g_{ik}(x, \dot{x}) = g_{\alpha\beta}(u, \dot{u})\, B_i^\alpha(x, \dot{x})\, B_k^\beta(x, \dot{x}) + N_{ik}(x, \dot{x}) \,, \tag{2.20}$$

together with

$$g^{ik}(x, y) = g^{\alpha\beta}(u, v)\, B_{\alpha\beta}^{ik} + N^{ik}(x, \dot{x}) \,. \tag{2.21}$$

Furthermore, differentiating (2.6) with respect to $\dot{u}^\gamma$ and taking into account (2.4a), we obtain

$$\frac{1}{2}\frac{\partial g_{\alpha\beta}(u, \dot{u})}{\partial \dot{u}^\gamma} \equiv C_{\alpha\beta\gamma}(u, \dot{u}) = C_{ijk}(x, \dot{x})\, B_{\alpha\beta\gamma}^{ijk} \,. \tag{2.22}$$

Also, we shall require the directional derivatives of the B_i^α. From (2.14) and (2.22) we have

$$\frac{\partial B_i^\alpha}{\partial \dot{u}^\gamma} = 2 g^{\alpha\delta}\{ C_{ijh} B_{\gamma\delta}^{hj} - C_{rjh} B_{\delta\gamma\beta i}^{rhj\beta}\} \,.$$

To the last term in this expression we apply (2.18), and on simplification this relation reduces to the form

$$\frac{\partial B_i^\alpha}{\partial \dot{u}^\gamma} = 2 C_{jh}^k N_i^j B_k^\alpha B_\gamma^h \,, \tag{2.23}$$

or, alternatively,

$$\frac{\partial B_i^\alpha}{\partial u^\gamma} = 2 g^{\alpha\delta} B_{\gamma\delta}^{hk} C_{jkh} N_i^j \,. \tag{2.23a}$$

From this result and (2.18) we deduce that

$$\frac{\partial N_j^i}{\partial \dot{u}^\gamma} = -2 g^{\alpha\delta} B_{\alpha\gamma\delta}^{ihk} C_{rkh} N_j^r \,. \tag{2.24}$$

We note that as a result of (1.3.5) and (2.4) we have

$$\frac{\partial B_i^\alpha}{\partial \dot{u}^\gamma}\, \dot{u}^\gamma = \frac{\partial N_j^i}{\partial u^\gamma}\, \dot{u}^\gamma = 0 \,. \tag{2.25}$$

§ 3. The Induced Connection Parameters

Consider a vector field $X^\alpha(u^\beta)$ defined over the subspace F_m. The components X^i of this field with respect to the surrounding space F_n are given by (2.4), so that we may form the covariant derivatives with respect to F_n in the usual manner. However, even though the field X^i is tangent to F_m, this is no longer true for DX^i or δX^i. Furthermore, it is evident that we require a process of covariant differentiation in F_m itself, and such a process may be defined in two distinct ways. Firstly, we may construct the covariant differential of X^i with respect to F_n, and then project the vector thus obtained onto F_m by means of the projection parameters discussed in § 2. This method would yield a vector tangent to F_m, which might be regarded as the *induced covariant derivative* of X^i. Secondly, the metric tensor $g_{\alpha\beta}$ of F_m [defined by (2.6)] together with its derivatives may be used to construct connection parameters according to the pattern of Chapters II and III. These parameters — called the *intrinsic parameters* — would yield an *intrinsic covariant derivative* of X^i.

Unfortunately, these alternative processes of differentiation do not in general coincide. The first alternative leads to a simpler theory than the second, so that for the purpose of the present discussion our choice falls on the former. In order not to interrupt our analysis we shall defer our discussion of the intrinsic covariant derivative to the last section of this chapter.

Corresponding to a displacement dx^i between two points x^i and $x^i + dx^i$ on F_m we may form Cartan's covariant differential DX^i as given by (3.1.2) with respect to an element of support $\dot{x}^i$, which we shall still consider to be tangent to F_m. According to the definition given above, we shall define the *induced* differential $\overline{D}X^\alpha$ by the equation

$$\overline{D}X^\alpha = B_i^\alpha(x, \dot{x})\, DX^i . \tag{3.1}$$

If du^γ denotes the displacement dx^k with respect to the coordinates u^α of F_m, it follows by differentiation of (2.4) that

$$dX^i = B_\beta^i\, dX^\beta + B_{\beta\gamma}^i X^\beta\, du^\gamma , \tag{3.2}$$

where we have written

$$B_{\beta\gamma}^i = \frac{\partial B_\beta^i}{\partial u^\gamma} = \frac{\partial^2 x^i}{\partial u^\beta \partial u^\gamma} , \tag{3.3}$$

while we have a similar equation for the corresponding change $d\dot{x}^k$ of the element of support $\dot{x}^k$. On substituting these relations together with (1.4) in (3.1.2), we find

$$DX^i = B_\beta^i\, dX^\beta + (B_{\beta\gamma}^i + C_{hk}^i B_\beta^h B_{\varepsilon\gamma}^k \dot{u}^\varepsilon + \Gamma_{hk}^i B_{\beta\gamma}^{h\,k})\, X^\beta\, du^\gamma +$$
$$+ C_{hk}^i B_{\beta\gamma}^{h\,k} X^\beta\, d\dot{u}^\gamma . \tag{3.4}$$

Thus in virtue of (3.1), (2.11), (2.14), (2.22) the definition (3.1) may be written in the form

$$\overline{D} X^{\alpha} = d X^{\alpha} + B_i^{\alpha} (B_{\beta\gamma}^i + C_{kh}^i B_{\beta}^h B_{\varepsilon\gamma}^k \dot{u}^{\varepsilon} + \Gamma_{hk}^i B_{\beta\gamma}^{hk}) X^{\beta} d u^{\gamma} + \\ + C_{\beta\gamma}^{\alpha} X^{\beta} d \dot{u}^{\gamma} . \tag{3.5}$$

But in analogy with the definition (3.1.2) we shall write the covariant differential in the form

$$\overline{D} X^{\alpha} = d X^{\alpha} + C_{\beta\gamma}^{\alpha} X^{\beta} d \dot{u}^{\gamma} + \Gamma_{\beta\gamma}^{\alpha} X^{\beta} d u^{\gamma} , \tag{3.6}$$

which defines the induced connection parameters $\Gamma_{\beta\gamma}^{\alpha}$. Comparison with (3.5) yields the explicit expression

$$\Gamma_{\beta\gamma}^{\alpha} = B_i^{\alpha} (B_{\beta\gamma}^i + C_{kh}^i B_{\beta}^h B_{\varepsilon\gamma}^k \dot{u}^{\varepsilon} + \Gamma_{hk}^i B_{\beta\gamma}^{hk}) . \tag{3.7}$$

Similarly, if we apply the δ-process to the field X^i, and define the induced δ-derivatives and parameters by

$$\bar{\delta} X^{\alpha} = B_i^{\alpha} \delta X^i , \tag{3.8}$$

and

$$\bar{\delta} X^{\alpha} = d X^{\alpha} + \Gamma_{\beta\gamma}^{*\alpha} X^{\beta} d u^{\gamma} , \tag{3.9}$$

it follows that these parameters are expressible as follows:

$$\Gamma_{\beta\gamma}^{*\alpha} = B_i^{\alpha} (B_{\beta\gamma}^i + \Gamma_{hk}^{*i} B_{\beta\gamma}^{hk}) , \tag{3.10}$$

or, alternatively, in view of (2.14),

$$\Gamma_{\alpha\gamma\beta}^{*} = g_{ij} B_{\gamma}^j (B_{\alpha\beta}^i + \Gamma_{hk}^{*i} B_{\alpha\beta}^{hk}) . \tag{3.10a}$$

We note that as a result of (1.3.5), (2.4), (3.7) and (3.10) we have

$$\Gamma_{\beta\gamma}^{\alpha} \dot{u}^{\beta} = \Gamma_{\beta\gamma}^{*\alpha} \dot{u}^{\beta} \tag{3.11}$$

for the induced connection parameters.

Clearly the induced connection parameters define a parallel displacement in F_m, and we shall now inquire as to whether the geodesics of F_m are autoparallel with respect to this form of parallelism. By definition, the geodesics of F_m are those curves of F_m which minimise integrals whose integrand is of the form $[g_{\alpha\beta}(u, du) d u^{\alpha} d u^{\beta}]^{1/2}$, and since we have seen that the metric of F_m satisfies the same conditions as that of F_n, we may immediately deduce from the analysis of Chapter II that the geodesics of F_m are the solutions of the differential equations

$$\frac{d^2 u^{\gamma}}{d s^2} + 2 G^{\gamma} \left(u^{\varepsilon}, \frac{d u^{\varepsilon}}{d s} \right) = 0 , \quad 2 G^{\gamma} \equiv \gamma_{\alpha}{}^{\gamma}{}_{\beta} \frac{d u^{\alpha}}{d s} \frac{d u^{\beta}}{d s} , \tag{3.12}$$

where s is the arc-length and the $\gamma_{\alpha}{}^{\gamma}{}_{\beta}$ are the Christoffel symbols of the second kind formed with respect to the tensor $g_{\alpha\beta}$. It is now a simple matter to express the equations (3.12) in terms of the parameters (3.10).

In order to do so, let us differentiate equations (2.6) with respect to u^{γ} along a curve C of F_m, the directional arguments of the tensors occurring in these equations being the tangent vectors to C. We thus obtain

$$\frac{\partial g_{\alpha\beta}}{\partial u^\gamma} = \frac{\partial g_{ij}}{\partial x^k}\, B^{ijk}_{\alpha\beta\gamma} + 2 C_{ijk} B^k_{\varepsilon\gamma}\, \dot{u}^\varepsilon\, B^{ij}_{\alpha\beta} + \\ + g_{ij} B^i_{\alpha\gamma} B^j_\beta + g_{ij} B^i_\alpha B^j_{\beta\gamma}\,. \tag{3.13}$$

By performing a cyclic interchange of α, β, γ in this equation, we obtain two similar relations, from the sum of which we subtract the one above. Interchange of indices and simplification leads to the following equation of transformation:

$$\gamma_{\alpha\gamma\beta} = g_{ij} B^j_\gamma (B^i_{\alpha\beta} + \gamma_h{}^i{}_k B^{hk}_{\alpha\beta}) + C_{ijk}(B^{ij}_{\beta\gamma} B^k_{\varepsilon\alpha} + B^{ij}_{\gamma\alpha} B^k_{\varepsilon\beta} - \\ - B^{ij}_{\alpha\beta} B^k_{\varepsilon\gamma})\, \dot{u}^\varepsilon\,. \tag{3.13a}$$

In particular we have as a result of (1.3.5):

$$\gamma_{\alpha\gamma\beta}\, \dot{u}^\alpha\, \dot{u}^\beta = g_{ij} B^j_\gamma (B^i_{\alpha\beta} + \gamma_h{}^i{}_k B^{hk}_{\alpha\beta})\, \dot{u}^\alpha\, \dot{u}^\beta\,, \tag{3.13b}$$

or, using (3.1.27'), (2.14), (3.12)

$$2 G^\alpha = B^\alpha_i (B^i_{\beta\gamma} + \Gamma^{*i}_{hk} B^{hk}_{\beta\gamma})\, \dot{u}^\beta\, \dot{u}^\gamma\,. \tag{3.14}$$

From (3.10) we therefore deduce that

$$2 G^\alpha = \Gamma^{*\alpha}_{\beta\gamma}\, \dot{u}^\beta\, \dot{u}^\gamma\,. \tag{3.15}$$

Applying equation (3.9) to the tangent vector x'^α of the geodesic (3.12), it follows from (3.15) that the geodesics of F_m satisfy the equations

$$\frac{\bar{\delta}\, x'^\alpha}{\delta s} = 0\,. \tag{3.16}$$

Thus the geodesics of the subspace are indeed the autoparallel curves with respect to the induced connection. Obviously, this is true also for the intrinsic connection.

Furthermore, if we multiply equations (3.1) and (3.8) by $g_{\alpha\gamma}$, we see that in virtue of (2.14) these equations may be written in the form

$$g_{ij} B^j_\gamma\, D X^i = g_{\alpha\gamma}\, \bar{D} X^\alpha\,, \tag{3.17}$$

and

$$g_{ij} B^j_\gamma\, \delta X^i = g_{\alpha\gamma}\, \bar{\delta} X^\alpha\,. \tag{3.17a}$$

The following very simple but useful theorems[1] result from these relations:

1. If a curve C lies in a subspace F_m of F_n, and a vector field in F_m is parallel along C with respect to F_n, it is also parallel with respect to the induced connection of F_m.

2. If a curve is a geodesic of F_n it is also a geodesic of any subspace F_m in which it is contained.

---

[1] These theorems are generalisations of well-known theorems of Riemannian geometry. Cf. EISENHART [1], p. 75. It is to be noted, however, that in Riemannian geometry the induced and intrinsic connections coincide automatically.

3. A necessary and sufficient condition that a vector field in F_m be parallel with respect to the induced connection of F_m along a curve C, while not being parallel in F_n, is that the covariant differentials $g_{ij}\,\delta X^j$ of this field with respect to F_n be normal to F_m [equation (2.15)].

4. *When a geodesic of F_m is not a geodesic of F_n, its principal normal* $r\,(\delta x'^i/\delta s)$ *is normal to F_m* (the normal being defined with respect to a line-element tangential to the geodesic).

Finally, we note that there exists a connection for vectors attached to F_m but normal to F_m in the sense of equation (2.15). This connection is defined in a manner analogous to equation (3.1). Let Y^i be normal to F_m (but defined over F_m), so that it is expressible in the form $Y^i = \sum_\mu N^i_\mu Y^\mu$. The covariant differential $\overline{D}\,Y^\mu$ is defined to be the projection of $D\,Y^i$ onto the $(n-m)$-dimensional direction normal to F_m, i. e.

$$\overline{D}\,Y^\mu = g_{ij}\,N^j_\mu\,D\,Y^i\,. \tag{3.18}$$

If we write [1]

$$\overline{D}\,Y^\mu = d\,Y^\mu + \overline{C}^\mu_{\nu\gamma}\,Y^\nu\,d\dot{u}^\gamma + \overline{\lambda}^\mu_{\nu\gamma}\,Y^\nu\,du^\gamma\,, \quad (\mu,\nu = m+1,\dots,n)\,, \tag{3.19}$$

it is easily verified (as above) that in virtue of (3.18)

$$\overline{C}^\mu_{\nu\gamma} = g_{ij}\,N^i_\mu\,(C^j_{hk}\,N^h_\nu\,B^k_\gamma)\,, \tag{3.20}$$

and [2]

$$\overline{\lambda}^\mu_{\nu\gamma} = g_{ij}\,N^j_\mu\left(\frac{\partial N^i_\nu}{\partial u^\gamma} + \Gamma^i_{hk}\,N^h_\nu\,B^k_\gamma + C^i_{hk}\,N^h_\nu\,B^k_{\varepsilon\gamma}\,\dot{u}^\varepsilon\right). \tag{3.21}$$

Alternatively, we may write (3.19) in the form

$$\overline{D}\,Y^\mu = d\,Y^\mu + \lambda^\mu_{\nu\gamma}\,Y^\nu\,du^\gamma + \overline{A}^\mu_{\nu\gamma}\,Y^\nu\,\overline{D}\,l^\gamma\,, \tag{3.22}$$

(l^α being, as usual, the unit vector in the direction of the element of support), where

$$\lambda^\mu_{\nu\gamma} = \overline{\lambda}^\mu_{\nu\gamma} - \overline{C}^\mu_{\nu\delta}\,\Gamma^\delta_{\varepsilon\gamma}\,\dot{u}^\varepsilon$$

$$= g_{ij}\,N^j_\mu\left\{\frac{\partial N^i_\nu}{\partial u^\gamma} + \Gamma^i_{hk}\,N^h_\nu\,B^k_\gamma + C^i_{hk}\,N^h_\nu\,(B^k_{\varepsilon\gamma} - \Gamma^\delta_{\varepsilon\gamma}\,B^k_\delta)\,\dot{u}^\varepsilon\right\}. \tag{3.23}$$

---

[1] In the subsequent equations repeated indices μ, ν, σ will imply summation over the range from $m+1$ to n.

[2] Here we have assumed that $g_{ij}\,N^i_\mu\,\dfrac{\partial N^j_\nu}{\partial \dot{u}^\gamma} = 0$.

This is permissible if we assume that the N^i are so chosen that $\partial N_i/\partial \dot{u}^\gamma_\nu = 0$, which is justifiable geometrically in virtue of equation (2.15a) (since the $B^i_\alpha{}^\mu$ are independent of direction). Compare DAVIES [2], pp. 22 and 25. The method outlined above follows that of DAVIES closely.

In conclusion we remark that the induced connection (3.10) is not metric, i. e. the corresponding covariant derivatives of the $g_{\alpha\beta}$ do not vanish. In fact, writing

$$g_{\alpha\beta|\gamma} = \frac{\partial g_{\alpha\beta}}{\partial u^\gamma} - \frac{\partial g_{\alpha\beta}}{\partial \dot u^\lambda}\, \Gamma^\lambda_{\varepsilon\gamma}\dot u^\varepsilon - g_{\alpha\delta}\Gamma^{*\,\delta}_{\beta\gamma} - g_{\beta\delta}\Gamma^{*\,\delta}_{\alpha\gamma}, \qquad (3.24)$$

we find on substitution from (3.13) and (3.10)

$$g_{\alpha\beta|\gamma} = \left(\frac{\partial g_{ij}}{\partial x^k} - \Gamma^*_{ijk} - \Gamma^*_{jik}\right) B^{ijk}_{\alpha\beta\gamma} + 2A_{ijk}(B^k_{\varepsilon\gamma} - \Gamma^\lambda_{\varepsilon\gamma}B^k_\lambda)\, l^\varepsilon\, B^{ij}_{\alpha\beta}.$$

In this equation we replace the Γ^*_{ijk} by Γ_{ijk}, using (3.1.25). As a result of (3.1.7) the first three terms will then vanish, leaving us with the relation

$$g_{\alpha\beta|\gamma} = 2A_{ijk}\{(B^k_{\varepsilon\gamma} - B^k_\lambda\Gamma^\lambda_{\varepsilon\gamma})\, l^\varepsilon + \Gamma^k_{rh}l^r\, B^h_\gamma\}\, B^{ij}_{\alpha\beta}. \qquad (3.25)$$

§ 4. Fundamental Aspects of the Theory of Subspaces Based on the Euclidean Connection

1°. The Normal Curvature and Associated Tensors

Hitherto we have been able to derive from a fairly unified point of view a set of formulae of which some are of importance to the theory of subspaces, irrespective of whether such a theory is to be based on the euclidean connection or whether the locally Minkowskian metric is to be used. However, in the application of these formulae, i. e. in the theory of subspaces proper, this distinction leads to two entirely different geometrical theories. We shall thus discuss the theory based on the euclidean connection in the present section, postponing the alternative treatment until § 7.

The discussion of the present section is based chiefly on the work of HOMBU and DAVIES[1]. The paper of HOMBU contains a more detailed description of the analytical tools to be used in the sequel, while the treatment of DAVIES is less formal.

We shall now endeavour to find some analytical expression for the normal curvature of the subspace. Consider a curve $C:\ x^i = x^i(s)$ [or $u^\alpha = u^\alpha(s)$] of F_m, referred to its arc-length as parameter. According to § 1, the length of the vector Dx'^i/Ds gives us a measure of the curvature of C, regarded as a curve of F_n, while the length of $\overline{D}u'^\alpha/Ds$ gives us a measure of the curvature of C, regarded as a curve of F_m. Clearly a representation of the normal curvature of F_m at a point P of C in the direction x'^i (tangent to C) will be given by the vector $Dx'^i - B^i_\alpha \overline{D}u'^\alpha$,

---

[1] HOMBU [3], DAVIES [2]. The more restricted theory of hypersurfaces is discussed in its various aspects by CARTAN [1], pp. 19—29; HAIMOVICI [1, 3, 4]; WEGENER [2]; NAZIM [1]. For the non-linear connection (see Ch. III) a theory of hypersurfaces may also be constructed (BARTHEL [4, 5]).

which itself is normal to F_m. However, instead of evaluating this expression explicitly, let us consider the vector $DX^i - B^i_\alpha \overline{D} X^\alpha$, where $X^\alpha(u, \dot u)$ is a vector field tangent to F_m defined along C (so that we shall not lose terms which would otherwise drop out due to homogeneity). We assume that $X^\alpha(u, \dot u)$ is homogeneous of degree zero with respect to $\dot u^\alpha$. The element of support is to be tangent to C: and since it will naturally enter our expressions, we immediately observe that

$$Dl^i - B^i_\alpha \overline{D} l^\alpha = H^i_\gamma(u, \dot u)\, du^\gamma\,, \tag{4.1}$$

where we have put

$$H^i_\gamma = (B^i_{\beta\gamma} - B^i_\alpha \Gamma^\alpha_{\beta\gamma})\, l^\beta + B^k_\gamma \Gamma^i_{hk}\, l^h\,, \tag{4.2}$$

having made use of (2.4) and (3.2).

It is often useful to express the induced covariant differential as defined by (3.1) in terms of the covariant differential $\overline{D} l^\alpha$ [analogously to the relation (3.2.7)]. For we then have

$$DX^i - B^i_\alpha \overline{D} X^\alpha = \left(F\, \frac{\partial X^i}{\partial \dot x^h} + A^i_{hk} X^k\right) Dl^h + X^i_{|h}\, d x^h - $$
$$- B^i_\alpha \left(F\, \frac{\partial X^\alpha}{\partial \dot u^\gamma} + A^\alpha_{\beta\gamma} X^\beta\right) \overline{D} l^\gamma - B^i_\alpha X^\alpha_{|\gamma}\, du^\gamma\,, \tag{4.3}$$

where we have written

$$X^\alpha_{|\gamma} = \frac{\partial X^\alpha}{\partial u^\gamma} - \frac{\partial X^\alpha}{\partial \dot u^\lambda}\, \Gamma^\lambda_{\varepsilon\gamma}\, \dot u^\varepsilon + \overline{\Gamma}^{*\,\alpha}_{\beta\gamma}\, X^\beta\,. \tag{4.4}$$

But at this stage great care must be exercised: for the $\overline{\Gamma}^{*\,\alpha}_{\beta\gamma}$ which appear in this equation are *not* in general identical with the $\Gamma^{*\,\alpha}_{\beta\gamma}$ defined by (3.10). In fact, as for equation (3.1.25), a comparison of (4.4) with (3.6) yields[1]

$$\overline{\Gamma}^{*\,\alpha}_{\beta\gamma} = \Gamma^\alpha_{\beta\gamma} - A^\alpha_{\beta\delta}\, \Gamma^\delta_{\varepsilon\gamma}\, l^\varepsilon\,. \tag{4.5}$$

In order to evaluate (4.3) we note, firstly, that in view of (4.1) and (2.4) we have

$$F\left\{\frac{\partial X^i}{\partial \dot x^h}\, Dl^h - B^i_\alpha\, \frac{\partial X^\alpha}{\partial \dot u^\gamma}\, \overline{D} l^\gamma\right\} = F\, \frac{\partial X^i}{\partial \dot x^j}\, H^j_\gamma\, du^\gamma\,. \tag{4.6}$$

Secondly, from (2.22) and (2.14) we deduce that

$$A^\alpha_{\beta\gamma} = B^\alpha_j A^j_{hk} B^{h\,k}_{\beta\gamma}\,. \tag{4.7}$$

With the aid of this equation together with (4.1) and (2.18) we thus find

$$A^i_{hk} X^k Dl^h - B^i_\alpha A^\alpha_{\beta\gamma} X^\beta \overline{D} l^\gamma = [A^i_{hk} H^h_\gamma\, du^\gamma + N^i_j A^j_{hk} B^h_\gamma \overline{D} l^\gamma]\, X^k\,. \tag{4.8}$$

Thirdly, differentiation of (2.4) with respect to u^γ gives

$$\frac{\partial X^i}{\partial x^h}\, B^h_\gamma + \frac{\partial X^i}{\partial \dot x^j}\, B^j_{\varepsilon\gamma}\, \dot u^\varepsilon = B^i_{\beta\gamma} X^\beta + B^i_\alpha\, \frac{\partial X^\alpha}{\partial u^\gamma}\,,$$

---

[1] We remark that the $\overline{\Gamma}^{*\,\alpha}_{\beta\gamma}$ are not even symmetric in β and γ. See equations (4.33) and (4.34) below.

and hence we deduce by direct expansion

$$X^i_{|h}\, dx^h - B^i_\alpha X^\alpha_{|\gamma}\, du^\gamma = (B^i_{\beta\gamma} - B^i_\alpha \overline{\Gamma}^{*\,\alpha}_{\beta\gamma} + \Gamma^{*\,i}_{hk} B^{hk}_{\beta\gamma})\, X^\beta\, du^\gamma -$$
$$- \frac{\partial X^i}{\partial \dot{x}^j}\, (B^j_{\varepsilon\gamma}\, \dot{u}^\varepsilon + \Gamma^j_{rh}\, \dot{x}^r\, B^h_\gamma - \Gamma^\delta_{\varepsilon\gamma}\, \dot{u}^\varepsilon\, B^j_\delta)\, du^\gamma \tag{4.9}$$
$$= (B^i_{\beta\gamma} - B^i_\alpha \overline{\Gamma}^{*\,\alpha}_{\beta\gamma} + \Gamma^{*\,i}_{hk} B^{hk}_{\beta\gamma})\, X^\beta\, du^\gamma - F\, \frac{\partial X^i}{\partial \dot{x}^j}\, H^j_\gamma\, du^\gamma,$$

where we have used (4.2) in the second step. We now substitute (4.6), (4.8) and (4.9) in (4.3), and on collecting terms we see that terms involving $\partial X^i/\partial \dot{x}^j$ cancel, leaving us with

$$DX^i - B^i_\alpha\, \overline{D}X^\alpha = (B^i_{\beta\gamma} - B^i_\alpha \overline{\Gamma}^{*\,\alpha}_{\beta\gamma} + \Gamma^{*\,i}_{hk} B^{hk}_{\beta\gamma} +$$
$$+ A^i_{hk} B^h_\beta H^k_\gamma)\, X^\beta\, du^\gamma + N^i_j (A^j_{hk} B^{hk}_{\beta\gamma})\, X^\beta\, \overline{D}l^\gamma\,. \tag{4.10}$$

Let us apply equation (4.10) to the tangent vector x'^i of C (this vector being the element of support). The last term on the right-hand side vanishes identically in virtue of (2.4) and (1.3.5). Thus, if we write

$$\overset{0}{H}{}^i_{\gamma\beta} = B^i_{\beta\gamma} - B^i_\alpha \overline{\Gamma}^{*\,\alpha}_{\beta\gamma} + (\Gamma^{*\,i}_{hk} B^k_\gamma + A^i_{hk} H^k_\gamma)\, B^h_\beta\,, \tag{4.11}$$

so that

$$\overset{0}{H}{}^i_{\gamma\beta}\, l^\beta = H^i_\gamma\,, \tag{4.11a}$$

equation (4.10) will in this case reduce to

$$D x'^i - B^i_\alpha\, \overline{D}u'^\alpha = \overset{0}{H}{}^i_{\gamma\beta}\, u'^\beta\, du^\gamma\,, \tag{4.12}$$

in agreement with (4.1) and (4.2). The right-hand side of this equation represents that part of the vector curvature of C which is normal to F_m. It is clear that this expression depends only on the direction x'^i of C at the point P under consideration and it is the same for all curves tangent to C at that point. We shall therefore call $\overset{0}{H}{}^i_{\gamma\beta}$ the *normal* or *Eulerian* curvature tensor of F_m at P in the direction u'^α [1].

In particular, if C is a geodesic of F_m, then $\overline{D}u'^\alpha = 0$, and it follows from (4.12) *that the magnitude of the normal curvature vector* $\overset{0}{H}{}^i_{\alpha\beta}\, u'^\alpha\, u'^\beta$ (*regarded as a vector of* F_n) *is equal to the curvature of the geodesic* (*regarded as a curve of* F_n) *of* F_m *through* P *in the direction* u'^α.

It is convenient to introduce "mixed" connection coefficients as follows: we define [2]

$$\Gamma^i_{h\gamma} = \Gamma^{*\,i}_{hk} B^k_\gamma + A^i_{hk} H^k_\gamma\,. \tag{4.13}$$

---

[1] The notation (4.11) is the same as that of HOMBU [3], p. 77 and DAVIES [2], p. 27. These authors introduce this tensor by a process of differentiation (as we shall verify almost immediately), thus generalising the formal methods by means of which it is introduced in Riemannian geometry. See, for instance, SCHOUTEN [1], p. 256.

[2] This definition is due to HOMBU [3], p. 83.

We may now express (4.11) in the form

$$\overset{0}{H}{}^{i}_{\gamma\beta} = B^{i}_{\beta\gamma} + \Gamma^{i}_{h\gamma} B^{h}_{\beta} - \Gamma^{*\alpha}_{\beta\gamma} B^{i}_{\alpha}. \tag{4.14}$$

Clearly this expression represents a mixed type of covariant derivative of the projection parameters B^{i}_{β}, this differentiation being carried out partly with respect to the metric of F_n and partly with respect to the metric of F_m. Thus in order to be able to appreciate the full significance of (4.14) it will be necessary for us to discuss briefly the general process involved.

2°. The D-symbolism[1]

Let X^i represent an arbitrary vector field defined at points of F_m. We may represent its covariant differential in the form

$$DX^i = \overset{0}{V}_h X^i\, dx^h + \overset{1}{V}_h X^i\, Dl^h, \tag{4.15}$$

where, in accordance with (3.2.7), we have written

$$\overset{0}{V}_h X^i = X^i{}_{|h}, \tag{4.16}$$

and

$$\overset{1}{V}_h X^i = \left(F\frac{\partial X^i}{\partial \dot{x}^h} + A^{i}_{hk} X^k\right). \tag{4.17}$$

Thus if we wish to express the covariant differential (4.15) in the form

$$DX^i = \overset{0}{D}_\gamma X^i\, du^\nu + \overset{1}{D}_\gamma X^i\, \overline{D}l^\gamma, \tag{4.18}$$

the operators $\overset{0}{D}, \overset{1}{D}$ must be defined by

$$\overset{0}{D}_\gamma X^i = B^{h}_\gamma \overset{0}{V}_h X^i + H^{h}_\gamma \overset{1}{V}_h X^i, \tag{4.19}$$

and

$$\overset{1}{D}_\gamma X^i = B^{h}_\gamma \overset{1}{V}_h X^i, \tag{4.20}$$

by virtue of equation (4.1).

The D-operators may be expressed in terms of the connection parameters. For on expanding (4.19) according to (4.16) and (4.17) we have

$$\overset{0}{D}_\gamma X^i = B^{h}_\gamma \left(\frac{\partial X^i}{\partial x^h} - \frac{\partial X^i}{\partial \dot{x}^k}\Gamma^{k}_{rh}\dot{x}^r\right) + (B^{h}_\gamma \Gamma^{*i}_{hk} + H^{h}_\gamma A^{i}_{hk}) X^k +$$

$$+ F\frac{\partial X^i}{\partial \dot{x}^k} H^{k}_\gamma.$$

---

[1] The process of covariant differentiation with respect to both of the connection parameters of an enveloping space and of a subspace, which now goes under the name of D-symbolism ("D-Symbolik"), was introduced in Riemannian geometry by BORTOLOTTI [2] and VAN DER WAERDEN [1]. Further details are given by BOMPIANI [3] and TUCKER [1]. For a general account of this method the reader should consult SCHOUTEN [1], p. 254 et seq. The method was generalised to apply to the covariant derivative of CARTAN in Finsler spaces by HOMBU [3] and applied extensively by the latter author and DAVIES [2].

Using (2.4), (4.2) and (4.13) we may reduce this relation to the form

$$\overset{0}{D}_\gamma X^i = \frac{\partial X^i}{\partial u^\nu} - F\,\frac{\partial X^i}{\partial \dot u^\delta}\,\Gamma^\delta_{\varepsilon\gamma}\,l^\varepsilon + \Gamma^i_{k\gamma}\,X^k\,. \tag{4.21}$$

Similarly, if V^α is a vector field tangent to F_m and $Y^i = \sum_\mu N^i_\mu Y^\mu$ a field normal to F_m, we define the corresponding $\overset{0}{D}$-operators by the equations

$$\overset{0}{D}_\gamma V^\alpha = \frac{\partial V^\alpha}{\partial u^\nu} - \frac{\partial V^\alpha}{\partial \dot u^\delta}\,\Gamma^\delta_{\varepsilon\gamma}\,\dot u^\varepsilon + \overset{*}{\Gamma}{}^{\alpha}_{\beta\gamma}\,V^\beta\,, \tag{4.22}$$

and

$$\overset{0}{D}_\gamma Y^\mu = \frac{\partial Y^\mu}{\partial u^\nu} - \frac{\partial Y^\mu}{\partial \dot u^\delta}\,\Gamma^\delta_{\varepsilon\gamma}\,\dot u^\varepsilon + \lambda^\mu_{\nu\gamma}\,Y^\nu\,, \tag{4.23}$$

so that for a mixed tensor involving three kinds of indices we would write, for instance,

$$\overset{0}{D}_\gamma T^{i\mu}_\alpha = \frac{\partial T^{i\mu}_\alpha}{\partial u^\nu} - \frac{\partial T^{i\mu}_\alpha}{\partial \dot u^\delta}\,\Gamma^\delta_{\varepsilon\gamma}\,\dot u^\varepsilon + \Gamma^i_{k\gamma}\,T^{k\mu}_\alpha - \overset{*}{\Gamma}{}^{\varepsilon}_{\alpha\gamma}\,T^{i\mu}_\varepsilon + \lambda^\mu_{\nu\gamma}\,T^{i\nu}_\alpha\,. \tag{4.24}$$

In connection with the latter two definitions we may note that the coefficients (3.23) may be expressed in a simpler form as a result of our new notation. Using (3.1.29) and substituting from (4.2) and (4.13) we find

$$\lambda^\mu_{\nu\gamma} = g_{ij}\,N^j_\mu \left(\frac{\partial N^i_\nu}{\partial u^\nu} + N^h_\nu\,\Gamma^i_{h\gamma}\right)\,. \tag{4.25}$$

Naturally[1] the $\overset{1}{D}$-operator may be applied to the vector fields V^α, Y^μ:

$$\overset{1}{D}_\gamma V^\alpha = B^\alpha_i\,B^h_\gamma\,\overset{1}{\nabla}_h V^i\,, \tag{4.26}$$

and

$$\overset{1}{D}_\gamma Y^\mu = g_{ij}\,N^j_\mu\,B^h_\gamma\,\overset{1}{\nabla}_h Y^i\,. \tag{4.27}$$

As a special application let us consider the $\overset{0}{D}$-derivative of the projection parameters B^i_β. From the above definitions we have

$$\overset{0}{D}_\gamma B^i_\beta = B^i_{\beta\gamma} + \Gamma^i_{h\gamma}\,B^h_\beta - \overset{*}{\Gamma}{}^{\alpha}_{\beta\gamma}\,B^i_\alpha = \overset{0}{H}{}^{i}_{\gamma\beta} \tag{4.28}$$

in accordance with (4.14). Similarly, we may introduce a further tensor connected with the curvature of F_m by writing

$$\overset{1}{H}{}^{i}_{\gamma\beta} = \overset{1}{D}_\gamma B^i_\beta = B^h_\gamma\,B^j_\beta\,\overset{1}{\nabla}_h(B^i_\alpha B^\alpha_j)\,.$$

Using (4.17), the evaluation of this tensor is a straight-forward application of (2.23). The result may be simplified by means of (2.15a), (2.17) and

____

[1] Hombu [3], p. 76. We shall not pursue the study of the $\overset{1}{D}$-operator to any extent, the reader being referred to this paper for further details, especially as regards the commutation formulae.

(2.18), yielding the expression

$$\overset{1}{H}{}^i_{\gamma\beta} = N^i_j A^j_{hk} B^{hk}_{\gamma\beta}\,. \tag{4.28a}$$

In a similar manner a *second normal curvature tensor* of F_m[1] at a point P of F_m in a given direction u'^α may be introduced by defining

$$\overset{0}{L}{}^i_{\gamma\mu} \equiv \overset{0}{D}_\gamma N^i = \frac{\partial \overset{\mu}{N}{}^i}{\partial u^\gamma} - \frac{\partial \overset{\mu}{N}{}^i}{\partial \dot u^\delta}\, \Gamma^\delta_{\varepsilon\gamma}\dot u^\varepsilon - \lambda^\nu_{\underset{\nu}{\mu\gamma}} N^i + \Gamma^i_{\underset{\mu}{k\gamma}} N^k\,. \tag{4.29}$$

Regarded as the components of a vector of F_n, the quantities (4.29) *define a vector tangent to F_m*, or, these quantities "lie with the index i in F_m". In order to verify this statement we observe that the $\overset{0}{D}$-operation is "metric", i. e. $\overset{0}{D}_\gamma g_{ij} = 0$[2], which follows directly from the definition of $\overset{0}{D}_\gamma g_{ij}$ and the application of equations (3.1.7) and (3.1.25) together with the homogeneity properties of the g_{ij}. The required result follows immediately if the $\overset{0}{D}$-operator is applied to equation (2.16).

Thus we shall write

$$B^\alpha_i \overset{0}{L}{}^i_{\gamma\mu} = \overset{0}{L}{}^\alpha_{\gamma\mu}\,, \tag{4.30}$$

so that

$$\overset{0}{L}{}^i_{\gamma\mu} = B^i_\alpha \overset{0}{L}{}^\alpha_{\gamma\mu}\,. \tag{4.30a}$$

Similarly, since $\overset{0}{H}{}^i_{\gamma\beta}$ is normal to F_m, we may put

$$N_{\underset{\mu}{i}} \overset{0}{H}{}^i_{\gamma\beta} = \overset{0}{H}{}^\mu_{\gamma\beta}\,, \tag{4.31}$$

so that

$$\overset{0}{H}{}^i_{\gamma\beta} = N^i_{\underset{\mu}{}} \overset{0}{H}{}^\mu_{\gamma\beta}\,. \tag{4.31a}$$

Hence, if we apply the $\overset{0}{D}_\gamma$-derivative to equation (2.15a), it follows from these equations together with (4.28) and (4.29) that

$$\overset{0}{L}{}^\alpha_{\gamma\mu} = -\, g^{\alpha\delta} \overset{0}{H}{}^\mu_{\gamma\delta}\,. \tag{4.32}$$

Thus the first and second tensors of Eulerian curvature are not independent.

We have noted above that the $\Gamma^{*\alpha}_{\beta\gamma}$ as defined by (4.5) are not, in general, equivalent to the $\Gamma^{*\alpha}_{\beta\gamma}$ as given by (3.10). The difference between these parameters is easily calculated as follows: on expanding (4.5) we find, using (3.7) and (4.7)

$$\Gamma^{*\alpha}_{\beta\gamma} = B^\alpha_i \{ B^i_{\beta\gamma} + \Gamma^i_{hk} B^{hk}_{\beta\gamma} + A^i_{hk} B^h_\beta (B^k_{\varepsilon\gamma} - B^k_\delta \Gamma^\delta_{\varepsilon\gamma})\, l^\varepsilon \}\,.$$

---

[1] For a discussion of the corresponding tensor in Riemannian geometry the reader is referred to SCHOUTEN [1], p. 256 and BOMPIANI [2].

[2] HOMBU [3], p. 80. The simple calculation involved in this assertion is left to the reader.

In this result we substitute from (3.1.29) and (3.10), thus obtaining

$$\Gamma^{*\,\alpha}_{\beta\,\gamma} = \Gamma^{*\,\alpha}_{\beta\,\gamma} + B^{\alpha}_{i} A^{i}_{h\,k} B^{h}_{\beta} \{ (B^{k}_{\varepsilon\,\gamma} - B^{k}_{\delta} \Gamma^{\delta}_{\varepsilon\,\gamma}) \, l^{\varepsilon} + \Gamma^{k}_{m\,j} \, l^{m} B^{j}_{\gamma} \} \, .$$

This, together with (4.2) yields

$$\overline{\Gamma}^{*\,\alpha}_{\beta\,\gamma} = \Gamma^{*\,\alpha}_{\beta\,\gamma} + B^{\alpha}_{i} A^{i}_{h\,k} B^{h}_{\beta} H^{k}_{\gamma} \, . \tag{4.33}$$

On the other hand, the $\overline{\Gamma}^{*\,\alpha}_{\beta\,\gamma}$ yield a metric connection which, as we have seen above, is not true for the $\Gamma^{*\,\alpha}_{\beta\,\gamma}$. In fact, in view of (4.2) we may write equations (3.25) in the form

$$g_{\alpha\beta\,|\,\gamma} = 2 A_{i\,j\,k} B^{i\,j}_{\alpha\,\beta} H^{k}_{\gamma} \, .$$

But from (3.24) and the definition of the $\overset{0}{D}$-operator we have

$$\overset{0}{D}_{\gamma} g_{\alpha\beta} = g_{\alpha\beta\,|\,\gamma} - g_{\alpha\delta} (\overline{\Gamma}^{*\,\delta}_{\beta\,\gamma} - \Gamma^{*\,\delta}_{\beta\,\gamma}) - g_{\delta\beta} (\overline{\Gamma}^{*\,\delta}_{\alpha\,\gamma} - \Gamma^{*\,\delta}_{\alpha\,\gamma}) \, ,$$

so that if we substitute from (4.33) we obtain

$$\overset{0}{D}_{\gamma} g_{\alpha\beta} = 2 A_{i\,j\,k} B^{i\,j}_{\alpha\,\beta} H^{k}_{\gamma} - g_{\alpha\delta} B^{\delta}_{i} A^{i}_{h\,k} B^{h}_{\beta} H^{k}_{\gamma} - g_{\delta\beta} B^{\delta}_{i} A^{i}_{h\,k} B^{h}_{\alpha} H^{k}_{\gamma} = 0 \, , \tag{4.33a}$$

as a result of (2.14).

Since the $\Gamma^{*\,\alpha}_{\beta\,\gamma}$ are symmetric in β, γ it follows from (4.33) that

$$\overline{\Gamma}^{*\,\alpha}_{\beta\,\gamma} - \overline{\Gamma}^{*\,\alpha}_{\gamma\,\beta} = B^{\alpha}_{i} A^{i}_{h\,k} (B^{h}_{\beta} H^{k}_{\gamma} - B^{h}_{\gamma} H^{k}_{\beta}) \, , \tag{4.34}$$

and from (3.11) and (4.33) we deduce that

$$\overline{\Gamma}^{*\,\alpha}_{\beta\,\gamma} \, \dot{u}^{\beta} = \Gamma^{*\,\alpha}_{\beta\,\gamma} \, \dot{u}^{\beta} = \Gamma^{\alpha}_{\beta\,\gamma} \, \dot{u}^{\beta} \, . \tag{4.34a}$$

Also, from the definition (4.28) it is clear that the $\overset{0}{H}{}^{i}_{\gamma\beta}$ are not, in general, symmetric. In fact, from (4.11) we have

$$\overset{0}{H}{}^{i}_{\gamma\beta} - \overset{0}{H}{}^{i}_{\beta\gamma} = A^{i}_{h\,k} (B^{h}_{\beta} H^{k}_{\gamma} - B^{h}_{\gamma} H^{k}_{\beta}) - B^{i}_{\alpha} (\overline{\Gamma}^{*\,\alpha}_{\beta\,\gamma} - \overline{\Gamma}^{*\,\alpha}_{\gamma\,\beta}) \, ,$$

and, hence, using (4.34) and applying (2.18) we finally find

$$\overset{0}{H}{}^{i}_{\gamma\beta} - \overset{0}{H}{}^{i}_{\beta\gamma} = N^{i}_{j} A^{j}_{h\,k} (B^{h}_{\beta} H^{k}_{\gamma} - B^{h}_{\gamma} H^{k}_{\beta}) \, . \tag{4.34b}$$

3°. The Generalised Equations of Gauss, Codazzi and Kühne

The analytical apparatus developed above will enable us to derive in a very simple manner the relevant generalisations of the equations of Gauss, Codazzi and Kühne of classical differential geometry. For this purpose it is necessary to derive suitable commutation formulae for the $\overset{0}{D}$-operator. Since the calculations which this process involves are straight-forward but a little lengthy, we shall only give a few intermediate steps.

We apply the $\overset{0}{D}$-operator to equation (4.28), obtaining

$$
\begin{aligned}
\overset{0}{D}_\alpha \overset{0}{H}{}^i_{\gamma\beta} =& \left(\frac{\partial \Gamma^i_{h\gamma}}{\partial u^\alpha} - \frac{\partial \Gamma^i_{h\gamma}}{\partial \dot{u}^\delta} \Gamma^\delta_{\lambda\alpha} \dot{u}^\lambda + \Gamma^i_{k\alpha}\Gamma^k_{h\gamma} \right) B^h_\beta - \\
&- \left(\frac{\partial \overline{\Gamma}^{*\varepsilon}_{\beta\gamma}}{\partial u^\alpha} - \frac{\partial \overline{\Gamma}^{*\varepsilon}_{\beta\gamma}}{\partial \dot{u}^\delta} \Gamma^\delta_{\lambda\alpha} \dot{u}^\lambda - \overline{\Gamma}^{*\delta}_{\beta\alpha} \overline{\Gamma}^{*\varepsilon}_{\delta\gamma} \right) B^i_\varepsilon - \qquad (4.35) \\
&- \overline{\Gamma}^{*\delta}_{\gamma\alpha} \overset{0}{H}{}^i_{\delta\beta} + \cdots ,
\end{aligned}
$$

where we have not written out explicitly a large number of terms which are symmetric in the indices α and γ.

If, in accordance with (4.1.7) we define "mixed" curvature tensors for F_m by writing

$$
\begin{aligned}
K^i_{h\gamma\alpha} =& \frac{\partial \Gamma^i_{h\gamma}}{\partial u^\alpha} - \frac{\partial \Gamma^i_{h\gamma}}{\partial \dot{u}^\delta} \Gamma^\delta_{\lambda\alpha} \dot{u}^\lambda + \Gamma^i_{k\alpha}\Gamma^k_{h\gamma} - \\
&- \frac{\partial \Gamma^i_{h\alpha}}{\partial u^\gamma} + \frac{\partial \Gamma^i_{h\alpha}}{\partial \dot{u}^\delta} \Gamma^\delta_{\lambda\gamma} \dot{u}^\lambda - \Gamma^i_{k\gamma}\Gamma^k_{h\alpha} ,
\end{aligned} \qquad (4.36)
$$

together with

$$
\begin{aligned}
K^\varepsilon_{\beta\gamma\alpha} =& \frac{\partial \overline{\Gamma}^{*\varepsilon}_{\beta\gamma}}{\partial u^\alpha} - \frac{\partial \overline{\Gamma}^{*\varepsilon}_{\beta\gamma}}{\partial \dot{u}^\delta} \Gamma^\delta_{\lambda\alpha} \dot{u}^\lambda - \overline{\Gamma}^{*\varepsilon}_{\delta\gamma} \overline{\Gamma}^{*\delta}_{\beta\alpha} - \\
&- \frac{\partial \overline{\Gamma}^{*\varepsilon}_{\beta\alpha}}{\partial u^\gamma} + \frac{\partial \overline{\Gamma}^{*\varepsilon}_{\beta\alpha}}{\partial \dot{u}^\delta} \Gamma^\delta_{\lambda\gamma} \dot{u}^\lambda + \overline{\Gamma}^{*\varepsilon}_{\delta\alpha} \overline{\Gamma}^{*\delta}_{\beta\gamma} ,
\end{aligned} \qquad (4.37)
$$

we deduce from (4.35) that

$$
\overset{0}{D}_\alpha \overset{0}{H}{}^i_{\gamma\beta} - \overset{0}{D}_\gamma \overset{0}{H}{}^i_{\alpha\beta} = B^h_\beta K^i_{h\gamma\alpha} - B^i_\varepsilon K^\varepsilon_{\beta\gamma\alpha} + \overset{0}{H}{}^i_{\delta\beta}(\overline{\Gamma}^{*\delta}_{\alpha\gamma} - \overline{\Gamma}^{*\delta}_{\gamma\alpha}) . \quad (4.38)
$$

Similarly, we have from (4.29)

$$
\begin{aligned}
\overset{0}{D}_\alpha \overset{0}{L}{}^i_{\gamma\mu} =& -N^i_{\ v} \left(\frac{\partial \lambda^v_{\mu\gamma}}{\partial u^\alpha} - \frac{\partial \lambda^v_{\mu\gamma}}{\partial \dot{u}^\varepsilon} \Gamma^\varepsilon_{\lambda\alpha} \dot{u}^\lambda - \lambda^v_{\sigma\gamma} \lambda^\sigma_{\mu\alpha} \right) + \\
&+ N^k_{\ \mu} \left(\frac{\partial \Gamma^i_{k\gamma}}{\partial u^\alpha} - \frac{\partial \Gamma^i_{k\gamma}}{\partial \dot{u}^\varepsilon} \Gamma^\varepsilon_{\lambda\alpha} \dot{u}^\lambda + \Gamma^i_{l\alpha}\Gamma^l_{k\gamma} \right) - \qquad (4.39) \\
&- \frac{\partial N^i_{\ \mu}}{\partial \dot{u}^\delta} \left\{ \frac{\partial}{\partial u^\alpha}(\Gamma^\delta_{\lambda\gamma}\dot{u}^\lambda) - \frac{\partial}{\partial \dot{u}^\varepsilon}(\Gamma^\delta_{\lambda\gamma}\dot{u}^\lambda)\Gamma^\varepsilon_{\beta\alpha}\dot{u}^\beta \right\} - \overline{\Gamma}^{*\delta}_{\gamma\alpha} \overset{0}{L}{}^i_{\delta\mu} + \cdots ,
\end{aligned}
$$

where again those terms which are symmetric in α and γ have not been written out. Furthermore, we note that the coefficient of $\partial N^i/\partial \dot{u}^\delta$ in μ

(4.39) may be expanded by means of (4.34a) to the form

$$
\frac{\partial \overline{\Gamma}^{*\delta}_{\varepsilon\gamma}}{\partial u^\alpha} \dot{u}^\varepsilon - \frac{\partial \overline{\Gamma}^{*\delta}_{\varepsilon\gamma}}{\partial \dot{u}^\lambda} \dot{u}^\varepsilon \Gamma^\lambda_{\beta\alpha}\dot{u}^\beta - \overline{\Gamma}^{*\delta}_{\lambda\gamma}\overline{\Gamma}^{*\lambda}_{\beta\alpha}\dot{u}^\beta ,
$$

or, on suitable interchange of indices,

$$
F \left(\frac{\partial \overline{\Gamma}^{*\delta}_{\varepsilon\gamma}}{\partial u^\alpha} - \frac{\partial \overline{\Gamma}^{*\delta}_{\varepsilon\gamma}}{\partial \dot{u}^\lambda} \Gamma^\lambda_{\beta\alpha}\dot{u}^\beta - \overline{\Gamma}^{*\delta}_{\lambda\gamma}\overline{\Gamma}^{*\lambda}_{\varepsilon\alpha} \right) l^\varepsilon .
$$

A third curvature tensor is defined by writing

$$K^{\nu}_{\mu\gamma\alpha} = \frac{\partial \lambda^{\nu}_{\mu\gamma}}{\partial u^{\alpha}} - \frac{\partial \lambda^{\nu}_{\mu\gamma}}{\partial \ddot{u}^{\varepsilon}} \, \Gamma^{\varepsilon}_{\lambda\alpha} \ddot{u}^{\lambda} + \lambda^{\nu}_{\sigma\alpha} \lambda^{\sigma}_{\mu\gamma} -$$
$$- \frac{\partial \lambda^{\nu}_{\mu\alpha}}{\partial u^{\gamma}} + \frac{\partial \lambda^{\nu}_{\mu\alpha}}{\partial \ddot{u}^{\varepsilon}} \, \Gamma^{\varepsilon}_{\lambda\gamma} \ddot{u}^{\lambda} - \lambda^{\nu}_{\sigma\gamma} \lambda^{\sigma}_{\mu\alpha} . \tag{4.40}$$

Thus, in consequence of (4.37) and (4.40) it follows from (4.39) that we have

$$\overset{0}{D}_{\alpha} \overset{0}{L}{}^{i}_{\gamma\mu} - \overset{0}{D}_{\gamma} \overset{0}{L}{}^{i}_{\alpha\mu} = N^{k} K^{i}_{\underset{\mu}{k\gamma\alpha}} - N^{i} K^{\nu}_{\underset{\nu}{\gamma\alpha}} -$$
$$- F \frac{\partial N^{i}}{\partial \ddot{u}^{\delta}} \overset{\mu}{} K^{\delta}_{\varepsilon\gamma\alpha} l^{\varepsilon} - \overset{0}{L}{}^{i}_{\delta\mu} (\overline{\Gamma}^{*\,\delta}_{\gamma\,\alpha} - \overline{\Gamma}^{*\,\delta}_{\alpha\,\gamma}) . \tag{4.41}$$

We now apply the $\overset{0}{D}$-operator to (4.31 a) and observe (4.29). In this way we obtain

$$\overset{0}{D}_{\gamma} \overset{0}{H}{}^{i}_{\alpha\beta} = B^{i}_{\delta} \overset{0}{L}{}^{\delta}_{\gamma\mu} \overset{0}{H}{}^{\mu}_{\alpha\beta} + N^{i} \overset{0}{D}_{\gamma} \overset{0}{H}{}^{\mu}_{\underset{\mu}{\alpha\beta}} . \tag{4.42}$$

On substituting this result in (4.38) we find

$$B^{i}_{\delta} (\overset{0}{L}{}^{\delta}_{\alpha\mu} \overset{0}{H}{}^{\mu}_{\gamma\beta} - \overset{0}{L}{}^{\delta}_{\gamma\mu} \overset{0}{H}{}^{\mu}_{\alpha\beta}) + N^{i} (\overset{0}{D}_{\alpha} \overset{0}{H}{}^{\mu}_{\gamma\beta} - \overset{0}{D}_{\gamma} \overset{0}{H}{}^{\mu}_{\underset{\mu}{\alpha\beta}})$$
$$= B^{h}_{\beta} K^{i}_{h\gamma\alpha} - B^{i}_{\varepsilon} K^{\varepsilon}_{\beta\gamma\alpha} + \overset{0}{H}{}^{i}_{\delta\beta} B^{\delta}_{j} A^{j}_{hk} (B^{h}_{\alpha} H^{k}_{\gamma} - B_{\gamma} H^{k}_{\alpha}) , \tag{4.43}$$

where we have applied (4.34) to the last term on the right-hand side. On multiplying this result with B^{ε}_{i}, one obtains in view of (2.17) and (4.31 a)

$$\overset{0}{L}{}^{\varepsilon}_{\alpha\mu} \overset{0}{H}{}^{\mu}_{\gamma\beta} - \overset{0}{L}{}^{\varepsilon}_{\gamma\mu} \overset{0}{H}{}^{\mu}_{\alpha\beta} = K^{i}_{h\gamma\alpha} B^{h} B^{\varepsilon}_{i} - K^{\varepsilon}_{\beta\gamma\alpha} . \tag{4.44}$$

Similarly, multiplication of (4.43) with $N_{\underset{\nu}{i}}$ yields

$$\overset{0}{D}_{\alpha} \overset{0}{H}{}^{\nu}_{\gamma\beta} - \overset{0}{D}_{\gamma} \overset{0}{H}{}^{\nu}_{\alpha\beta} = K^{i}_{h\gamma\alpha} B^{h}_{\beta} N_{\underset{\nu}{i}} +$$
$$+ \overset{0}{H}{}^{i}_{\delta\beta} B^{\delta}_{j} A^{j}_{hk} (B^{h}_{\alpha} H^{k}_{\gamma} - B^{h}_{\gamma} H^{k}_{\alpha}) N_{\underset{\nu}{i}} . \tag{4.45}$$

Again, let us apply the $\overset{0}{D}$-operator to (4.30 a), observing (4.28) and (4.31 a), obtaining

$$\overset{0}{D}_{\gamma} \overset{0}{L}{}^{i}_{\beta\mu} = N^{i} \overset{0}{H}{}^{\nu}_{\gamma\delta} \overset{0}{L}{}^{\delta}_{\beta\mu} + B^{i}_{\delta} \overset{0}{D}_{\gamma} \overset{0}{L}{}^{\delta}_{\beta\mu} . \tag{4.46}$$

This result is substituted in (4.41), giving

$$N^{i} (\overset{0}{H}{}^{\nu}_{\alpha\delta} \overset{0}{L}{}^{\delta}_{\gamma\mu} - \overset{0}{H}{}^{\nu}_{\gamma\delta} \overset{0}{L}{}^{\delta}_{\alpha\mu}) + B^{i}_{\delta} (\overset{0}{D}_{\alpha} \overset{0}{L}{}^{\delta}_{\gamma\mu} - \overset{0}{D}_{\gamma} \overset{0}{L}{}^{\delta}_{\alpha\mu})$$
$$= N^{k} K^{i}_{\underset{\mu}{k\gamma\alpha}} - N^{i} K^{\nu}_{\underset{\nu}{\mu\gamma\alpha}} - F \frac{\partial N^{i}}{\partial \ddot{u}^{\delta}} K^{\delta}_{\varepsilon\gamma\alpha} l^{\varepsilon} - \overset{0}{L}{}^{i}_{\delta\mu} (\overline{\Gamma}^{*\,\delta}_{\gamma\,\alpha} - \overline{\Gamma}^{*\,\delta}_{\alpha\,\gamma}) . \tag{4.47}$$

On multiplying this result by B_i^ε we find as above

$$\overset{0}{D}_\alpha \overset{0}{L}{}^\varepsilon_{\gamma\mu} - \overset{0}{D}_\gamma \overset{0}{L}{}^\varepsilon_{\alpha\mu} = \underset{\mu}{K}{}^i_{k\gamma\alpha} N^k B_i^\varepsilon - F B_i^\varepsilon \frac{\partial \underset{\mu}{N}{}^i}{\partial \dot u^\delta} K^\delta_{\beta\gamma\alpha} l^\beta - \tag{4.48}$$
$$- B_i^\varepsilon \overset{0}{L}{}^i_{\delta\mu} B_j^\delta A^j_{hk} (B^h_\gamma H^k_\alpha - B^h_\alpha H^k_\gamma) ,$$

where we have applied (3.34) to the last term on the right-hand side.

This formula may be modified to some extent. If, as before, we assume that the normal vectors N^i are so chosen that $\partial \underset{\mu}{N}_i / \partial \dot u^\nu = 0$, it follows from (2.16) and (2.24) that

$$\frac{\partial \underset{\mu}{N}{}^i}{\partial \dot u^\delta} = - 2 B_j^\varrho B_\varrho^i B_\delta^h \underset{\mu}{C}{}^j_{hk} N^k , \tag{4.49}$$

and

$$\underset{\nu}{N}_i \frac{\partial \underset{\mu}{N}{}^i}{\partial \dot u^\delta} = 0 . \tag{4.50}$$

Hence (4.48) may easily be reduced to the form

$$\overset{0}{D}_\alpha \overset{0}{L}{}^\varepsilon_{\gamma\mu} - \overset{0}{D}_\gamma \overset{0}{L}{}^\varepsilon_{\alpha\mu} = \underset{\mu}{K}{}^i_{k\gamma\alpha} N^k B_i^\varepsilon +$$
$$+ 2 B_j^\varepsilon B_\delta^h A^j_{hk} N^k \underset{\mu}{K}{}^\delta_{\varrho\gamma\alpha} l^\varrho - \overset{0}{L}{}^\varepsilon_{\delta\mu} B_i^\delta A^i_{hk} (B^h_\gamma H^k_\alpha - B^h_\alpha H^k_\gamma) . \tag{4.51}$$

Finally, multiplication of (4.47) with $\underset{\nu}{N}_i$ yields in view of (4.50) the relation

$$\overset{0}{H}{}^\nu_{\alpha\delta} \overset{0}{L}{}^\delta_{\gamma\mu} - \overset{0}{H}{}^\nu_{\gamma\delta} \overset{0}{L}{}^\delta_{\alpha\mu} = \underset{\mu}{K}{}^i_{k\gamma\alpha} N^k \underset{\nu}{N}_i - K^\nu_{\mu\gamma\alpha} . \tag{4.52}$$

Equations (4.44) represent the generalisation of the equations of GAUSS, (4.45) and (4.51) of the equations of CODAZZI, while (4.52) is the generalisation of the equations of KÜHNE[1].

§ 5. The Lie Derivative and its Application to the Theory of Subspaces

In the theory of small deformations of subspaces of a Riemannian space the Lie derivative has proved itself to be an extremely useful and powerful tool. In the present section we shall discuss the generalisation of the

---

[1] In essence these equations and their derivation are due to DAVIES [2], p. 29. As has already been mentioned above, the treatment of HOMBU involves commutation formulae with respect to both the $\overset{1}{D}$- and the $\overset{0}{D}$-operators. In this way additional formulae are obtained which may well be regarded as further generalisations of the equations of GAUSS, CODAZZI and KÜHNE. Furthermore, HOMBU expresses the latter equations in terms of CARTAN's curvature tensors $R^\alpha_{\beta\gamma\varepsilon}$, $P^\alpha_{\beta\gamma\varepsilon}$, $S^\alpha_{\beta\gamma\varepsilon}$ (Ch. IV) with respect to F_m. Naturally such formulae are numerous and also extremely complicated and we refer the reader to HOMBU [3] for further details. The curvature tensors (4.36), (4.37), (4.40) were introduced by DAVIES [2]. See also RAPCSÁK [1] and VARGA [5].

Lie derivative to Finsler spaces, which is due to DAVIES [1], and its application to the theory of subspaces. The most significant result which this investigation will yield is the fact that the Eulerian curvature tensor taken by itself does not possess all the geometrical properties exhibited by its counterpart in Riemannian geometry. Since the implications of this constitute the main object of the present section, the discussion of the Lie derivative proper in F_n will of necessity be brief[1].

Let $v^i(x)$ be a vector field of class C^2 defined over a region R of F_n. With this field we may associate an infinitesimal transformation of the type

$$\bar{x}^i = x^i + v^i(x)\, d\tau\,, \tag{5.1}$$

where $d\tau$ is to be regarded as an infinitesimal constant. We may interpret (5.1) by assigning to each point x^i of F_n a shift or displacement $d x^i = v^i(x)\, d\tau$, while it is natural to stipulate that the corresponding variation of the components $\dot{x}^i$ of the element of support is represented by

$$\dot{\bar{x}}^i = \dot{x}^i + \left(\frac{\partial v^i}{\partial x^h}\, \dot{x}^h\right) d\tau\,. \tag{5.2}$$

If $X^i(x, \dot{x})$ is a vector field defined over R, where we assume that $X^i(x, \dot{x})$ is homogeneous of degree zero with respect to $\dot{x}^k$, this field will be affected by the variations (5.1) and (5.2); in fact, if we denote the variation arising from (5.1) and (5.2) by $\overset{v}{d}X^i$, we shall have

$$\overset{v}{d}X^i = \frac{\partial X^i}{\partial x^k}\, v^k\, d\tau + \frac{\partial X^i}{\partial \dot{x}^h}\left(\frac{\partial v^h}{\partial x^k}\, \dot{x}^k\right) d\tau\,,$$

or, if we use (3.1.27′), adding and subtracting the same term, this may be written as

$$\overset{v}{d}X^i = \left(\frac{\partial X^i}{\partial x^k} - \frac{\partial X^i}{\partial \dot{x}^h}\, \frac{\partial G^h}{\partial \dot{x}^k}\right) v^k\, d\tau + \frac{\partial X^i}{\partial \dot{x}^h}\, (v^h{}_{|k}\, \dot{x}^k)\, d\tau\,, \tag{5.3}$$

where we have used the fact that the field $v^i(x)$ is independent of direction.

However, if we interpret (5.1) not as a general shift, but merely as an infinitesimal coordinate transformation [with which (5.2) would be consistent], and if we denote by $\overline{X}^i$ the components of the field X^i in the

---

[1] For the geometrical background and the motivation of the introduction of the Lie derivative the reader is referred to SCHOUTEN [1], Ch. II, § 10, and SCHOUTEN and VAN DER KULK [1], Ch. II, § 13. Apart from the work of DAVIES [1, 2], LAPTEW [1] also defines a Lie derivative for Finsler spaces.

Again, infinitesimal transformations more general than (5.1) may be defined in the sense that the vector v^i may be a function of direction as well as of position: $v^i = v^i(x, \dot{x})$, where $\dot{x}^k$ represents the direction of the previously defined element of support. In this manner the results of DAVIES concerning geodesic deviation are extended by SU [1].

new coordinate system, we would have

$$\overline{X}^i = \frac{\partial \overline{x}^i}{\partial x^j} X^j = \left(\delta_j^i + \frac{\partial v^i}{\partial x^j} d\tau\right) X^j \,.$$

We shall call $\overline{X}^i$ the vector X^i "displaced"[1] from $(x, \dot{x})$ to $(\overline{x}, \dot{\overline{x}})$, and we write

$$\overset{m}{d}X^i = \overline{X}^i - X^i = \frac{\partial v^i}{\partial x^j} X^j \, d\tau \,. \tag{5.4}$$

The *Lie derivative* of the vector field X^i in the Finsler space F_n may now be defined by

$$\underset{L}{D} X^i = \frac{\overset{v}{d}X^i - \overset{m}{d}X^i}{d\tau} \,. \tag{5.5}$$

On substituting (5.3) and (5.4) in this expression — again adding and subtracting the same term — we find

$$\underset{L}{D} X^i = \left(\frac{\partial X^i}{\partial x^k} - \frac{\partial X^i}{\partial \dot{x}^h} \frac{\partial G^h}{\partial \dot{x}^k} + \Gamma_{jk}^{*i} X^j\right) v^k -$$

$$- \left(\frac{\partial v^i}{\partial x^j} + \Gamma_{kj}^{*i} v^k\right) X^j + \frac{\partial X^i}{\partial \dot{x}^h} (v_{|k}^h \dot{x}^k) \,,$$

or,

$$\underset{L}{D} X^i = X_{|k}^i v^k - v_{|k}^i X^k + \frac{\partial X^i}{\partial \dot{x}^h} (v_{|k}^h \dot{x}^k) \,. \tag{5.6}$$

Regarding this relation as the definition of the Lie derivative of a contravariant vector field X^i, we may define the Lie derivative of an arbitrary tensor field $T^{i_1 \cdots i_r}{}_{j_1 \ldots j_s}$ as follows:

$$\underset{L}{D} T^{i_1 \cdots i_r}{}_{j_1 \ldots j_s} = v^k \, T^{i_1 \cdots i_r}{}_{j_1 \ldots j_s | k} + \frac{\partial T^{i_1 \cdots i_r}{}_{j_1 \ldots j_s}}{\partial \dot{x}^h} (v_{|k}^h \dot{x}^k) -$$

$$- \sum_\nu T^{i_1 \cdots i_{\nu-1} k \, i_{\nu+1} \cdots i_r}{}_{j_1 \ldots j_s} v^{i_\nu}{}_{|k} + \tag{5.7}$$

$$+ \sum_\mu T^{i_1 \cdots i_r}{}_{j_1 \ldots j_{\mu-1} k \, j_{\mu+1} \ldots j_s} v^k{}_{|j_\mu} \,.$$

In our applications of the Lie derivative we shall require in particular the Lie derivatives of the metric tensor and of the connection coefficients of F_n. The former is easily evaluated directly from (5.7):

$$\underset{L}{D} g_{ij} = g_{ij|k} v^k + \frac{\partial g_{ij}}{\partial \dot{x}^h} (v_{|k}^h \dot{x}^k) + g_{kj} v_{|i}^k + g_{ik} v_{|j}^k \,;$$

but since $g_{ij|k} = 0$ identically, this reduces to

$$\underset{L}{D} g_{ij} = 2 A_{ijh} v_{|k}^h l^k + v_{j|i} + v_{i|j} \,, \tag{5.8}$$

---

[1] The original term used by SCHOUTEN and VAN KAMPEN [1] is "mitgeschleppt".

[2] In addition to (5.3) and (5.4) we may introduce a further difference $d X^i$, by considering the vector obtained by transporting X^i by parallelism from $(x, \dot{x})$ to $(\overline{x}, \dot{\overline{x}})$. DAVIES [1], p. 266.

where we have made use of (3.2.4). In passing, we note that the conditions

$$\underset{L}{D} g_{ij} = 0 \tag{5.9}$$

are the generalisations to Finsler spaces of the Killing equations of Riemannian geometry, giving the necessary and sufficient conditions for a motion in such spaces[1].

In order to find the Lie derivatives of the Γ^{*i}_{hk} we cannot apply (5.7) directly, since the Γ^{*i}_{hk} do not form the components of a tensor, so that we have to revert to the definition (5.5). Firstly, we note that

$$\overset{v}{d}\Gamma^{*i}_{jk} = \left(\frac{\partial \Gamma^{*i}_{jk}}{\partial x^h} v^h + \frac{\partial \Gamma^{*i}_{jk}}{\partial \dot{x}^r} \frac{\partial v^r}{\partial x^h} \dot{x}^h \right) d\tau ,$$

and if we use the same method as before in order to introduce the covariant derivatives $v^r_{|h}$ of the field v^r, we find

$$\overset{v}{d}\Gamma^{*i}_{jk} = \left\{ \left(\frac{\partial \Gamma^{*i}_{jk}}{\partial x^h} - \frac{\partial \Gamma^{*i}_{jk}}{\partial \dot{x}^l} \Gamma^l_{rh} \dot{x}^r \right) v^h + \frac{\partial \Gamma^{*i}_{jk}}{\partial \dot{x}^l} (v^l_{|r} \dot{x}^r) \right\} d\tau . \tag{5.10}$$

Secondly, we remark that the law of transformation (2.4.6a) of the Γ^{*i}_{hk} may be written in the form

$$\Gamma^{*i}_{jk} = \frac{\partial x^i}{\partial \bar{x}^r} \left(\frac{\partial^2 \bar{x}^r}{\partial x^j \partial x^k} + \bar{\Gamma}^{*r}_{st} \frac{\partial \bar{x}^s}{\partial x^j} \frac{\partial \bar{x}^t}{\partial x^k} \right) ,$$

so that for

$$\frac{\partial \bar{x}^i}{\partial x^r} = \delta^i_r + \frac{\partial v^i}{\partial x^r} d\tau ,$$

we obtain after some simplification that

$$\overset{m}{d}\Gamma^{*i}_{jk} = \bar{\Gamma}^{*i}_{jk} - \Gamma^{*i}_{jk}$$
$$= -\left(\frac{\partial^2 v^i}{\partial x^j \partial x^k} - \frac{\partial v^i}{\partial x^r} \Gamma^{*r}_{jk} + \frac{\partial v^r}{\partial x^k} \Gamma^{*i}_{jr} + \frac{\partial v^r}{\partial x^j} \Gamma^{*i}_{rk} \right) d\tau . \tag{5.11}$$

Thus according to (5.5), (5.10) and (5.11) we have

$$\underset{L}{D}\Gamma^{*i}_{jk} = \frac{\partial^2 v^i}{\partial x^j \partial x^k} - \frac{\partial v^i}{\partial x^r} \Gamma^{*r}_{jk} + \frac{\partial v^r}{\partial x^k} \Gamma^{*i}_{jr} + \frac{\partial v^r}{\partial x^j} \Gamma^{*i}_{rk} +$$
$$+ \left(\frac{\partial \Gamma^{*i}_{jk}}{\partial x^h} - \frac{\partial \Gamma^{*i}_{jk}}{\partial \dot{x}^l} \Gamma^l_{rh} \dot{x}^r \right) v^h + \frac{\partial \Gamma^{*i}_{jk}}{\partial \dot{x}^l} (v^l_{|r} \dot{x}^r) .$$

But if we consider the expansion

$$v^i_{|jk} = \frac{\partial^2 v^i}{\partial x^j \partial x^k} + \frac{\partial \Gamma^{*i}_{hj}}{\partial x^k} v^h - \frac{\partial \Gamma^{*i}_{hj}}{\partial \dot{x}^l} \Gamma^{*l}_{rk} \dot{x}^r v^h +$$
$$+ \Gamma^{*i}_{hj} \frac{\partial v^h}{\partial x^k} + \Gamma^{*i}_{hk} \frac{\partial v^h}{\partial x^j} - \Gamma^{*h}_{jk} \frac{\partial v^i}{\partial x^h} + (\Gamma^{*i}_{rk}\Gamma^{*r}_{hj} - \Gamma^{*i}_{hr}\Gamma^{*r}_{jk}) v^h ,$$

---

[1] EISENHART [1], p. 234. A similar generalisation of the Killing equations is due to KNEBELMAN [1], p. 557. We shall return to this point in Chapter VI.

it follows from (4.1.7) that we have[1]

$$\underset{L}{D}\, \Gamma^{*i}_{jk} = v^i_{|jk} + K^i_{jkh}\, v^h + \frac{\partial \Gamma^{*i}_{jk}}{\partial \dot{x}^h}\,(v^h_{|r}\, \dot{x}^r)\,. \tag{5.12}$$

After these preliminary considerations we may now turn to the study of small deformations of F_m resulting from the application of (5.1) to each point of F_m. In this manner we obtain a new subspace $\overline{F}_m$ of F_n, whose fundamental quantities differ by an infinitesimal amount from the corresponding quantities of F_m. We shall suppose that the new element of support remains tangential to F_m [which imposes certain restrictions on the derivatives of the field $v^i(x)$]. In order to find the variation of the Eulerian curvature tensor $\overset{0}{H}{}^i_{\gamma\beta}$ we shall have to find the variation of the more elementary quantities of F_m. We notice immediately that if we differentiate (5.1) with respect to u^α:

$$\overline{B}^i_\alpha = B^i_\alpha + \frac{\partial v^i}{\partial u^\alpha}\, d\tau\,,$$

so that

$$\overset{v}{d}\, B^i_\alpha = \frac{\partial v^i}{\partial u^\alpha}\, d\tau = \frac{\partial v^i}{\partial x^k}\, B^k_\alpha\, d\tau\,. \tag{5.13}$$

But this is also the "displaced" value $\overset{m}{d}\, B^i_\alpha$, as is obvious when we regard (5.1) as a coordinate transformation, since the coordinate system of F_m is "taken on" in the transformation. Thus if we define a further operator $\overset{a}{D}$ by $\overset{a}{D}\, d\tau = \overset{v}{d} - \overset{m}{d}$ (so that $\overset{a}{D} = \underset{L}{D}$ for quantities of F_n), we have

$$\overset{a}{D}\, B^i_\alpha = 0\,. \tag{5.14}$$

As a result of this relation and (2.6) it follows that

$$\overset{a}{D}\, g_{\alpha\beta} = \left(\underset{L}{D}\, g_{ij}\right) B^{ij}_{\alpha\beta}\,, \tag{5.15}$$

the right-hand side being given by (5.8). Also, since $g_{\alpha\beta}\, g^{\beta\varepsilon} = \delta^\varepsilon_\alpha$, it is obvious that

$$\overset{a}{D}\, g^{\alpha\beta} = -g^{\alpha\varepsilon}\, g^{\delta\beta}\, \overset{a}{D}\, g_{\varepsilon\delta}\,. \tag{5.16}$$

---

[1] The formulae (5.8) and (5.12) are given by DAVIES [1] [equations (35) and (30) respectively]. In this paper the increments $\overset{v}{d},\ \overset{2}{d},\ \overset{m}{d}$ are denoted by $\overset{1}{d},\ \overset{2}{d},\ \overset{3}{d}$ and invariant derivatives $\overset{(r,s)}{D}$ are defined by

$$\overset{(r,s)}{D} = \frac{\overset{r}{d} - \overset{s}{d}}{d\tau}\,,$$

so that $\overset{(1,3)}{D} \equiv \underset{L}{D}$. The commutation formulae corresponding to these operators are developed completely and are applied to the derivation of the equations of geodesic deviation as well as to the generalised Frenet formulae.

Hence we have, using (2.14) and (2.18)

$$\overset{a}{D} B_i^\alpha = g^{\alpha\varepsilon} B_\varepsilon^j N_i^h \left(\underset{L}{D} g_{hj}\right), \tag{5.17}$$

and further, from (2.18),

$$\overset{a}{D} N_i^j = - B_\alpha^j \overset{a}{D} B_i^\alpha = - g^{\alpha\varepsilon} B_{\varepsilon\alpha}^{kj} N_i^h \left(\underset{L}{D} g_{hk}\right). \tag{5.18}$$

Also, as for (5.11), it may be shown that

$$\overset{m}{d} g_{ij} = - \left(g_{ik}\frac{\partial v^k}{\partial x^j} + g_{kj}\frac{\partial v^k}{\partial x^i}\right) d\tau,$$

and thus, from (2.14) and (5.13), we deduce that

$$\overset{m}{d} B_j^\alpha = - B_k^\alpha \frac{\partial v^k}{\partial x^j} d\tau.$$

This result is applied to (2.18), which yields

$$\overset{v}{d} N_j^i = \overset{a}{D} N_j^i \, d\tau + \overset{m}{d} N_j^i = \left\{\overset{a}{D} N_j^i + \left(N_j^k\frac{\partial v^i}{\partial x^k} - N_k^i\frac{\partial v^k}{\partial x^j}\right)\right\} d\tau. \tag{5.19}$$

With the aid of the above formulae we may now proceed to the evaluation of $\overset{a}{D} H_\gamma^i$. We have seen that $\overset{0}{H}_{\gamma\beta}^i$ is normal to F_m, and it therefore follows from (4.11a) that this is true also for H_γ^i. Thus we may deduce from (2.17) and (4.2) that

$$H_\gamma^i = N_k^i H_\gamma^k = N_k^i (B_{\beta\gamma}^k + \Gamma_{jh}^{*k} B_{\beta\gamma}^{jh}) \, l^\beta, \tag{5.20}$$

the reduction of the last term being possible in view of (2.15a). Taking the $\overset{v}{d}$-differential of this equation, noting (5.13) and (5.19) together with the fact that

$$\overset{v}{d} \Gamma_{jh}^{*k} = D\Gamma_{jh}^{*k} \, d\tau + \overset{m}{d}\Gamma_{jh}^{*k},$$

one obtains

$$\overset{v}{d} H_\gamma^i = \left\{\overset{a}{D} N_k^i + \left(N_k^h\frac{\partial v^i}{\partial x^h} - N_h^i\frac{\partial v^h}{\partial x^k}\right)\right\} (B_{\beta\gamma}^k + \Gamma_{jh}^{*k} B_{\beta\gamma}^{jh}) \, l^\beta \, d\tau +$$

$$+ N_k^i \left\{\frac{\partial^2 v^k}{\partial u^\beta \partial u^\gamma} d\tau + \left(\underset{L}{D}\Gamma_{jh}^{*k} \, d\tau + \overset{m}{d}\Gamma_{jh}^{*k}\right) B_{\beta\gamma}^{jh} + \right.$$

$$\left. + \Gamma_{jh}^{*k}\left(\frac{\partial v^j}{\partial u^\beta} B_\gamma^h + \frac{\partial v^h}{\partial u^\gamma} B_\beta^j\right) d\tau\right\} l^\beta.$$

In this equation the expression for $\overset{m}{d}\Gamma_{jh}^{*k}$ is substituted from (5.11), and after some simplification and further application of (5.20) this gives

$$\overset{v}{d} H_\gamma^i = \left\{\overset{a}{D} N_k^i H_\gamma^k + \frac{\partial v^i}{\partial x^h} H_\gamma^h + N_k^i \left(\underset{L}{D}\Gamma_{jh}^{*k}\right) B_{\beta\gamma}^{jh} l^\beta\right\} d\tau.$$

Hence
$$\overset{a}{D} H^i_\gamma = \left(\frac{\overset{v}{d} - \overset{m}{d}}{d\tau}\right) H^i_\gamma = \overset{a}{D} N^i_k H^k_\gamma + N^i_k \left(\underset{L}{D\Gamma^{*k}_{jh}}\right) B^{jh}_{\beta\gamma} l^\beta . \tag{5.21}$$

This result may now be substituted in the expression for the variation of the connection parameters $\Gamma^h_{j\gamma}$:

$$\overset{a}{D} \Gamma^h_{j\gamma} = \left(\underset{L}{D\Gamma^{*h}_{ji}}\right) B^i_\gamma + \left(\underset{L}{D A^h_{ji}}\right) H^i_\gamma + A^h_{ji} \left(\overset{a}{D} H^i_\gamma\right), \tag{5.22}$$

which follows directly from (4.13) and (5.14).

In order to calculate the corresponding derivative of the $\Gamma^{*\alpha}_{\beta\gamma}$ we observe that if we substitute (3.10) in (4.33) and use the abbreviation (4.13), we have
$$\Gamma^{*\alpha}_{\beta\gamma} = B^\alpha_i (B^i_{\beta\gamma} + \Gamma^i_{h\gamma} B^h_\beta) . \tag{5.23}$$

Thus
$$\overset{v}{d} \Gamma^{*\alpha}_{\beta\gamma} = \overline{B}^\alpha_i (\overline{B}^i_{\beta\gamma} + \Gamma^i_{h\gamma} \overline{B}^h_\beta) - \Gamma^{*\alpha}_{\beta\gamma} ,$$

the barred quantities referring to $\overline{F}_m$ [with the exception of $\Gamma^{*\alpha}_{\beta\gamma}$, which refers to F_m according to definition (4.5)]; and, in virtue of (5.14), we also find
$$\overset{a}{D} \Gamma^{*\alpha}_{\beta\gamma} = \overset{a}{D} B^\alpha_i (B^i_{\beta\gamma} + \Gamma^i_{h\gamma} B^h_\beta) + \left(\overset{a}{D} \Gamma^i_{h\gamma}\right) B^h_\beta B^\alpha_j . \tag{5.24}$$

We may substitute for $\overset{a}{D} B^\alpha_i$ from (5.17), and taking into account (2.15) we see that the first term on the right-hand side may be written in the form
$$g^{\alpha\varepsilon} B^j_\varepsilon \left(\underset{L}{D g_{kj}}\right) N^k_i (B^i_{\beta\gamma} + \Gamma^i_{h\gamma} B^h_\beta - B^i_\delta \Gamma^{*\delta}_{\beta\gamma}) = \overset{a}{D} B^\alpha_i \overset{0}{H}^i_{\gamma\beta}$$

in view of (4.14). Consequently (5.24) becomes
$$\overset{a}{D} \Gamma^{*\alpha}_{\beta\gamma} = \overset{0}{H}^i_{\gamma\beta} \overset{a}{D} B^\alpha_i + \left(\overset{a}{D} \Gamma^i_{h\gamma}\right) B^h_\beta B^\alpha_j . \tag{5.25}$$

The explicit expression for the variation of $\Gamma^{*\alpha}_{\beta\gamma}$ is then given by substitution of (5.21) and (5.22) in the relevant terms of (5.25), thus giving a final result in terms of the Lie derivatives (5.8) and (5.12).

We are now in a position to evaluate the variation of $\overset{0}{H}^i_{\beta\gamma}$. From (4.14) we have, using (5.25) and taking into account (5.14):
$$\overset{a}{D} \overset{0}{H}^i_{\gamma\beta} = \left(\overset{a}{D} \Gamma^i_{h\gamma}\right) B^h_\beta - B^i_\alpha \left[\left(\overset{a}{D} B^\alpha_j\right) \overset{0}{H}^j_{\gamma\beta} + \left(\overset{a}{D} \Gamma^i_{h\gamma}\right) B^h_\beta B^\alpha_j\right].$$

By means of (2.18) and (5.18) this is easily reduced to the form
$$\overset{a}{D} \overset{0}{H}^i_{\gamma\beta} = \overset{a}{D} N^i_j \overset{0}{H}^j_{\gamma\beta} + N^i_j \left(\overset{a}{D} \Gamma^j_{h\gamma}\right) B^h_\beta , \tag{5.26}$$

the last term being given by (5.22)[1].

---

[1] The above formulae, giving the variation of the fundamental quantities of F_m with respect to (5.1), are essentially those given by Davies [2], § 6. In this paper equation (5.26) is derived from commutation formulae involving the operators $\overset{0}{D}$ and $\overset{a}{D}$.

We shall apply the above equations to the following geometrical problem. In Riemannian geometry the second fundamental form may be introduced by two distinct methods. The first, due to BOMPIANI[1], depends on transporting a vector of F_m by parallel displacement from a point $P(x^i)$ of F_m to a neighbouring point $Q(x^i + dx^i)$ of F_m, firstly as a vector of the enveloping space F_n and secondly as a vector of the subspace F_m. The vectorial difference between the vectors thus obtained is expressed by means of the second fundamental form. The second method, due to BIANCHI[2], involves the calculation of the variation of the first fundamental form as we pass from the given subspace F_m to an infinitely near subspace parallel to F_m. In Riemannian geometry both methods yield the same second fundamental form[3], *but we shall see that this is not the case for Finsler spaces.*

Let us consider the first method. Suppose that X^i is a vector tangent to F_m at P, and denote by $X^i(P|Q)$ the vector obtained by transporting X^i by parallel displacement with respect to F_n from P to Q, so that

$$X^i(P|Q) = X^i + d^*X^i = X^i - \Gamma^{*i}_{hk}X^h\,dx^k - A^i_{hk}X^h\,Dl^k . \quad (5.27)$$

Similarly, if the components of X^i with respect to F_m are X^α, and if we denote by $X^\alpha(P|Q)$ the vector obtained by parallel displacement of X^α with respect to F_m from P to Q, we have

$$X^\alpha(P|Q) = X^\alpha + d^*X^\alpha = X^\alpha - \overline{\Gamma}^{*\,\alpha}_{\beta\,\gamma}X^\beta\,du^\gamma - A^\alpha_{\beta\,\gamma}X^\beta\,\overline{D}l^\gamma . \quad (5.28)$$

The vectorial difference V^i between these vectors is given by $V^i = X^i(P|Q) - B^i_\alpha X^\alpha(P|Q)$, where the B^i_α have to be given their values at Q. Since $X^i = B^i_\alpha X^\alpha$ at P, it follows from (5.27) and (5.28) that

$$V^i = d^*X^i - B^i_\alpha\,d^*X^\alpha - B^i_{\beta\,\gamma}X^\beta\,du^\gamma \qquad (5.29)$$

to a first approximation.

Taking the values of d^*X^i, d^*X^α as indicated by (5.27) and (5.28) and using (4.8) we find

$$d^*X^i - B^i_\alpha\,d^*X^\alpha = -N^i_j A^i_{hk} B^h_\beta X^\beta B^k_\gamma \overline{D}l^\gamma - A^i_{hk} H^h_\gamma B^k_\beta X^\beta\,du^\gamma +$$
$$+ (B^i_\alpha \overline{\Gamma}^{*\,\alpha}_{\beta\,\gamma} - \Gamma^{*i}_{hk} B^h_\beta B^k_\gamma)\,X^\beta\,du^\gamma ,$$

or, using (4.13) and (4.28a),

$$d^*X^i - B^i_\alpha\,d^*X^\alpha = -\overset{1}{H}{}^i_{\gamma\,\beta}X^\beta\,\overline{D}l^\gamma - (\Gamma^i_{h\,\gamma} B^h_\beta - \overline{\Gamma}^{*\,\alpha}_{\beta\,\gamma} B^i_\alpha)\,X^\beta\,du^\gamma .$$

On substituting from (4.14) we see that this reduces to

$$d^*X^i - B^i_\alpha\,d^*X^\alpha = -\overset{1}{H}{}^i_{\gamma\,\beta}X^\beta\,\overline{D}l^\gamma + \left(B^i_{\beta\,\gamma} - \overset{0}{H}{}^i_{\gamma\,\beta}\right)X^\beta\,du^\gamma .$$

---

[1] BOMPIANI [4], Ch. I.
[2] BIANCHI [1], Vol. II, Part II, p. 450.
[3] DAVIES [4], p. 291.

Thus equation (5.29) finally becomes

$$V^i = -\overset{1}{H^i_{\gamma\beta}} X^\beta \overline{D} l^\nu - \overset{0}{H^i_{\gamma\beta}} X^\beta\, du^\nu\,. \tag{5.30}$$

This equation represents the geometrical interpretation of the two quadratic forms whose coefficients are $\overset{1}{H^i_{\gamma\beta}}$, $\overset{0}{H^i_{\gamma\beta}}$. The quadratic form $\overset{0}{H^i_{\gamma\beta}}\, du^\nu\, du^\beta$ could well be regarded as a second fundamental form: for instance, two directions du^ν, δu^ν could be described as being *conjugate* provided that $\overset{0}{H^i_{\gamma\beta}}\, du^\nu\, \delta u^\beta = 0$. It is evident, however, that the *lack of symmetry of the* $\overset{0}{H^i_{\gamma\beta}}$ *would prevent this relation of conjugacy from being reciprocal*[1].

Let us now consider the second, i. e. Bianchi's approach to the second fundamental form. In the expression (5.15) for the variation of $g_{\alpha\beta}$ we substitute the values of the Lie derivative of the g_{ij} according to (5.8), thus obtaining

$$\overset{a}{D} g_{\alpha\beta} = B^{ij}_{\alpha\beta}(g_{ih} v^h_{|j} + g_{jh} v^h_{|i} + 2 A_{ijh} v^h_{|k} l^k)\,. \tag{5.31}$$

But from (4.13) and (4.21) we have

$$\overset{0}{D_\beta} v^h = \frac{\partial v^h}{\partial u^\beta} + \Gamma^{*h}_{kr} B^r_\beta v^k + A^h_{kr} H^r_\beta v^k\,,$$

so that

$$g_{ih} B^j_\beta v^h_{|j} = g_{ih}\left(\frac{\partial v^h}{\partial u^\beta} + \Gamma^{*h}_{kr} B^r_\beta v^k\right) = g_{ih}\left(\overset{0}{D_\beta} v^h - A^h_{kr} H^r_\beta v^k\right).$$

Hence (5.31) may be written in the form

$$\overset{a}{D} g_{\alpha\beta} = g_{jh}\left\{ B^j_\alpha \overset{0}{D_\beta} v^h + B^j_\beta \overset{0}{D_\alpha} v^h - \right.$$
$$\left. - A^h_{kr} v^k (B^j_\alpha H^r_\beta + B^j_\beta H^r_\alpha) + 2 B^{ij}_{\alpha\beta} A^h_{ir} v^r_{|k} l^k \right\}. \tag{5.32}$$

Now, following the method of BIANCHI, let us suppose that the displacement corresponding to (5.1) is of such a nature that v^i is normal to F_m, i. e. such that

$$g_{jh} B^j_\alpha v^h = 0\,.$$

Taking the $\overset{0}{D}$-derivative of this equation and noting (4.28) it follows that as a result of this assumption

$$g_{jh} B^j_\alpha \overset{0}{D_\beta} v^h = - g_{jh} \overset{0}{H^j_{\beta\alpha}} v^h\,.$$

On substituting this relation in (5.32), the latter equation becomes, after some rearrangement

$$\overset{a}{D} g_{\alpha\beta} = - v_j\left\{\left(\overset{0}{H^j_{\alpha\beta}} + \overset{0}{H^j_{\beta\alpha}}\right) + A^j_{ik}(B^k_\alpha H^i_\beta + B^k_\beta H^i_\alpha)\right\} + 2 A_{ijh} v^h_{|k} l^k B^{ij}_{\alpha\beta}\,. \tag{5.33}$$

---

[1] Equation (5.30) with the relevant observations is due to DAVIES [2], § 7.

This equation represents the second geometrical interpretation of the $\overset{0}{H}{}^{i}_{\alpha\beta}$ and is the generalisation of a similar equation given by BIANCHI *for the case of a Riemannian hypersurface*[1].

§ 6. Surfaces Imbedded in an F_3

Before proceeding to the theory of subspaces of a Finsler space from the point of view of locally Minkowskian geometry, we should treat the more special case of an F_2 imbedded in an F_3. Our reason for doing so is twofold: firstly, because the geometrical picture is very simple, and secondly because a theory, distinct from the one discussed above, has been suggested by BERWALD[2] for this case. This theory is very similar to the one proposed by FINSLER and CARTAN[3], and is remarkable for its geometrical clarity and simplicity. Furthermore, an alternative second fundamental form may be introduced in such a manner that its coefficients are symmetrical, in contrast with the tensors $\overset{0}{H}{}^{i}_{\alpha\beta}$ of § 5.

Let us consider then a three-dimensional Finsler space F_3 endowed with a local coordinate system $x^i \, (i = 1, 2, 3)$, and suppose that this system is chosen such that the surface F_2 is represented by the equation $x^3 = 0$. The coordinates $u^\alpha \, (\alpha, \, \beta = 1, 2)$ of F_2 may be taken to be $u^1 = x^1$, $u^2 = x^2$. At an arbitrary point P of F_2 let l^α represent the unit

---

[1] BIANCHI [1], loc. cit. equation (32). Equation (5.33) is due to DAVIES [2], § 7. In Riemannian geometry the right-hand side of (5.33) reduces to the single term

$$-2\,v_j \overset{0}{H}{}^{j}_{\alpha\beta} = -2 g_{ij}\,v^i \overset{0}{H}{}^{j}_{\alpha\beta}\,.$$

Since v^j is normal to F_m, we may write

$$v^j = \underset{\mu}{\Sigma}\, \underset{\mu}{N^j}\, v^\mu\,,$$

and if we assume in particular that v^μ has only one component of magnitude $\varepsilon/d\tau$ along the normal $\underset{\mu}{N^j}$, equation (5.33) becomes

$$\overset{a}{D}\, g_{\alpha\beta}\, d\tau = -2\varepsilon N_j \overset{0}{H}{}^{j}_{\alpha\beta} = -2\varepsilon \overset{0}{H}{}^{\mu}_{\alpha\beta}\,.$$

Hence if we denote by $\overline{ds^2}$ the fundamental metric form of the $\overline{F}_m$ resulting from F_m by such a displacement,

$$\overline{ds^2} = \left(g_{\alpha\beta} + \overset{a}{D} g_{\alpha\beta}\, d\tau \right) du^\alpha\, du^\beta = ds^2 + \overset{a}{D} g_{\alpha\beta}\, du^\alpha\, du^\beta\, d\tau\,,$$

so that

$$\frac{\overline{ds^2} - ds^2}{-2\varepsilon} = \overset{0}{H}{}^{\mu}_{\alpha\beta}\, du^\alpha\, du^\beta\,.$$

This relation (of Riemannian geometry) illustrates the geometrical significance of $\overset{a}{D} g_{\alpha\beta}$. The corresponding relation (in Finsler geometry) resulting from (5.33) would naturally be more complicated.

[2] BERWALD [8].

[3] FINSLER [1], Ch. XV; CARTAN [1], Ch. IX.

vector in the direction of the (tangential) element of support $\dot{x}^i$. We may then define a second vector with covariant components m_α tangential to F_2 at P by putting $m_1 = -\sqrt{\bar{g}}\, l^2$, $m_2 = +\sqrt{\bar{g}}\, l^1$, where $\bar{g} = \det |g_{\alpha\beta}|$. Clearly m^α is normal to l^α, for it follows from the definition that $l^\alpha m_\alpha = 0$. Also, since $l_\alpha l^\alpha = 1$, we have

$$l_1 m_2 - l_2 m_1 = \sqrt{\bar{g}}\, , \quad \text{or} \quad l^1 m^2 - l^2 m^1 = \frac{1}{\sqrt{\bar{g}}}\, , \tag{6.1}$$

and hence $m_\alpha m^\alpha = 1$, so that m_α is a unit vector. The metric tensor $g_{\alpha\beta}$ of F_2 may be decomposed with respect to the pair of orthogonal vectors l^α, m^α; it is easily seen that this decomposition leads to

$$g_{\alpha\beta} = l_\alpha l_\beta + m_\alpha m_\beta\, , \quad g^{\alpha\beta} = l^\alpha l^\beta + m^\alpha m^\beta\, . \tag{6.2}$$

At P we may also introduce a vector n^i normal to the surface whose covariant components are $(0,\, 0,\, (g^{33})^{-1/2})$, oriented in such a manner that

$$\sqrt{\bar{g}}\, \begin{vmatrix} l^1 & l^2 & l^3 \\ m^1 & m^2 & m^3 \\ n^1 & n^2 & n^3 \end{vmatrix} = 1\, , \tag{6.3}$$

which is consistent with (6.1) since $\bar{g} = g\, g^{33}$ by construction[1].

We are thus equipped with a set of three mutually perpendicular unit vectors defined at each point P of F_2. It is necessary to evaluate the covariant derivatives of these vectors when the direction of the element of support $\dot{x}^i$ at P undergoes a small change, becoming $\dot{x}^i + d\dot{x}^i$. From the general expression (3.1.2) for the covariant derivative, we have in this case (since $du^\alpha = 0$) by virtue of (1.3.5),

$$\bar{D} l^\alpha = dl^\alpha\, , \quad \bar{D} m^\alpha = dm^\alpha + \frac{1}{F} A^\alpha_{\beta\gamma}\, m^\beta\, d\dot{u}^\gamma\, . \tag{6.4}$$

But since l^α, m^α are unit vectors, $l_\alpha \bar{D} l^\alpha = 0$, and $m_\alpha \bar{D} m^\alpha = 0$, and we may write

$$\bar{D} l^\alpha = d\varphi \cdot m^\alpha\, , \quad \bar{D} m^\alpha = d\psi \cdot l^\alpha\, , \tag{6.5}$$

where

$$d\varphi = m_\alpha \bar{D} l^\alpha = -l^\alpha \bar{D} m_\alpha = -l_\alpha \bar{D} m^\alpha = -d\psi\, , \tag{6.6}$$

so that (6.5) and (6.4) may be written in the form

$$dl^\alpha = d\varphi \cdot m^\alpha\, ; \quad dm^\alpha = -d\varphi \cdot l^\alpha - \frac{1}{F} A^\alpha_{\beta\gamma}\, m^\beta\, d\dot{u}^\gamma\, . \tag{6.7}$$

---

[1] This construction represents a special case of the following perfectly general theorem: Let $(-1)^{i-1} p_i$ denote the determinant obtained by deleting the ith column of the matrix $||B^i_\alpha||$. Then the determinants $\bar{g} = \det |g_{\alpha\beta}|$ and $g = \det |g_{ij}|$ are related to each other by the formula

$$\bar{g}(u,\, \dot{u}) = g(x,\, \dot{x})\, g^{ij}(x,\, \dot{x})\, p_i\, p_j\, ,$$

where $(u,\, \dot{u})$, $(x,\, \dot{x})$ refer to the same tangential direction. The proof of this theorem (which is left to the reader) depends on equations (2.6) and the fact that

$$p_i\, B^i_\alpha = 0$$

by construction.

Similarly

$$dl_\alpha = d\varphi \cdot m_\alpha , \quad dm_\alpha = -d\varphi \cdot l_\alpha + \frac{1}{F} A_{\alpha\beta\gamma} m^\beta \, d\dot{u}^\gamma . \tag{6.7a}$$

The factor of proportionality $d\varphi$ has a definite geometrical significance as the increment of an angle. In fact, we have from (6.5) and (6.6)

$$d\varphi = m_\alpha \, dl^\alpha = m_\alpha \, d\left(\frac{\dot{u}^\alpha}{F}\right) = \frac{1}{F} m_\alpha \, d\dot{u}^\alpha ,$$

having used the fact that $m_\alpha \dot{u}^\alpha = 0$. From the definition of m_α it then follows that

$$\varphi = \int \frac{\sqrt{\bar{g}}}{F^2} (\dot{u}^1 \, d\dot{u}^2 - \dot{u}^2 \, d\dot{u}^1) , \tag{6.8}$$

which is essentially equivalent to the definition (1.7.12) of LANDSBERG for the notion of angle.

The theories of BERWALD and FINSLER depend to a great extent on derivatives with respect to φ. These involve the scalar J defined by the equation

$$J = A_{\alpha\beta\gamma} m^\alpha m^\beta m^\gamma . \tag{6.9}$$

If we consider the decomposition on F_2:

$$A_{\alpha\beta\gamma} m^\alpha m^\beta = J_1 l_\gamma + J_2 m_\gamma ,$$

it follows by multiplication of this equation with l^γ, m^γ that $J_1 = 0$ (since $A_{\beta\alpha\gamma} l^\gamma = 0$), and $J_2 = J$. Hence we have[1]

$$A_{\alpha\beta\gamma} m^\alpha m^\beta = J \, m_\gamma . \tag{6.9a}$$

Noting that

$$\frac{d\dot{u}^\gamma}{F} = dl^\gamma + \frac{\dot{u}^\gamma}{F^2} dF ,$$

it follows from (6.9) that we may write (6.7) in the form

$$dl^\alpha = d\varphi \cdot m^\alpha ; \quad dm^\alpha = -d\varphi \cdot l^\alpha - J \, d\varphi \cdot m^\alpha ,$$

or

$$\frac{\partial l^\alpha}{\partial \varphi} = m_\alpha ; \quad \frac{\partial m^\alpha}{\partial \varphi} = -l^\alpha - J \, m^\alpha . \tag{6.10}$$

Similarly,

$$\frac{\partial l_\alpha}{\partial \varphi} = m^\alpha ; \quad \frac{\partial m_\alpha}{\partial \varphi} = -l_\alpha + J \, m_\alpha . \tag{6.10a}$$

After these preliminaries we may now discuss the normal curvature of a curve $C: x^i = x^i(s)$ of F_2 tangential to l^i. As in euclidean geometry we define the latter by the scalar product

$$N(x, \dot{x}) = \frac{Dl^i}{Ds} n_i = -\frac{Dn^i}{Ds} l_i . \tag{6.11}$$

---

[1] The scalar $\frac{1}{2} J$ was introduced by BERWALD [8], p. 6, where it is called the "Hauptskalar".

Using (3.1.16), (3.1.27') and the fact that $G^i(x, \dot{x})$ is homogeneous of the second degree in $\dot{x}^i$, it is evident that this may be written in the form[1]

$$N(x, \dot{x}) = n_i \left(\frac{dl^i}{ds} + 2F^{-2}G^i \right) ,$$

or, since $l^3 \equiv 0$, $n_1 \equiv n_2 \equiv 0$,

$$F^2(x, \dot{x}) N(x, \dot{x}) = 2n_3 G^3(x, \dot{x}) . \tag{6.12}$$

Thus $N(x, \dot{x})$ is homogeneous of degree zero in $\dot{x}^i$, and we observe that since it is independent of the derivatives of the tangent vector $\dot{x}^i$ of C, it must be the same for all curves of F_2 tangent to C at P. We shall therefore regard $N(x, \dot{x})$ as the *normal curvature of the surface F_2 with respect to the line element* $(x, \dot{x})$. This independence of higher derivatives implies *that Meusnier's theorem also holds for surfaces imbedded in Finsler spaces*[2].

In the Minkowskian tangent space $T_3(P)$ of F_3 at P we now consider the set of vectors $\dot{u}^\alpha$ tangent to F_2 at P for which the right-hand side of (6.12) has a fixed value, namely unity. Clearly these vectors define a curve

$$F^2(u, \dot{u}) N(u, \dot{u}) = 1 , \tag{6.13}$$

in $T_2(P)$ (the u^α being fixed, while the $\dot{u}^\alpha$ are variable). This locus is obviously the generalisation of the *Dupin indicatrix* of classical differential geometry. The length of any vector $\dot{u}^\alpha$ satisfying (6.13) is given by

$$F(u, \dot{u}) = \frac{1}{\sqrt{N(u, \dot{u})}} ,$$

which implies

$$\dot{u}^\alpha = \frac{l^\alpha}{\sqrt{N(u, \dot{u})}} ,$$

so that the Dupin indicatrix represents the locus of end-points of vectors of $T_2(P)$ attached to P whose length is inversely proportional to the square root of the normal curvature. Again, in general (6.13) will not represent a quadric; *hence corresponding to each direction $\dot{u}^\alpha$ of F_2 at P we construct the osculating Dupin indicatrix*, which is defined by the equation

$$\Omega_{\alpha\beta}(u, \dot{u}) v^\alpha v^\beta = 1 , \qquad (u^\alpha, \dot{u}^\beta \text{ fixed}) , \tag{6.14}$$

in which the v^α are the running variables, where the coefficients $\Omega_{\alpha\beta}$ are defined by[3]

$$\Omega_{\alpha\beta} = \frac{1}{2} \frac{\partial^2(F^2 N)}{\partial \dot{u}^\alpha \partial \dot{u}^\beta} . \tag{6.15}$$

It is natural to regard the differential form $\Omega_{\alpha\beta}(u, \dot{u}) \, du^\alpha \, du^\beta$ as the *second fundamental form of F_2 with respect to the line element* $(u, \dot{u})$. It is

---

[1] CARTAN [1], p. 21. A slightly more general definition of normal curvature is studied by NAGATA [1].

[2] FINSLER [3], CARTAN [1], loc. cit. See also § 7.

[3] This construction is due to BERWALD [8], p. 9.

obvious from the definition (6.15) that the coefficients $\Omega_{\alpha\beta}$ of this form are symmetric in α and β. The orthogonal invariants (i. e. the trace and the discriminant) of this form may be used in order to define the *mean* and the *Gaussian* curvatures of F_2.

In order to be able to give explicit expressions for these invariants, it is necessary to evaluate the coefficients (6.15). Carrying out the differentiations indicated by this equation we obtain

$$\Omega_{\alpha\beta} = g_{\alpha\beta} N + F l_\beta \frac{\partial N}{\partial \dot{u}^\alpha} + F l_\alpha \frac{\partial N}{\partial \dot{u}^\beta} + \frac{1}{2} F^2 \frac{\partial^2 N}{\partial \dot{u}^\alpha \partial \dot{u}^\beta} . \tag{6.16}$$

The derivatives occurring in this relation must be transformed into derivatives involving φ. Firstly, we observe that in view of (6.10)

$$\frac{\partial \dot{u}^\beta}{\partial \varphi} = \frac{\partial}{\partial \varphi} (F l^\beta) = \frac{\partial F}{\partial \varphi} l^\beta + F m^\beta , \tag{6.17}$$

so that

$$\frac{\partial N}{\partial \varphi} = \frac{\partial N}{\partial \dot{u}^\beta} \frac{\partial \dot{u}^\beta}{\partial \varphi} = F \frac{\partial N}{\partial \dot{u}^\beta} m^\beta , \tag{6.18}$$

where we have used the fact that N is homogeneous of degree zero in the $\dot{u}^\alpha$. Decomposing the vector $\partial N/\partial \dot{u}^\beta$ with respect to l_β and m_β it follows immediately from the orthogonality of these vectors and equation (6.18) that

$$\frac{\partial N}{\partial \dot{u}^\beta} = \frac{1}{F} \frac{\partial N}{\partial \varphi} m_\beta . \tag{6.19}$$

Differentiating (6.18) once more with respect to φ and substituting from (6.10) and (6.17), we obtain

$$\frac{\partial^2 N}{\partial \varphi^2} = \frac{\partial F}{\partial \varphi} \frac{\partial N}{\partial \dot{u}^\beta} m^\beta + F \frac{\partial^2 N}{\partial \dot{u}^\beta \partial \dot{u}^\gamma} \left(\frac{\partial F}{\partial \varphi} l^\gamma + F m^\gamma \right) m^\beta -$$
$$- F \frac{\partial N}{\partial \dot{u}^\beta} (l^\beta + J m^\beta) ,$$

or, observing (6.18) and the relevant homogeneity properties of N,

$$\frac{\partial^2 N}{\partial \varphi^2} = F^2 \frac{\partial^2 N}{\partial \dot{u}^\beta \partial \dot{u}^\gamma} m^\beta m^\gamma - J \frac{\partial N}{\partial \varphi} . \tag{6.20}$$

Again we decompose the first term on the right-hand side of this equation as indicated above, and it is easily verified that (6.20) is equivalent to

$$F^2 \frac{\partial^2 N}{\partial \dot{u}^\alpha \partial \dot{u}^\beta} = - \frac{\partial N}{\partial \varphi} (l_\alpha m_\beta + l_\beta m_\alpha) + \left(\frac{\partial^2 N}{\partial \varphi^2} + J \frac{\partial N}{\partial \varphi} \right) m_\alpha m_\beta . \tag{6.21}$$

We may now substitute (6.19) and (6.21) in the expression (6.16) for $\Omega_{\alpha\beta}$, at the same time inserting (6.2) for the $g_{\alpha\beta}$. After some slight simplification we find

$$\Omega_{\alpha\beta} = N l_\alpha l_\beta + \frac{1}{2} \frac{\partial N}{\partial \varphi} (l_\alpha m_\beta + l_\beta m_\alpha) +$$
$$+ \left(N + \frac{1}{2} J \frac{\partial N}{\partial \varphi} + \frac{1}{2} \frac{\partial^2 N}{\partial \varphi^2} \right) m_\alpha m_\beta . \tag{6.22}$$

If we now represent the variable vector v^α of equation (6.14) by means of its components X, Y with respect to the fixed vectors l^α, m^α, i. e. if we put

$$v^\alpha = X l^\alpha + Y m^\alpha , \qquad (6.23)$$

so that

$$X = l_\alpha v^\alpha ; \qquad Y = m_\alpha v^\alpha , \qquad (6.23\text{a})$$

the equation (6.14) of the osculating Dupin indicatrix with respect to the line-element $(u, \dot u)$ reads

$$N X^2 + \frac{\partial N}{\partial \varphi} X Y + \left(N + \frac{1}{2} J \frac{\partial N}{\partial \varphi} + \frac{1}{2} \frac{\partial^2 N}{\partial \varphi^2} \right) Y^2 = 1 . \qquad (6.24)$$

Thus the *mean curvature* for the line-element $(u, \dot u)$ may be defined to be trace of the left-hand side of (6.24):

$$2 H \equiv \frac{1}{R_1} + \frac{1}{R_2} = 2 N + \frac{1}{2} J \frac{\partial N}{\partial \varphi} + \frac{1}{2} \frac{\partial^2 N}{\partial \varphi^2} , \qquad (6.25)$$

while the *Gaussian curvature* for the line-element $(u, \dot u)$ is defined by the discriminant of (6.24):

$$K \equiv \frac{1}{R_1 R_2} = N^2 - \frac{1}{4} \left(\frac{\partial N}{\partial \varphi} \right)^2 + \frac{1}{2} N \frac{\partial^2 N}{\partial \varphi^2} + \frac{1}{2} J N \frac{\partial N}{\partial \varphi} . \qquad (6.26)$$

These invariants[1] have properties very similar to those of their euclidean counterparts. For instance, it is immediately seen that if we multiply (6.22) by $g^{\alpha\beta}$, summing over α and β:

$$\frac{1}{R_1} + \frac{1}{R_2} = g^{\alpha\beta} \Omega_{\alpha\beta} , \qquad (6.27)$$

while

$$\frac{1}{R_1 R_2} = \frac{\Omega}{g} , \qquad (6.27\text{a})$$

where $\Omega = \det(\Omega_{\alpha\beta})$. From these equations it is obvious that the principal radii of curvature R_1, R_2 of F_2 at P with respect to the line-element $(u, \dot u)$ are simply the extreme values assumed by the quotient

$$\frac{1}{R} = \frac{\Omega_{\alpha\beta}(u, \dot u) \, v^\alpha v^\beta}{g_{\alpha\beta}(u, \dot u) \, v^\alpha v^\beta} , \qquad (u, \dot u \text{ fixed}) . \qquad (6.28)$$

For these extreme values are the roots of the equation

$$|g_{\alpha\beta} - R \, \Omega_{\alpha\beta}| = 0 , \qquad (6.29)$$

which is in accordance with equations (6.27) and (6.27a). Furthermore, we note that as an immediate consequence of (6.25) and (6.26) we have the relation

$$\left(\frac{1}{R_1} - N \right) \left(\frac{1}{R_2} - N \right) = - \frac{1}{4} \left(\frac{\partial N}{\partial \varphi} \right)^2 . \qquad (6.30)$$

On the basis of the above construction it is possible to introduce further notions of classical differential geometry, such as asymptotic curves, points of elliptic, hyperbolic or parabolic curvature. We shall

---

[1] BERWALD [8], § 2.

restrict ourselves to a brief discussion of the notion of geodesic torsion, for it was pointed out by CARTAN[1] that *in general lines of curvature do not correspond to those curves of F_2 whose geodesic torsion vanishes*. In fact, this observation is contained in an interesting generalisation of the theorem of BELTRAMI and ENNEPER[2].

As in euclidean differential geometry, we define the *geodesic torsion* of C by the scalar product

$$\frac{1}{\tau_g} = - m^i \frac{D n_i}{D s} = n_i \frac{D m^i}{D s} , \tag{6.31}$$

and since $n_1 = 0$, $n_2 = 0$, $m^3 = 0$, it follows from the expression (3.2.7) for the covariant derivative of m^i that[3]

$$\frac{1}{\tau_g} = n_3 A^3_{kh} m^h \frac{D l^k}{D s} + n_3 \Gamma^{*3}_{hk} \frac{d x^k}{d s} m^h . \tag{6.31a}$$

Since $l_i (D l^i / D s) = 0$, we may decompose $D l^k / D s$ by virtue of (6.11) as follows:

$$\frac{D l^k}{D s} = N n^k + \frac{m^k}{\varrho_g} , \tag{6.32}$$

where $1/\varrho_g$ is the *geodesic curvature* of C (being the length of $\overline{D} l^\alpha / D s$). Following CARTAN[4], we introduce the invariants

$$A = n_i A^i_{kh} m^k m^h , \qquad B = n_i A^i_{kh} m^k n^h . \tag{6.33}$$

It then follows from (3.1.27'), (6.31a), (6.32) and (6.33) that the geodesic torsion may be expressed in the form

$$\frac{1}{\tau_g} = \frac{A}{\varrho_g} + B N + \frac{n_i m^h}{F} \frac{\partial G^i}{\partial \dot{x}^h} . \tag{6.34}$$

In order to arrive at the generalisation of the above-mentioned theorem, we require an expression more explicit than (6.18) for $\partial N / \partial \varphi$. Differentiating (6.12) with respect to $\dot{x}^k$, we have

$$2 F l_k N + F^2 \frac{\partial N}{\partial \dot{x}^k} = 2 \left[n_3 \frac{\partial G^3}{\partial \dot{x}^k} + G^3 \frac{\partial n_3}{\partial \dot{x}^k} \right] ,$$

or, since $m^k l_k = 0$,

$$F^2 \frac{\partial N}{\partial \dot{x}^k} m^k = 2 n_3 \frac{\partial G^3}{\partial \dot{x}^k} m^k + N F^2 \frac{\partial}{\partial \dot{x}^k} (\log n_3) m^k . \tag{6.35}$$

Remembering that $n_3 = (g^{33})^{-1/2}$, it is easily verified that

$$\frac{1}{n_3} \frac{\partial n_3}{\partial \dot{x}^k} = - \frac{1}{2} (n_3)^2 \frac{\partial g^{33}}{\partial \dot{x}^k} = n^h n^i C_{hik} .$$

---

[1] CARTAN [1], p. 21.

[2] BLASCHKE [4], p. 113.

[3] Here the elements of support are taken to be the tangent vectors of C.

[4] CARTAN [1], p. 22.

Substituting this result in (6.35) we have

$$F \frac{\partial N}{\partial \dot{x}^k} m^k = \frac{2 n_i m^k}{F} \frac{\partial G^i}{\partial \dot{x}^k} + N n^h n^i A_{hik} m^k ,$$

and hence, from (6.18) and (6.33)

$$\frac{\partial N}{\partial \varphi} = \frac{2 n_i m^k}{F} \frac{\partial G^i}{\partial \dot{x}^k} + B N . \tag{6.36}$$

Comparing this result with (6.34), we finally obtain

$$\frac{\partial N}{\partial \varphi} = 2 \left(\frac{1}{\tau_g} - \frac{A}{\varrho_g} \right) - B N . \tag{6.37}$$

Now suppose that the direction $\dot{u}^\alpha$ at P of F_2 is an *asymptotic* direction, i. e. $N(u, \dot{u}) = 0$. It then follows from (6.26) that in this case

$$\frac{\partial N}{\partial \varphi} = 2 \sqrt{-K} ,$$

or, using (6.37),

$$\left(\frac{1}{\tau_g} - \frac{A}{\varrho_g} \right)^2 = -K . \tag{6.38}$$

This geometrical interpretation of K *represents the generalisation of the theorem of* BELTRAMI *and* ENNEPER.

The above-mentioned remark of CARTAN concerning lines of curvature follows from equation (6.37), for the condition $\partial N/\partial \varphi = 0$ (corresponding to a stationary value of N) does not imply $(\tau_g)^{-1} = 0$ as in euclidean differential geometry.

Note: The above conclusions are in essence due to CARTAN [1] and FINSLER [1]. It is remarkable, however, that the mean and Gaussian curvatures used by these authors are not equivalent to the concepts used in the present text. In defining $2H$ and K, FINSLER[1] considers the equation of EULER of euclidean differential geometry:

$$N = \frac{1}{R_1} \sin^2 \varphi + \frac{1}{R_2} \cos^2 \varphi .$$

Writing this in the form

$$N = \frac{1}{2} \left(\frac{1}{R_1} + \frac{1}{R_2} \right) + \frac{1}{2} \left(\frac{1}{R_2} - \frac{1}{R_1} \right) \cos 2\varphi ,$$

successive differentiations with respect to φ yield

$$\frac{\partial N}{\partial \varphi} = \left(\frac{1}{R_1} - \frac{1}{R_2} \right) \sin 2\varphi , \qquad \frac{\partial^2 N}{\partial \varphi^2} = 2 \left(\frac{1}{R_1} - \frac{1}{R_2} \right) \cos 2\varphi ,$$

so that

$$2H' \equiv \frac{1}{R_1} + \frac{1}{R_2} = 2 N + \frac{1}{2} \frac{\partial^2 N}{\partial \varphi^2} , \tag{6.39}$$

and

$$K' \equiv \frac{1}{R_1 R_2} = N^2 + \frac{1}{2} N \frac{\partial^2 N}{\partial \varphi^2} - \frac{1}{4} \left(\frac{\partial N}{\partial \varphi} \right)^2 . \tag{6.40}$$

The latter equations are regarded as the *definitions* of the mean and Gaussian curvatures of F_2 respectively by both FINSLER and CARTAN. Comparing these

---

[1] FINSLER [1], § 70 et seq., where further geometrical consequences of equation (6.37) are discussed.

equations with (6.25) and (6.26), we see that the definitions used in the above treatment have each included an additional term involving the invariant J. This, however, does not affect the above conclusions since these depend on the identity (6.30), which is equally true for both definitions.

Some aspects of the above theory may be generalised to hypersurfaces of an n-dimensional Finsler space; how this may be done is indicated by BERWALD[1].

§ 7. Fundamental Aspects of the Theory of Subspaces from the Point of View of the Locally Minkowskian Metric

As we have remarked at the beginning of the present chapter, the theory of subspaces of a Finsler space presents a completely different picture if we dispense with the notion of element of support and the euclidean connection of CARTAN. From an analytical point of view the difficulties that occur if we abandon these concepts result from the fact that the covariant derivative of the metric tensor does not vanish, and also because we cannot construct fields of mutually orthogonal unit vectors. From the first of these remarks it follows that in contrast to classical differential geometry the covariant derivative of the unit normal vector of a hypersurface is not, in general, tangential to the latter, which leads to a new geometrical picture with peculiar difficulties; while from the second remark it is evident that we shall have to revise somewhat our ideas concerning the normals themselves. We shall see that these factors force us to introduce two distinct second fundamental forms, which between each other share the properties of the second fundamental form of classical differential geometry. A further feature of basic importance is the fact that due to lack of linearity it is generally impossible to specify the number of principal directions at a point.

For the sake of geometrical clarity, we shall first develop the theory of hypersurfaces F_{n-1} of the Finsler space F_n: towards the end of this section we shall indicate the corresponding generalisations for the case of an F_m imbedded in an F_n.

1°. Normal Curvature

For the case of a hypersurface F_{n-1} equation (2.15) will specify a unique normal n^i, provided we normalise n^i such that its Minkowskian length is unity:

$$n_i B_\alpha^i = 0 , \qquad n^i = g^{ij}(x, n) n_j , \qquad F(x^j, n^j) = 1 . \tag{7.1}$$

---

[1] BERWALD [8], § 4. A detailed discussion of the notion of mean curvature in Finsler spaces is given by ZHANG [1, 2]. Totally geodesic hypersurfaces are studied by KIKUCHI [1], HAIMOVICI [7], RAPCSÁK [4]. See also NASU [1]. Using a class of non-linear connections BARTHEL [5, 6] investigates the properties of minimal hypersurfaces, the measure of area involved being that of CHOQUET and BUSEMANN (Ch. I, § 8). In these latter investigations the partial arbitrariness of the connection parameters proves to be most advantageous.

Nevertheless, in the course of our analysis it will become evident that however useful n^i might be, it does not possess all the properties that we shall require of a normal. For instance, let $C: x^i = x^i(s)$ be a curve of F_{n-1}, passing through a point P of F_{n-1}. Since C is referred to its arc-length, its tangent vector $x'^i = dx^i/ds$ is a unit vector. If we write equation (3.17a) in the form

$$g_{ij}(x, x')\, B_\alpha^j(x)\, \frac{\delta x'^i}{\delta s} = g_{\alpha\delta}(u, u')\, \frac{\delta u'^\delta}{\delta s}\,, \tag{7.2}$$

it is clear that the principal normal vector $r(\delta x'^i/\delta s)$ of the geodesic of F_{n-1} tangent to C at P does *not* coincide with the normal of F_{n-1} defined by (7.1) as in classical differential geometry; instead its direction coincides with the unit vector $n^{*i}(x, x')$ defined by the equations

$$g_{ij}(x, x')\, B_\alpha^j(x)\, n^{*i}(x, x') = 0\,, \qquad F(x^j, n^{*j}(x, x')) = 1\,. \tag{7.3}$$

To each direction x'^i tangent to F_{n-1} at P there corresponds a geodesic of F_{n-1} and hence such a vector: the totality of these vectors generates a cone with vertex at P in the Minkowskian tangent space $T_n(P)$. We shall call this cone the *normal cone* of F_{n-1} at P (since in euclidean geometry this cone would consist simply of the unique normal to F_{n-1} at P)[1].

The following identities involving n^i and n^{*i} will be found useful in the sequel. For the sake of brevity let us write[2]

$$n_i^*(x, x') = g_{ij}(x, x')\, n^{*j}(x, x')\,. \tag{7.4}$$

From (7.1) and (7.3) it then follows that $n_i^*(x, x')$ and n_i are proportional; and using the definition (1.6.4) of the Minkowskian cosine and the fact that both n^i and n^{*i} are unit vectors it is easily seen that

$$n_i^*(x, x') = \frac{\psi(x, x')}{\cos(n, n^*)}\, n_i\,, \tag{7.5}$$

where we have written

$$\psi(x, x') = g_{ij}(x, x')\, n^{*i}(x, x')\, n^{*j}(x, x')\,. \tag{7.6}$$

Furthermore, from (2.14) and (7.3) we have

$$B_i^\alpha(x, x')\, n^{*i}(x, x') = 0\,, \tag{7.7}$$

---

[1] The geometrical distinction between the vector n^i and a generator $n^{*i}(x, x')$ of the normal cone is easily described as follows. The $(n-1)$ vectors B_α^i define a hyperplane T_{n-1} in $T_n(P)$; there exist two points P_1, P_2 on the indicatrix in $T_n(P)$ whose tangent hyperplanes are parallel to T_{n-1} as a result of the convexity of the indicatrix. The vectors PP_1, PP_2 both satisfy (7.1); by choice of a suitable orientation, one of these vectors is taken as n^i. Let P_3 be a corresponding point on the hyperellipsoid $g_{ij}(x, x'_{(0)})\, x'^i x'^j = 1$ $(x'_{(0)}$ fixed) whose tangent hyperplane is parallel to T_{n-1}, and denote by P_4 the point at which the vector PP_3 (produced if necessary) intersects the indicatrix. Then $n^{*i}(x, x'_{(0)})$ is the vector PP_4.

[2] It should be noted that n_i^* does not represent the covariant components of n^{*i}.

and it is evident that we must find a second set of projection parameters b_i^α inverse to B_α^i such that

$$b_i^\alpha(x)\, n^i(x) = 0 \,. \tag{7.7a}$$

This is easily done as follows: in analogy with (2.6) we define a hyper-surface tensor independent of direction by putting

$$\gamma_{\alpha\beta}(u) = g_{ij}(x, n)\, B_{\alpha\beta}^{ij}\,, \qquad \gamma^{\alpha\beta}\, \gamma_{\alpha\gamma} = \delta_\gamma^\beta \,. \tag{7.8}$$

Writing

$$b_i^\alpha(u) = \gamma^{\alpha\delta}(u)\, g_{ij}(x, n)\, B_\delta^j(x) \,, \tag{7.8a}$$

it is obvious that (7.7a) is satisfied, while

$$b_i^\alpha\, B_\beta^i = \delta_\beta^\alpha\,, \qquad \text{together with} \qquad B_\beta^i\, B_i^\alpha = \delta_\beta^\alpha \,. \tag{7.9}$$

It is easily verified that corresponding to (2.18) we now have

$$B_\beta^k(x)\, B_j^\beta(x, x') = \delta_j^k - \frac{1}{\psi}\, n^{*k}(x, x')\, n_j^*(x, x') \,, \tag{7.10}$$

together with

$$B_\beta^k(x)\, b_j^\beta(x) = \delta_j^k - n^k(x)\, n_j(x) \,. \tag{7.10a}$$

We shall now define the normal curvature of F_{n-1} by means of the following geometrical construction. Consider two neighbouring points $P(x^i)$, $Q(x^i + dx^i)$ of a curve C on F_{n-1}. At P we have

$$n_i\, x'^i = n_i\, B_\alpha^i\, u'^\alpha = 0 \tag{7.11}$$

for the tangent vector x'^i of C in virtue of (7.1); thus, by (1.6.4) $\cos(n, x') = 0$ at P. Let us now transport by δ-parallelism (Ch. II) the vector n_i from P to Q, thus obtaining a vector

$$n_i + d^*n_i = n_i + P_{ik}^h(x, x')\, n_h\, dx^k \,. \tag{7.12}$$

At Q let us evaluate the cosine

$$dq \equiv \cos(n + d^*n, x' + dx') = \frac{(n_i + d^*n_i)(x'^i + dx'^i)}{H(x + dx, n + d^*n)\, F(x + dx, x' + dx')}, \tag{7.13}$$

where $x'^i + dx'^i$ is the unit tangent vector to C at Q. Clearly dq will indicate to some extent the change in the direction of n_i as we pass from P to Q, and hence dq will represent some measure of the curvature of F_{n-1} at P in the direction PQ.

In view of (7.11) the numerator on the right-hand side of (7.13) may be written as $n_i\, dx'^i + d^*n_i\, x'^i$. But since (7.11) holds generally along C, we have by differentiation: $n_i\, dx'^i = -dn_i\, x'^i$. Also, the denominator on the right-hand side of (7.13) differs from unity only by terms of the first order of smallness; hence neglecting second order terms we find

$$\frac{dq}{ds} = \frac{x'^i(d^*n_i - dn_i)}{ds} \,, \tag{7.14}$$

noting that $n^i + dn^i$ is the unit normal to F_{n-1} at Q. But clearly this expression is independent of the derivatives of the tangent vector x'^i

of C and is therefore the same for all curves of F_{n-1} with common tangent x'^i at P. Thus (7.14) represents a property of the hypersurface, and we shall regard it as the *normal curvature* of F_{n-1} at P in the direction x'^i, to be denoted by $[R(x, x')]^{-1}$.

It is obvious that this definition coincides with that of classical differential geometry. For if we substitute from (7.12) in (7.14) we have

$$\frac{1}{R(x, x')} = \frac{dq}{ds} = -\left(\frac{dn_i}{ds} - P^h_{ik}(x, x') n_h \frac{dx^k}{ds}\right) x'^i,$$

or, by (2.3.17) and (7.11)

$$\frac{1}{R(x, x')} = -\frac{\delta n_i}{\delta s} x'^i = n_i \frac{\delta x'^i}{\delta s}. \tag{7.15}$$

Also, regarding the curve C as a curve of F_n, and using the definition (1.8a) for its curvature $\frac{1}{r}$, it follows from (7.15) and the definition (1.6.4) of the Minkowskian cosine that

$$\frac{\cos\left(n^i, \dfrac{\delta x'^i}{\delta s}\right)}{r} = \frac{1}{R(x, x')}. \tag{7.16}$$

Since $\delta x'^i/\delta s$ coincides in direction with the principal normal of C with respect to F_n, it is clear that this result represents the *generalisation of Meusnier's theorem*.

As a special application of this theorem let us examine the case when C is a geodesic of F_{n-1}. Denoting the curvature of this geodesic (regarded as a curve of F_n) by $(R^*(x, x'))^{-1}$, we have from (1.8a), (7.2) and (7.3)

$$\frac{\delta x'^i}{\delta s} = n^{*i}(x, x')\,(R^*(x, x'))^{-1},$$

and hence, since the cosine is a function homogeneous of degree zero in its directional arguments, equation (7.16) becomes

$$\frac{\cos(n, n^*)}{R^*(x, x')} = \frac{1}{R(x, x')}. \tag{7.17}$$

Thus the normal curvature $R^{-1}(x, x')$ *does not in general coincide with the curvature of the geodesic of F_{n-1}* (regarded as a curve of F_n) *tangent to x'^i at x^i*. In fact, it follows from (7.16) that $R^{-1}(x, x')$ is the curvature of that curve of F_{n-1} tangent to x'^i at x^i whose principal normal coincides in direction with n^i. However, the above construction clearly indicates that both R and R^* are invariants of fundamental importance, which have to be considered on their respective merits, and which naturally lead to two distinct second fundamental forms. We shall call $(R^*(x, x'))^{-1}$ the *secondary normal curvature*[1] of F_{n-1} in the direction x'^i.

---

[1] RUND [10, II], p. 202. We observe also that $R^*(x, x')$ cannot be interpreted geometrically by the variation of the unit normal $n^{*i}(x, x')$ in a manner similar to the construction leading to (7.14).

2°. The Two Second Fundamental Forms

Again, let us consider a vector field $X^i(s)$ tangent to F_{n-1}, defined along a curve C of F_{n-1}. From the definition (2.3.17) of the δ-derivative, together with (2.4.10) and (3.2) we have

$$\frac{\delta X^i}{\delta s} = B^i_{\beta\gamma} X^\beta \frac{du^\gamma}{ds} + B^i_\alpha \frac{dX^\alpha}{ds} + \Gamma^{*i}_{hk} B^{hk}_{\beta\gamma} X^\beta \frac{du^\gamma}{ds},$$

or, introducing the δ-derivative of X^α with respect to F_{n-1},

$$\frac{\delta X^i}{\delta s} = I^i_{\beta\gamma} X^\beta u'^\gamma + B^i_\alpha \frac{\overline{\delta} X^\alpha}{\delta s}, \tag{7.18}$$

where we have written

$$I^i_{\beta\gamma} = B^i_{\beta\gamma} - B^i_\delta \Gamma^{*\delta}_{\beta\gamma} + \Gamma^{*i}_{hk} B^{hk}_{\beta\gamma}. \tag{7.19}$$

It is immediately obvious that these quantities are components of a tensor obtained by a process of "mixed" differentiation of the type considered in § 4. In fact, one can develop an analogous "D-symbolism" for δ-differentiation, namely by defining in analogy with (4.24) for a tensor field $T^{i_1\cdots i_r}{}_{\alpha_1\ldots\alpha_s}$ the following derivative:

$$\overset{0}{\delta}_\gamma T^{i_1\cdots i_r}{}_{\alpha_1\ldots\alpha_s} = \partial_\gamma T^{i_1\cdots i_r}{}_{\alpha_1\ldots\alpha_s}$$
$$+ \sum_\mu T^{i_1\cdots i_{\mu-1} j i_{\mu+1}\cdots i_r}{}_{\alpha_1\ldots\alpha_s} \Gamma^{*i_\mu}_{jk} B^k_\gamma - \tag{7.20}$$
$$- \sum_\nu T^{i_1\cdots i_r}{}_{\alpha_1\ldots\alpha_{\nu-1}\beta\alpha_{\nu+1}\ldots\alpha_s} \Gamma^{*\beta}_{\alpha_\nu\gamma}.$$

The term $\partial_\gamma T^{i_1\cdots i_r}{}_{\alpha_1\ldots\alpha_s}$ on the right-hand side denotes differentiation of $T^{i_1\cdots i_r}{}_{\alpha_1\ldots\alpha_s}$ with respect to u^γ, taking into account the variation of the directional argument (if any) of $T^{i_1\cdots i_r}{}_{\alpha_1\ldots\alpha_s}$. In this connection it is important to observe whether these arguments are given in terms of the coordinate system of F_n or of F_{n-1}. In particular, we have over F_{n-1},

$$\overset{0}{\delta}_\gamma g_{ij}(x, \xi) = g_{ij;k}(x, \xi) B^k_\gamma.$$
$$\overset{1}{}$$

Clearly one might also define a $\overset{1}{\delta}_\gamma$-operator, but since this operator will not be used in the sequel, it will be ignored.

The following remarks, however, are significant. Firstly, it is clear that (7.19) represents a special case of (7.20) if we write

$$I^i_{\beta\gamma} = \overset{0}{\delta}_\gamma B^i_\beta, \tag{7.20a}$$

where it is to be noted that the B^i_β are independent of direction. In contrast to the Eulerian curvature tensor $\overset{0}{H}{}^i_{\beta\gamma}$ of § 4, which is defined in a similar manner, the tensor $I^i_{\beta\gamma}$ is symmetric in its lower indices. In fact, a simple comparison based on (4.33) shows that

$$\overset{0}{H}{}^i_{\gamma\beta} = I^i_{\gamma\beta} + N^i_j A^j_{hk} B^h_\beta H^k_\gamma$$

in the notation of § 4.

Secondly, a direct calculation involving equation (3.25) yields the relation

$$\overset{0}{\delta}_\gamma \, g_{\alpha\beta}(u, u') = B^{ijk}_{\alpha\beta\gamma} \, g_{ij;k}(x, x') = \overset{0}{\delta}_\gamma \, g_{ij} B^{ij}_{\alpha\beta} \,, \tag{7.21}$$

where the directional argument is tangent to F_{n-1}, and coincides with the direction of differentiation.

The significance of the $I^i_{\alpha\beta}$ depends on the fact that these quantities, regarded as components of a vector of F_n, represent a generator of the normal cone. For in virtue of (2.6) equation (3.10a) may be written in the form

$$g_{ij} B^{ji}_{\gamma\delta} \Gamma^{*\,\delta}_{\alpha\beta} = g_{ij} B^j_\gamma (B^i_{\alpha\beta} + \Gamma^{*\,i}_{hk} B^{hk}_{\alpha\beta}) \,,$$

which, by means of (7.19), reduces to

$$g_{ij}(x, x') \, B^j_\gamma I^i_{\alpha\beta}(x, x') = 0 \,. \tag{7.22}$$

Comparing this result with (7.3), it is clear that we may define a hypersurface tensor $\Omega^*_{\alpha\beta}(u, u')$ by the equation

$$I^i_{\alpha\beta}(x, x') = n^{*\,i}(x, x') \, \Omega^*_{\alpha\beta}(u, u') \,. \tag{7.23}$$

We are justified in regarding the components of this tensor as the coefficients of one of the second fundamental forms of F_{n-1}. For we may now write equation (7.18) in the form

$$\frac{\delta X^i}{\delta s} = n^{*\,i} \Omega^*_{\alpha\beta}(u, u') \, X^\alpha u'^\beta + B^i_\alpha \frac{\overline{\delta} X^\alpha}{\delta s} \,, \tag{7.24}$$

which, incidentally, represents the decomposition of the covariant derivative of X^i into a component tangential to the hypersurface and a component tangential to the normal cone. If we now apply this equation to the tangent vector x'^i of a geodesic of $F_{n-1} (\overline{\delta} u'^\alpha / \delta s = 0)$, we have

$$\frac{\delta x'^i}{\delta s} = n^{*\,i} \Omega^*_{\alpha\beta}(u, u') \, u'^\alpha u'^\beta \,,$$

and since $n^{*\,i}$ is a unit vector, it follows that the curvature of this geodesic (regarded as a curve of F_n) is represented by

$$\frac{1}{R^*(x, x')} = \Omega^*_{\alpha\beta}(u, u') \, u'^\alpha u'^\beta = \frac{\Omega^*_{\alpha\beta}(u, u') \, du^\alpha \, du^\beta}{g_{\alpha\beta}(u, u') \, du^\alpha \, du^\beta} \,, \tag{7.25}$$

as in classical differential geometry. We shall call the numerator of the second term on the right-hand side the *secondary* second fundamental form[1], since we shall have to distinguish it from another second fundamental form, which is introduced as follows.

As in equation (7.25) we wish to represent the normal curvature also in terms of "quadratic" forms. Substituting (7.24) in (7.15) we have in view of (7.1) and (7.17)

$$\frac{1}{R(x, x')} = n_i I^i_{\alpha\beta}(u, u') \, u'^\alpha u'^\beta \,. \tag{7.26}$$

---

[1] RUND [9], p. 493.

Hence we shall write

$$\Omega_{\alpha\beta}(u, u') = n_i I^i_{\alpha\beta}(u, u') ,\qquad(7.27)$$

and equation (7.26) becomes

$$\frac{1}{R(x, x')} = \frac{\Omega_{\alpha\beta}(u, u')\, du^\alpha\, du^\beta}{g_{\alpha\beta}(u, u')\, du^\alpha\, du^\beta} .\qquad(7.28)$$

The numerator of the expression on the right-hand side of this equation is to be regarded as the second fundamental form of F_{n-1}. From the symmetry of $I^i_{\alpha\beta}$ it follows that both sets of coefficients $\Omega_{\alpha\beta}$ and $\Omega^*_{\alpha\beta}$ are symmetric in α and β.

There exists a simple relationship between the $\Omega_{\alpha\beta}$ and the $\Omega^*_{\alpha\beta}$. If we multiply equation (7.23) by n_i, noting that both n^i and n^{*i} are unit vectors, it follows from the definition (1.6.4) of the Minkowskian cosine that

$$n_i I^i_{\alpha\beta} = \cos(n, n^*)\, \Omega^*_{\alpha\beta} ,\qquad(7.29)$$

and thus we have from (7.27) and (7.29)

$$\Omega_{\alpha\beta}(u, u') = \Omega^*_{\alpha\beta}(u, u') \cos(n, n^*) .\qquad(7.30)$$

The coefficients $\Omega_{\alpha\beta}$ of the second fundamental form may again be interpreted geometrically in terms of the variation of the unit normal of F_{n-1}[1]. For if we take the $\overset{0}{\delta}_\alpha$-derivative of (7.1), we have

$$\left(\overset{0}{\delta}_\alpha n_i\right) B^i_\beta = - n_i \left(\overset{0}{\delta}_\alpha B^i_\beta\right) ,$$

or, substituting from (7.20a) and (7.27),

$$\Omega_{\alpha\beta} = - B^i_\beta \overset{0}{\delta}_\alpha n_i = - \tfrac{1}{2}\left(B^i_\beta \overset{0}{\delta}_\alpha n_i + B^i_\alpha \overset{0}{\delta}_\beta n_i\right).\qquad(7.31)$$

This equation is useful also when we wish to express the variation of the unit normal in terms of the second fundamental form. This is done as follows: we multiply (7.31) by b^β_j, noting (7.10a), obtaining

$$\Omega_{\alpha\beta} b^\beta_j = - \overset{0}{\delta}_\alpha n_j + n_j \left(n^h \overset{0}{\delta}_\alpha n_h\right),$$

or, since n^i is unit,

$$g^{ij}(x, n)\, \Omega_{\alpha\beta} b^\beta_j = - g^{ij}(x, n)\, \overset{0}{\delta}_\alpha n_j - n^i \left(n_h \overset{0}{\delta}_\alpha n^h\right).\qquad(7.32)$$

But from (7.1) we have

$$g^{ij}(x, n)\, \overset{0}{\delta}_\alpha n_j = \overset{0}{\delta}_\alpha n^i + g^{ij}(x, n)\, c_{hj,\alpha}\, n^h ,\qquad(7.33)$$

where we have written

$$c_{jh,\alpha} = \overset{0}{\delta}_\alpha g_{jh}(x, n)\qquad(7.34)$$

---

[1] In fact, $\Omega_{\alpha\beta}$ was first defined by (7.31). RUND [10, II], p. 199.

for the sake of brevity. And further, again by (7.1)

$$n_h \overset{0}{\delta}_\alpha n^h = - \tfrac{1}{2} c_{hj,\alpha} n^h n^j \, . \tag{7.35}$$

Substituting (7.33) and (7.35) in (7.32) and using (7.8a) we find the following expression for the variation of n^i:

$$\overset{0}{\delta}_\alpha n^i = - B^i_\delta \gamma^{\beta\delta} \Omega_{\alpha\beta} - c_{hk,\alpha} n^h [g^{ik}(x, n) - \tfrac{1}{2} n^i n^k] \, . \tag{7.36}$$

In connection with this formula we observe that the term $c_{hk,\alpha} n^h$ does not contain the derivatives $\partial n^k / \partial u^\alpha$ as one might suspect at first sight, for if we expand this expression according to (7.34), we find that the term which involves these derivatives vanishes identically due to homogeneity. Thus the expression for $\overset{0}{\delta}_\alpha n^i$ depends only on positional coordinates and the direction x'^i along which we are differentiating, i. e. it is the same for all curves of F_{n-1} which have a common tangent x'^i at the point under consideration.

It is also possible to evaluate the $\overset{0}{\delta}_\alpha$-derivatives of the generators n^{*i} of the normal cone. By a process similar to the one described above, we find

$$\overset{0}{\delta}_\alpha n^{*i}(x, x') = - \psi(x, x') g^{\beta\delta}(u, u') B^i_\delta \Omega^*_{\alpha\beta}(u, u') -$$
$$- c^*_{hj,\alpha}(x, x') n^{*h}(x, x') [g^{ij}(x, x') - \tag{7.36a}$$
$$- \frac{1}{2\psi} n^{*i}(x, x') n^{*j}(x, x')] + \tfrac{1}{2} \overset{0}{\delta}_\alpha (\log \psi) n^{*i}(x, x') \, ,$$

where we have written

$$c^*_{hj,\alpha}(x, x') = \overset{0}{\delta}_\alpha g_{hj}(x, x') \, . \tag{7.37}$$

In contrast to (7.36), equation (7.37) suffers from the drawback that the right-hand side involves derivatives of x'^i, so that $\overset{0}{\delta}_\alpha n^{*i}(x, x')$ depends on the curve of F_{n-1} along which the process of differentiation takes place. Considering the geometrical background of $n^{*i}(x, x')$, this is to be expected.

3°. Principal Directions

Returning to equation (7.28), we see that a natural generalisation of the *Dupin indicatrix* of classical differential geometry is the $(n-2)$-dimensional locus

$$\Omega_{\alpha\beta}(u, u') u'^\alpha u'^\beta = 1 \tag{7.38}$$

in the hyperplane T_{n-1} spanned by the vectors B^i_α in T_n. In general, of course, (7.38) will not represent a quadric surface.

We shall now define *principal directions* to be those directions which are determined by points on (7.38) whose (Minkowskian) distance from the centre of the indicatrix assumes an extreme value relative to neighbouring points of (7.38). In other words, principal directions are given by extreme values of $g_{\alpha\beta}(u, u')\, u'^{\alpha}\, u'^{\beta}$ subject to the subsidiary condition $\Omega_{\alpha\beta}(u, u')\, u'^{\alpha}\, u'^{\beta} = 1$, where u^{α} is kept fixed.

In view of (7.28) this definition implies that *principal directions are those directions for which the normal curvature assumes extreme values.* In fact, it follows from (7.28) and (7.38) that the lengths of the radius vectors u'^{α} of the Dupin indicatrix are simply $\sqrt{R(u, u')}$.

According to the multiplier rule we therefore seek solutions of the equations

$$\frac{\partial}{\partial u'^{\gamma}} \{g_{\alpha\beta}(u, u')\, u'^{\alpha}\, u'^{\beta} + \lambda\,(\Omega_{\alpha\beta}(u, u')\, u'^{\alpha}\, u'^{\beta} - 1)\} = 0 .$$

In consequence of (1.3.5) these equations reduce to

$$2g_{\alpha\gamma}(u, u')\, u'^{\alpha} + 2\lambda\,\Omega_{\alpha\gamma}(u, u')\, u'^{\alpha} + \lambda\,\frac{\partial \Omega_{\alpha\beta}}{\partial u'^{\gamma}}\, u'^{\alpha}\, u'^{\beta} = 0 . \tag{7.39}$$

But as a result of (7.1) we may write (7.27) in the form

$$\Omega_{\alpha\beta} = n_i\,(B^{i}_{\alpha\beta} + \Gamma^{*\,i}_{h\,k}\, B^{h\,k}_{\alpha\beta}) ,$$

and on differentiating this equation with respect to u'^{γ}, we obtain

$$\frac{\partial \Omega_{\alpha\beta}}{\partial u'^{\gamma}} = n_i\,\frac{\partial \Gamma^{*\,i}_{h\,k}}{\partial u'^{\gamma}}\, B^{h\,k}_{\alpha\beta} ,$$

and hence, using (2.5.12), we have

$$\frac{\partial \Omega_{\alpha\beta}}{\partial u'^{\gamma}}\, u'^{\alpha}\, u'^{\beta} = 0 . \tag{7.40}$$

It therefore follows that (7.39) reduces to

$$g_{\alpha\gamma}(u, u')\, u'^{\alpha} = -\,\lambda\,\Omega_{\alpha\gamma}(u, u')\, u'^{\alpha} ,$$

and on multiplying this result by u'^{γ} and noting (7.28), we find

$$\lambda = -\,R(x, x') .$$

Thus principal directions satisfy the equations

$$g_{\alpha\gamma}(u, u')\, u'^{\alpha} = R(x, x')\,\Omega_{\alpha\gamma}(u, u')\, u'^{\alpha} , \tag{7.41}$$

where $R^{-1}(x, x')$ is the normal curvature corresponding to these directions.

Since this is not a linear problem, nothing can be said in general as regards the number of possible independent solutions, and therefore it is not feasible to define the Gaussian and the mean curvature as in the classical theory of surfaces. We shall see, however, that despite this difficulty the principal directions (7.41) possess many of the properties of their euclidean counterparts.

Firstly, let us assume for the moment that at least two independent solutions $u'^{\alpha}_{(1)}$, $u'^{\alpha}_{(2)}$ exist, this assumption being geometrically feasible. Writing down the two equations (7.41) corresponding to each of these solutions, and multiplying them by $u_{(2)}^{\prime\gamma}$, $u_{(1)}^{\prime\gamma}$ respectively, we have

$$g_{\alpha\gamma}(u,\,u'_{(1)})\,u_{(1)}^{\prime\alpha}\,u_{(2)}^{\prime\gamma} = R_{(1)}\,\Omega_{\alpha\gamma}(u,\,u'_{(1)})\,u_{(1)}^{\prime\alpha}\,u_{(2)}^{\prime\gamma}\,, \qquad (7.41\,\mathrm{a})$$

and

$$\dot{g}_{\alpha\gamma}(u,\,u'_{(2)})\,u_{(2)}^{\prime\alpha}\,u_{(1)}^{\prime\gamma} = R_{(2)}\,\Omega_{\alpha\gamma}(u,\,u'_{(2)})\,u_{(2)}^{\prime\alpha}\,u_{(1)}^{\prime\gamma}\,, \qquad (7.41\,\mathrm{b})$$

where $R_{(1)}^{-1}$ and $R_{(2)}^{-1}$ are the normal, i. e. principal curvatures corresponding to the directions $u_{(1)}^{\prime\alpha}$, $u_{(2)}^{\prime\alpha}$. Since the $\Omega_{\alpha\gamma}(u,\,u')$ are homogeneous of degree zero in the u'^{α}, we can normalise $u_{(1)}^{\prime\alpha}$, $u_{(2)}^{\prime\alpha}$ to be unit vectors; and recalling the definition (1.6.4) of the Minkowskian cosine it follows by subtraction of (7.41 b) from (7.41 a) that

$$\frac{\cos(u'_{(1)},\,u'_{(2)}) - \cos(u'_{(2)},\,u'_{(1)})}{R_{(1)}R_{(2)}} = \left[\frac{\Omega_{\alpha\gamma}(u,\,u'_{(1)})}{R_{(2)}} - \frac{\Omega_{\alpha\gamma}(u,\,u'_{(2)})}{R_{(1)}}\right] u_{(1)}^{\prime\alpha}\,u_{(2)}^{\prime\gamma}\,. \quad (7.42)$$

This result[1] represents the *generalisation of the orthogonality relation between principal directions* of the classical theory. For if the cosine were symmetric in its directional arguments, and if the coefficients $\Omega_{\alpha\gamma}$ were independent of direction, it would follow from (7.42) that principal directions would correspond to conjugate directions of the Dupin indicatrix, and hence either (7.41 a) or (7.41 b) would lead to the law of orthogonality.

Secondly, if we multiply (7.36) by u'^{α}, summing over α, it is easily seen that we may rewrite this equation in the following form

$$\frac{\delta n_j}{\delta s} = \left(\frac{\delta g_{jh}}{\delta s}\right) n^h - g_{ij}\,B_{\delta}^i\,\gamma^{\beta\delta}\,\Omega_{\alpha\beta}\,u'^{\alpha} - \left(\frac{\delta g_{hk}}{\delta s}\right) n^h\,[\delta_j^k - \tfrac{1}{2}\,n^k\,n_j]\,,$$

or, using (7.8 a) and simplifying,

$$\frac{\delta n_j}{\delta s} = -\Omega_{\alpha\beta}\,u'^{\alpha}\,b_j^{\beta} + \tfrac{1}{2}\,n_j\left(\frac{\delta g_{hk}}{\delta s}\right) n^h\,n^k\,. \qquad (7.43)$$

In particular, it then follows from (7.1) and (7.9) that

$$B_{\gamma}^j\,\frac{\delta n_j}{\delta s} = -\Omega_{\alpha\gamma}\,u'^{\alpha}\,. \qquad (7.44)$$

Now let us suppose that u'^{α} in this equation represents a principal direction. Denoting by $v_{\alpha} = g_{\alpha\beta}(u,\,u')\,u'^{\beta}$ the covariant components of u'^{α}, and taking into account equations (7.41) for the principal directions, we have as a result of (7.44)

$$B_{\alpha}^i\,\frac{\delta n_i}{\delta s} = -R^{-1}(x,\,x')\,v_{\alpha}\,. \qquad (7.45)$$

Thus the projection of the covariant derivative along a principal direction of the unit normal onto the hypersurface coincides with that principal direction[2]. This theorem is a generalisation of the classical formula of

---

[1] RUND [9], p. 496.

[2] A different formulation of similar results is given by RUND [10, II], p. 206.

RODRIGUES. It should be noted, however, that in contrast to the classical theory, the covariant differential of n_i has a normal component, in general non-vanishing.

Since, as we have seen above, the mean curvature cannot be defined directly by means of the Dupin indicatrix, we seek another analogy from classical differential geometry in our endeavour to arrive at a suitable definition. We recall the fact that the surface Laplacian $V^2\vec{r}$ of the position vector $\vec{r}$ of a given point of a surface imbedded in a 3-dimensional euclidean space is given by the formula $V^2\vec{r} = M\vec{n}$, where $\vec{n}$ is the unit normal to the surface at that point, while M is the mean curvature. Noting that in Finsler spaces the Laplacian operator has to be represented by $g^{\alpha\beta}\, \overset{0}{\delta}_\alpha\, \overset{0}{\delta}_\beta$, which depends upon direction, we have to consider the expression

$$g^{\alpha\beta}(u,\, u')\, \overset{0}{\delta}_\alpha \left(\overset{0}{\delta}_\beta\, x^i\right) = g^{\alpha\beta}(u,\, u')\, \overset{0}{\delta}_\alpha B^i_\beta = g^{\alpha\beta}(u,\, u')\, I^i_{\alpha\beta}\,,$$

the second relation being a consequence of (7.20a). But we have seen [equation (7.22)] that $I^i_{\alpha\beta}$ is, in fact, a generator of the normal cone, so that we are justified in defining a scalar $M^*(u,\, u')$ by the equation

$$g^{\alpha\beta}(u,\, u')\, \overset{0}{\delta}_\alpha \left(\overset{0}{\delta}_\beta\, x^i\right) = M^*(u,\, u')\, n^{*\,i}(x,\, x')\,,$$

in analogy with the theorem quoted above. Since n^{*i} is a unit vector, $M^*(u,\, u')$ *represents the magnitude of the surface Laplacian* and will therefore be regarded as the *mean curvature*. In view of (7.23) it follows from the last two equations that the mean curvature may be expressed in the form

$$M^*(u,\, u') = g^{\alpha\beta}(u,\, u')\, \Omega^*_{\alpha\beta}(u,\, u')\,. \tag{7.46}$$

Similarly an alternative definition of mean curvature corresponding to the second fundamental form $\Omega_{\alpha\beta}$ would read

$$M(u,\, u') = \gamma^{\alpha\beta}(u)\, \Omega_{\alpha\beta}(u,\, u')\,. \tag{7.46a}$$

This latter invariant has a very simple geometrical meaning. For using (7.31) and (7.8a) and the fact that n_i is independent of direction, we may write

$$M(u,\, u') = -\, g^{ij}(x,\, n)\, b^\alpha_j\, \overset{0}{\delta}_\alpha\, n_i\,,$$

and hence from (7.1) and (7.10a) we deduce

$$M(u,\, u') = -\, n^k_{;k} - g_{hi;k}\, n^h\, [g^{ik} - \tfrac{1}{2}\, n^i\, n^k]\,. \tag{7.47}$$

Thus apart from the second term on the right-hand side (which invariably appears since δn^i is not tangential to F_{n-1}), the mean curvature $M(u,\, u')$ represents the divergence of the unit normal vector field n^k of F_{n-1}.

A direction u'^α is said to be *asymptotic* if it satisfies the equation

$$\Omega_{\alpha\beta}(u,\, u')\, u'^\alpha\, u'^\beta = 0\,. \tag{7.48}$$

In Finsler geometry the properties of asymptotic directions do not differ to any great extent from those of classical geometry. This is demonstrated by the following theorems: *For asymptotic directions the normal as well as the secondary normal curvature of F_{n-1} vanishes.* This follows directly from (7.28) and (7.30). In virtue of (7.24) we have furthermore: *the principal normal $(r\,\delta x'^i/\delta s)$ of an asymptotic curve of F_{n-1} (regarded as a curve of F_n) is tangential to F_{n-1}.* In particular, *if the tangents to a geodesic Γ of F_{n-1} are asymptotic directions, Γ is also a geodesic of F_n.* Conversely, *geodesics of F_n which are also curves of F_{n-1} are geodesics of F_{n-1} and are tangential to asymptotic directions of F_{n-1}.* From equation (1.9) we may also deduce the fact that *if a geodesic of F_{n-1} is tangential to an asymptotic direction at P it has contact of at least the second order with the geodesic of F_n touching it at P.*

If all the geodesics of F_{n-1} are also geodesics of the enveloping F_n, the hypersurface (if such a hypersurface exists) is called *totally geodesic.* A necessary and sufficient condition for F_{n-1} to be totally geodesic is that (7.48) holds for all (u, u'). Differentiating (7.48) with respect to u'^ν and u'^δ, noting (7.40) and (3.3.15), *we find that totally geodesic hypersurfaces are characterised by the condition*

$$\Omega_{\delta\gamma}(u, u') + n_i\,C^i_{jk|h}(x, x')\,x'^h B^{kj}_{\gamma'\delta} = 0 \,. \tag{7.48a}$$

This equation, together with (7.36), indicates that in general the unit normals of totally geodesic hypersurfaces are *not* parallel in the enveloping space (in contrast to Riemannian geometry)[1].

In § 5 we remarked in connection with the Eulerian curvature tensor that the latter may be interpreted in terms of small deformations of the subspace, although the final result is not as simple or as direct as in Riemannian geometry [equation (5.33)]. The same observation is true for the present theory. Since the calculations relevant to this problem are very similar in both theories, we shall content ourselves by stating the final results. Corresponding to F_{n-1} we construct a new hypersurface $\overline{F}_{n-1}$ defined by the locus of points whose coordinates are $x^i + \varepsilon\,n^i$, n^i being the unit normal corresponding to a point x^i of F_{n-1}. If $P(x^i)$, $Q(x^i + dx^i)$ are two neighbouring points of F_{n-1}, a distance ds apart, these define points $\overline{P}$, $\overline{Q}$ on $\overline{F}_{n-1}$, a distance $d\bar{s}$ apart from each other. Neglecting higher powers of ε it is then easily shown that[2]

$$\left(\frac{d\bar{s}}{ds}\right)^2 - 1 = 2\,\varepsilon\,g_{ij}(x, dx)\,\frac{dx^i}{ds}\,\frac{\delta n^j}{\delta s} \,, \tag{7.49}$$

---

[1] A study of geodesic manifolds transversal to a given subspace (of arbitrarily many dimensions) at a given point in terms of normal coordinates is made by CAIRNS [1].

[2] RUND [9], pp. 499—500.

and using (7.36) one may deduce that this result may be written more explicitly in the form

$$\left(\frac{d\bar{s}}{ds}\right)^2 - 1 = -2\varepsilon\, g_{\beta\gamma}(u,\,u')\,\gamma^{\beta\varepsilon}\,[\Omega_{\alpha\varepsilon} +$$
$$+\, c_{i\,h,\alpha}\, n^h (B_\varepsilon^i + \tfrac{1}{2} B_j^\delta\, \gamma_{\varepsilon\delta}\, n^i\, n^j)]\, u'^\alpha\, u'^\gamma\,. \tag{7.50}$$

4°. The Equations of Gauss and Codazzi

In order to find the desired relations between the coefficients $\Omega_{\alpha\beta}$ of the second fundamental form of F_{n-1} and the curvature tensors of F_n, it is necessary to express the $I_{\alpha\beta}^i$ in terms of n^i in a manner analogous to (7.23). We therefore define a new set of quantities $\omega_{\alpha\beta}^i$ by the equations

$$I_{\alpha\beta}^i = n^i \Omega_{\alpha\beta} + \omega_{\alpha\beta}^i\,, \tag{7.51}$$

and we shall first find explicit expressions for the $\omega_{\alpha\beta}^i$. Using (7.23) and (7.30) we see that (7.51) may be written as

$$\Omega_{\alpha\beta}\,[\sec(n,\,n^*)\, n^{*i} - n^i] = \omega_{\alpha\beta}^i\,.$$

But from (1.6.4) it follows that

$$n^i = \sec(n,\,n^*)\, n^{*i} - \mu^\alpha\, B_\alpha^i\,,$$

where we have put

$$\mu^\alpha = -\, n^i\, B_i^\alpha\,. \tag{7.52}$$

Hence we have

$$\omega_{\alpha\beta}^i = \Omega_{\alpha\beta}\, \mu^\delta\, B_\delta^i\,, \tag{7.53}$$

and equations (7.51) become

$$I_{\alpha\beta}^i = n^i\, \Omega_{\alpha\beta} + B_\delta^i\, \mu^\delta\, \Omega_{\alpha\beta}\,. \tag{7.54}$$

We now suppose that the directional arguments occurring in these tensors are given by a vector field $\xi^i(x^k)$ tangent to F_{n-1}. We may then take the corresponding $\overset{0}{\delta}_\gamma$-derivative of (7.54), and a straight-forward calculation similar to previous deductions of commutation formulae shows that we have

$$\overset{0}{\delta}_\gamma I_{\alpha\beta}^i - \overset{0}{\delta}_\beta I_{\alpha\gamma}^i = B_{\alpha\beta\gamma}^{hkj}\, \tilde{K}_{hkj}^i - B_\delta^i\, \tilde{K}_{\alpha\beta\gamma}^\delta\,, \tag{7.55}$$

where the latter curvature tensor is defined with respect to the $\Gamma_{\beta\gamma}^{*\alpha}$ of F_{n-1} by a relation which corresponds exactly to (4.1.5). Applying (7.55) to (7.51) we find

$$-B_\delta^i\, \tilde{K}_{\alpha\beta\gamma}^\delta + \tilde{K}_{hkj}^i\, B_{\alpha\beta\gamma}^{hkj} = \overset{0}{\delta}_\gamma\, n^i \cdot \Omega_{\alpha\beta} - \overset{0}{\delta}_\beta\, n^i \cdot \Omega_{\alpha\gamma} +$$
$$+\, n^i (\Omega_{\alpha\beta;\gamma} - \Omega_{\alpha\gamma;\beta}) + \overset{0}{\delta}_\gamma\, \omega_{\alpha\beta}^i - \overset{0}{\delta}_\beta\, \omega_{\alpha\gamma}^i\,. \tag{7.56}$$

In this relation we substitute for the $\overset{0}{\delta}_\gamma\, n_i$ from equation (7.36), and after some simplification we obtain the following equation:

$$B_\delta^i\left[-\tilde{K}_{\alpha\beta\gamma}^{\ \delta}+\gamma^{\delta\varepsilon}\left(\Omega_{\gamma\varepsilon}\Omega_{\alpha\beta}-\Omega_{\beta\varepsilon}\Omega_{\alpha\gamma}\right)\right]$$

$$=-\tilde{K}_{hkj}^i B_{\alpha\beta\gamma}^{hkj}-n^i\left\{\Omega_{\alpha\gamma;\beta}-\Omega_{\alpha\beta;\gamma}+\tfrac{1}{2}\,n^k n^h\left[c_{hk,\beta}\Omega_{\alpha\gamma}-c_{hk,\gamma}\Omega_{\alpha\beta}\right]\right\}+$$

$$+\,g^{ik}(x,n)\,n^h\left(c_{hk,\beta}\Omega_{\alpha\gamma}-c_{hk,\gamma}\Omega_{\alpha\beta}\right)+\overset{0}{\delta}_\gamma\,\omega_{\alpha\beta}^i-\overset{0}{\delta}_\beta\,\omega_{\alpha\gamma}^i\,. \qquad (7.56\mathrm{a})$$

We multiply this equation by $g_{ij}(x,n)\,B_\lambda^j$, noting (7.1) and (7.8). This leads to

$$\gamma_{\delta\lambda}\,\tilde{K}_{\alpha\gamma\beta}^{\ \delta}+\left(\Omega_{\gamma\lambda}\Omega_{\alpha\beta}-\Omega_{\beta\lambda}\Omega_{\alpha\gamma}\right)$$

$$=g_{ij}(x,n)\,\tilde{K}_{hlk}^i B_{\alpha\beta\gamma\lambda}^{hkj}+n^h B_\lambda^k\left(c_{hk,\beta}\Omega_{\alpha\gamma}-c_{hk,\gamma}\Omega_{\alpha\beta}\right)+ \qquad (7.57)$$

$$+\,g_{ij}(x,n)\,B_\lambda^j\left(\overset{0}{\delta}_\gamma\,\omega_{\alpha\beta}^i-\overset{0}{\delta}_\beta\,\omega_{\alpha\gamma}^i\right).$$

On multiplying (7.56a) by $g_{ij}(x,n)\,n^j$ we obtain similarly after some simplification

$$g_{ij}(x,n)\,n^j\,\tilde{K}_{hkl}^i B_{\alpha\beta\gamma}^{hkl}=\left(\Omega_{\alpha\beta;\gamma}-\Omega_{\alpha\gamma;\beta}\right)+$$

$$+\tfrac{1}{2}\,n^h n^k\left(c_{hk,\beta}\Omega_{\alpha\gamma}-c_{hk,\gamma}\Omega_{\alpha\beta}\right)+g_{ij}(x,n)\,n^j\left(\overset{0}{\delta}_\gamma\,\omega_{\alpha\beta}^i-\overset{0}{\delta}_\beta\,\omega_{\alpha\gamma}^i\right). \qquad (7.58)$$

Equations (7.57) and (7.58) represent the *generalisations of the equations of* GAUSS *and* CODAZZI *of classical differential geometry*. On comparing these equations with the corresponding equations of Riemannian geometry, we see that the essential differences (apart from the impossibility of contracting terms with different directional arguments) lie in the additional terms involving the $\omega_{\alpha\beta}^i$ and the $c_{hk,\alpha}$. This, again, is due to the fact that the covariant derivatives of the unit normals are not tangential to the hypersurface.

It is possible, however, to eliminate from equations (7.57) and (7.58) the $\overset{0}{\delta}$-derivatives of the $\omega_{\alpha\beta}^i$. For from (7.53) and (7.20a) we find by differentiation

$$\overset{0}{\delta}_\gamma\,\omega_{\alpha\beta}^i-\overset{0}{\delta}_\beta\,\omega_{\alpha\gamma}^i=B_\delta^i\{(\Omega_{\alpha\beta;\gamma}-\Omega_{\alpha\gamma;\beta})\,\mu^\delta+$$

$$+\,(\Omega_{\alpha\beta}\,\mu_{;\gamma}^\delta-\Omega_{\alpha\gamma}\,\mu_{;\beta}^\delta)\}+\mu^\delta\,(I_{\delta\gamma}^i\,\Omega_{\alpha\beta}-I_{\delta\beta}^j\,\Omega_{\alpha\gamma})\,.$$

In this equation we substitute for the $I_{\delta\gamma}^i$ from (7.54), and it follows that this relation may then be written in the form

$$\overset{0}{\delta}_\gamma\,\omega_{\alpha\beta}^i-\overset{0}{\delta}_\beta\,\omega_{\alpha\gamma}^i=B_\delta^i\{(\Omega_{\alpha\beta;\gamma}-\Omega_{\alpha\gamma;\beta})\,\mu^\delta+(\Omega_{\alpha\beta}\,\mu_{;\gamma}^\delta-\Omega_{\alpha\gamma}\,\mu_{;\beta}^\delta)+$$

$$+\,(\Omega_{\varepsilon\gamma}\Omega_{\alpha\beta}-\Omega_{\varepsilon\beta}\Omega_{\alpha\gamma})\,\mu^\varepsilon\,\mu^\delta\}+n^i\,(\Omega_{\delta\gamma}\Omega_{\alpha\beta}-\Omega_{\delta\beta}\Omega_{\alpha\gamma})\,\mu^\delta\,. \qquad (7.59)$$

Multiplication of (7.59) by $g_{ij}(x,n)\,B_\lambda^j$ yields in view of (7.8) and (7.1),

$$g_{ij}(x,n)\,B_\lambda^j\left(\overset{0}{\delta}_\gamma\,\omega_{\alpha\beta}^i-\overset{0}{\delta}_\beta\,\omega_{\alpha\gamma}^i\right)=\gamma_{\delta\lambda}\{(\Omega_{\alpha\beta;\gamma}-\Omega_{\alpha\gamma;\beta})\,\mu^\delta+ \qquad (7.60)$$

$$+\,(\Omega_{\varepsilon\gamma}\Omega_{\alpha\beta}-\Omega_{\varepsilon\beta}\Omega_{\alpha\gamma})\,\mu^\varepsilon\,\mu^\delta+(\Omega_{\alpha\beta}\,\mu_{;\gamma}^\delta-\Omega_{\alpha\gamma}\,\mu_{;\beta}^\delta)\}\,;$$

and similarly, if we multiply (7.59) by $g_{ij}(x, n)\, n^j$, we find

$$g_{ij}(x, n)\, n^j \left(\overset{0}{\delta_\gamma}\, \omega^i_{\alpha\beta} - \overset{0}{\delta_\beta}\, \omega^i_{\alpha\gamma} \right) = (\Omega_{\delta\gamma}\, \Omega_{\alpha\beta} - \Omega_{\delta\beta}\, \Omega_{\alpha\gamma})\, \mu^\delta \,. \tag{7.60a}$$

Equations (7.60) and (7.60a) are substituted in (7.57) and (7.58) respectively. The final form[1] of the equations of Gauss and Codazzi is then as follows:

$$\begin{aligned}
\gamma_{\delta\lambda}\tilde{K}^\delta_{\alpha\gamma\beta} &+ (\Omega_{\gamma\lambda}\, \Omega_{\alpha\beta} - \Omega_{\beta\lambda}\, \Omega_{\alpha\gamma}) \\
&= g_{ij}(x, n)\tilde{K}^i_{hlk} B^{hk\,l\,j}_{\alpha\beta\gamma\lambda} + n^h B^k_\lambda (c_{hk,\beta}\, \Omega_{\alpha\gamma} - c_{hk,\gamma}\, \Omega_{\alpha\beta}) + \\
&\quad + \gamma_{\delta\lambda}\{ (\Omega_{\alpha\beta;\gamma} - \Omega_{\alpha\gamma;\beta})\, \mu^\delta + (\Omega_{\varepsilon\gamma}\, \Omega_{\alpha\beta} - \Omega_{\varepsilon\beta}\, \Omega_{\alpha\gamma})\, \mu^\varepsilon\, \mu^\delta + \\
&\quad + (\Omega_{\alpha\beta}\, \mu^\delta_{;\gamma} - \Omega_{\alpha\gamma}\, \mu^\delta_{;\beta})\} \,,
\end{aligned} \tag{7.61}$$

together with

$$\begin{aligned}
g_{ij}(x, n)\, n^j \tilde{K}^i_{hkl} B^{h\,kl}_{\alpha\beta\gamma} &= (\Omega_{\alpha\beta;\gamma} - \Omega_{\alpha\gamma;\beta}) + \\
&+ \tfrac{1}{2} n^h n^k (c_{hk,\beta}\, \Omega_{\alpha\gamma} - c_{hk,\gamma}\, \Omega_{\alpha\beta}) + (\Omega_{\delta\gamma}\, \Omega_{\alpha\beta} - \Omega_{\delta\beta}\, \Omega_{\alpha\gamma})\, \mu^\delta \,.
\end{aligned} \tag{7.62}$$

Clearly different forms of the equations of Gauss and Codazzi are obtained when one considers the coefficients $\Omega^*_{\alpha\beta}$ of the secondary second fundamental form together with a given generator $n^{*i}(x, x')$ of the normal cone. In order to derive the required relations we consider equation (7.23) instead of (7.54). The calculation proceeds along similar lines to the one outlined above (in fact, it is simpler, because there is no analogue of the terms $\omega^i_{\alpha\beta}$) and will therefore be omitted. *The alternative forms of the equations of* Gauss *and* Codazzi *read as follows:*

$$\begin{aligned}
\tilde{K}_{\alpha\lambda\gamma\beta}(u, u') &- \psi (\Omega^*_{\alpha\gamma}\Omega^*_{\lambda\beta} - \Omega^*_{\alpha\beta}\Omega^*_{\lambda\gamma}) \\
&= \tilde{K}_{hjlk}(x, x')\, B^{j\,hkl}_{\lambda\alpha\beta\gamma} - n^{*l} B^i_\lambda (c^*_{jl,\gamma}\, \Omega^*_{\alpha\beta} - c^*_{jl,\beta}\, \Omega^*_{\alpha\gamma}) \,,
\end{aligned} \tag{7.63}$$

together with

$$\begin{aligned}
\tilde{K}_{hjlk}(x, x')\, n^{*j} B^{hkl}_{\alpha\beta\gamma} &= \psi (\Omega^*_{\alpha\gamma;\beta} - \Omega^*_{\alpha\beta;\gamma}) + \\
&+ \tfrac{1}{2} n^{*h} n^{*j} (c^*_{jh,\gamma}\, \Omega^*_{\alpha\beta} - c^*_{jh,\beta}\, \Omega^*_{\alpha\gamma}) - \tfrac{1}{2} (\Omega^*_{\alpha\beta}\, \psi_{;\gamma} - \Omega^*_{\alpha\gamma}\, \psi_{;\beta}) \,,
\end{aligned} \tag{7.64}$$

where the $c^*_{ij,\alpha}$ are given by equation (7.37).

5°. Subspaces of Arbitrary Dimension

In our discussion of the theory of subspaces of a Finsler space from the point of view of the locally Minkowskian metric we have so far restricted ourselves to the case of a hypersurface F_{n-1} imbedded in an F_n. The above theory has been extended to the case of an F_m imbedded in an F_n $(m < n)$ by Eliopoulos[2]. This generalisation is not trivial, since one cannot construct systems of normals as in the classical theory. The final results are, however, similar to those described above, so that we shall

---

[1] Rund [9], p. 502.
[2] Eliopoulos [1], pp. 1—63; [2].

not give a detailed description of this theory, except to point out the chief differences which arise.

Again we have to consider $(n-m)$ normals n^i_μ, which are normalised as in (7.1)[1], together with the $(n-m)$ independent solutions $n^{*i}_\mu(x, x')$ of the equations $g_{ij}(x, x')\, B^j_\alpha\, n^{*i} = 0$. As regards the projection factors we note that we now have to put

$$b^\alpha_{i\,\mu} = g_{ij}\left(x, n_\mu\right) \gamma^{\alpha\beta}_\mu\, B^j_\beta \,, \tag{7.65}$$

where

$$\gamma_{\alpha\beta\,\mu} = g_{ij}\left(x, n_\mu\right) B^{ij}_{\alpha\beta} \,. \tag{7.66}$$

The vectors n^{*i}_μ can be chosen so as to satisfy relations of the type

$$g_{ij}(x, x')\, n^{*i}_\mu(x, x')\, n^{*j}_\nu(x, x') = \psi_\mu(x, x')\, \delta^\mu_\nu \,, \tag{7.67}$$

the functions $\psi_\mu(x, x')$ being the generalisations of the function defined by equation (7.6). The relation between the various normals is clearly exhibited by the formulae

$$n_{i\,\mu} = \sum_\nu p_{\mu\nu}\, n^*_{i\,\nu} \,, \tag{7.68}$$

where we have put

$$p_{\mu\nu} = \cos\left(n_\mu, n^*_\nu\right)(\psi_\nu)^{-1} \,. \tag{7.68a}$$

Again the induced connection parameters are to be used, and it is clear that the tensor $I^i_{\alpha\beta}$ is defined as before, and in particular continues to satisfy the orthogonality relation (7.22). Hence we define the coefficients $\Omega^*_{\alpha\beta\,\mu}$ by the equations

$$I^i_{\alpha\beta} = \sum_\mu \Omega^*_{\alpha\beta\,\mu}\, n^{*i}_\mu \,, \tag{7.69}$$

while the alternative coefficients $\Omega_{\alpha\beta\,\mu}$ result from the natural definition

$$\Omega_{\alpha\beta\,\nu} = \sum_\mu \Omega^*_{\alpha\beta\,\mu} \cos\left(n_\nu, n^*_\mu\right) \,, \tag{7.70}$$

which is the generalisation of (7.30). The covariant derivative of the tangent vector x'^i to a curve of F_m may then be expressed in the form

$$\frac{\delta x'^i}{\delta s} = \sum_\mu (\Omega^*_{\alpha\beta\,\mu}\, u'^\alpha\, u'^\beta)\, n^{*i}_\mu + B^i_\alpha\, \frac{\overline{\delta} u'^\alpha}{\delta s} \,, \tag{7.71}$$

and from (7.70) it then follows that

$$n_{i\,\nu}\, \frac{\delta x'^i}{\delta s} = \Omega_{\alpha\beta\,\nu}\, u'^\alpha\, u'^\beta \,. \tag{7.72}$$

---

[1] For the rest of this section $\mu, \nu, \ldots = m + 1, \ldots, n$. Summation with respect to these indices will be indicated explicitly.

Thus the quadratic form on the right-hand side serves to describe the normal curvature of F_m corresponding to the normal n^i_v. From equation (7.72) a form of Meusnier's theorem may be derived once more. The secondary second fundamental forms $\Omega^*_{\alpha\beta} u'^\alpha u'^\beta$ are related to the corresponding secondary normal curvatures of F_m in a similar manner. The generalised "orthogonality" relation (7.42) preserves its form, although the analogue of Rodrigues' formula (7.45) becomes considerably more complicated and loses its geometrical significance.

Naturally the Gauss-Codazzi equations also become a great deal more involved: corresponding to (7.56) one has the relation[1]

$$B^i_\delta \widetilde{K}^\delta_{\alpha\gamma\beta}(u, u') = - \widetilde{K}^i_{hkl}(x, x') B^{h\,k\,l}_{\alpha\beta\gamma} +$$
$$+ \sum_\mu \left(\Omega^*_{\alpha\beta} \overset{0}{\delta_\gamma}\, n^{*i} - \Omega^*_{\alpha\gamma} \overset{0}{\delta_\beta}\, n^{*i} \right) + \sum_\mu n^{*i} (\Omega^*_{\alpha\beta;\gamma} - \Omega^*_{\alpha\gamma;\beta}) . \qquad (7.73)$$

This equation is multiplied in turn by $g_{ij}(x, x')\, B^j_\lambda$ and by $g_{ij}(x, x')\, n^{*j}_\mu$, and the required relations are thus obtained. The final form of these equations becomes somewhat complicated when the explicit expressions for the derivatives of the unit normals are substituted in (7.73). As before, there are two sets of such relations corresponding to $\Omega_{\alpha\beta}$ and $\Omega^*_{\alpha\beta}$. For a full description of these results the reader is referred to the original work of ELIOPOULOS.

§ 8. The Differential Geometry of the Indicatrix and the Geometrical Significance of the Tensor S_{ijhk}

In the previous chapter we observed that in Cartan's theory of the curvature of a Finsler space three tensors make their appearance, one of which is denoted by S_{ijhk}. By studying the curvature properties of the indicatrix (which, after all, is a hypersurface in a Minkowskian space), VARGA [8] derived several theorems involving S_{ijhk}, as a result of which the geometrical significance of this tensor is clearly exhibited. The present section will be devoted to a description of these relationships, and with the use of the formalism of the previous sections we shall be able to follow VARGA's methods fairly closely.

Consider a Minkowskian space M_n, in which there exists a coordinate system x^i for which the Γ^{*i}_{hk} etc. vanish. The distance function is then of the form $F(x^i)$, and the hypersurface

$$F^2(x^i) = \frac{1}{k}, \qquad (k = \text{const.} > 0) , \qquad (8.1)$$

is homothetic to the indicatrix $F(x^i) = 1$. The same hypersurface (8.1) may be represented parametrically by

$$x^i = x^i(u^\alpha) , \quad (\alpha, \beta = 1, \ldots, n-1) , \qquad (8.2)$$

---

[1] ELIOPOULOS [1], p. 52.

and if these values of x^i are substituted in (8.1) we obtain an identity in the u^α, which yields by differentiation

$$\frac{\partial F^2}{\partial x^i}\, B^i_\alpha = 0,\qquad(8.3)$$

so that the vector $y_i = F\dfrac{\partial F}{\partial x^i} = g_{ij}(x)\, x^j$ is normal to the hypersurface (8.1) (which is in agreement with the construction of Ch. I, according to which the radius vector x^i of the indicatrix is normal to the tangent plane spanned by the B^i_α). If n^i is the unit normal vector, it then follows from (8.1) that $n^i = \sqrt{k}\, x^i$, $n_i = \sqrt{k}\, y_i$.

By means of equation (7.8) [in which the $g_{ij}(x, n)$ are now to be replaced by $g_{ij}(n) = g_{ij}(x)$] a positive definite Riemannian metric represented by the tensor $\gamma_{\alpha\beta}$ is induced on the hypersurface (8.1). If, in accordance with § 7, we now write

$$B^i_{\alpha\beta} = \omega_{\alpha\beta}\, n^i + B^i_\delta\, \lambda^\delta_{\alpha\beta},\qquad(8.4)$$

it follows from (7.1) and (7.8) that

$$\omega_{\alpha\beta} = n_i\, B^i_{\alpha\beta},\qquad g_{ij}(n)\, B^j_\gamma\, B^i_{\alpha\beta} = \gamma_{\gamma\delta}\, \lambda^\delta_{\alpha\beta}.\qquad(8.4\,\text{a})$$

But if we differentiate (8.3) with respect to u^β, noting (7.8) and (8.1), we find

$$\gamma_{\alpha\beta} = -\, y_i\, B^i_{\alpha\beta},$$

and hence, from (8.4 a),

$$\omega_{\alpha\beta} = \sqrt{k}\, y_i\, B^i_{\alpha\beta} = -\sqrt{k}\, \gamma_{\alpha\beta}.\qquad(8.5)$$

Also, differentiating (7.8) with respect to u^γ and taking into account the fact that the $g_{ij}(n)$ are homogeneous of degree zero in n^i, we find by a method similar to that leading up to equation (3.13a) that

$$g_{ij}\, B^i_{\alpha\beta}\, B^j_\gamma = \gamma_{\alpha\gamma\beta} - \overline{C}_{\alpha\gamma\beta}.\qquad(8.6)$$

Here the $\gamma_{\alpha\gamma\beta}$ are the Christoffel symbols of the first kind formed with respect to the $\gamma_{\alpha\beta}$, and the $\overline{C}_{\alpha\gamma\beta}$ are defined by

$$\overline{C}_{\alpha\gamma\beta}(u) = C_{ijk}(n)\, B^{ijk}_{\alpha\gamma\beta}.\qquad(8.7)$$

Thus in view of (8.4a), (8.5) and (8.6) equation (8.4) assumes the more explicit form

$$B^i_{\alpha\beta} = -k\, \gamma_{\alpha\beta}\, x^i + B^i_\delta(\gamma_\alpha{}^\delta{}_\beta - \overline{C}_\alpha{}^\delta{}_\beta).\qquad(8.8)$$

We may again introduce a $\overset{0}{\delta}_\gamma$-operator as in the previous sections: this operator is defined with respect to the $\gamma_\alpha{}^\delta{}_\beta$ of the hypersurface. It follows from the definition of the Christoffel symbols that $\overset{0}{\delta}_\varepsilon\gamma_{\alpha\beta} = 0$. A particular case of this operation is represented by

$$\overset{0}{\delta}_\beta B^i_\alpha = B^i_{\beta\alpha} - B^i_\delta\, \gamma_\alpha{}^\delta{}_\beta,$$

where we have used the fact that the x^i-coordinate system of the Min-kowskian space has been so chosen that the $\Gamma^{*\,i}_{h\,k}$ vanish identically. Equation (8.8) may now be written in the form

$$\overset{0}{\delta}_\beta B^i_\alpha = - k\,\gamma_{\alpha\beta}\,x^i - B^i_\delta\,\overline{C}^\delta_{\alpha\beta}\,, \tag{8.8a}$$

which gives in particular

$$n_i\,\overset{0}{\delta}_\beta B^i_\alpha = - k\,\gamma_{\alpha\beta}\,n_i\,x^i = - \sqrt{k}\,\,\gamma_{\alpha\beta}\,. \tag{8.8b}$$

However, in analogy with (7.20a) and (7.27) we are entitled to regard the left-hand side of (8.8b) as the coefficients of the second fundamental form: by (8.5) these are seen to be the $\omega_{\alpha\beta}$. Furthermore, if the normal curvature is defined as in (7.28), namely by putting

$$\frac{1}{R} = - \frac{\omega_{\alpha\beta}\,du^\alpha\,du^\beta}{\gamma_{\alpha\beta}\,du^\alpha\,du^\beta}\,,$$

it follows from (8.5) that $R^{-1} = \sqrt{k}$, which is not only independent of direction but is constant over the hypersurface (8.1). Thus the principal directions at each point of (8.1) are indeterminate — in fact, every point of (8.1) is a *umbilical* point[1]. We have therefore shown that all hyper-surfaces homothetic to the indicatrix possess the property that all their points are umbilical points.

On taking the $\overset{0}{\delta}_\gamma$-derivative of (8.8a) and substituting from the same equation we obtain

$$\overset{0}{\delta}_\gamma\,\overset{0}{\delta}_\beta B^i_\alpha = - k\,\gamma_{\alpha\beta}B^i_\gamma + \overline{C}^\delta_{\alpha\beta}(k\gamma_{\delta\gamma}\,x^i + B^i_\varepsilon\overline{C}^\varepsilon_{\delta\gamma}) - B^i_\delta\,\overset{0}{\delta}_\gamma\overline{C}^\delta_{\alpha\beta}\,.$$

In this equation we interchange the indices β and γ and subtract, noting that as before

$$\left(\overset{0}{\delta}_\gamma\,\overset{0}{\delta}_\beta - \overset{0}{\delta}_\beta\,\overset{0}{\delta}_\gamma\right)B^i_\alpha = + B^i_\delta K^\delta_{\alpha\gamma\beta} = - B^i_\delta K^\delta_{\alpha\beta\gamma}\,,$$

where $K^\delta_{\alpha\gamma\beta}$ represents the (Riemannian) curvature tensor defined with respect to the $\gamma_{\alpha\beta}$:

$$K^\delta_{\alpha\gamma\beta} = \frac{\partial}{\partial u^\beta}\left(\gamma_\alpha{}^\delta{}_\gamma\right) - \frac{\partial}{\partial u^\gamma}\left(\gamma_\alpha{}^\delta{}_\beta\right) + \gamma_\varepsilon{}^\delta{}_\beta\,\gamma_\alpha{}^\varepsilon{}_\gamma - \gamma_\varepsilon{}^\delta{}_\gamma\,\gamma_\alpha{}^\varepsilon{}_\beta\,.$$

In this manner we find after some simplification

$$B^i_\delta K^\delta_{\alpha\beta\gamma} = k(\gamma_{\alpha\beta}B^i_\gamma - \gamma_{\alpha\gamma}B^i_\beta) + B^i_\delta(\overline{C}^\delta_{\varepsilon\beta}\overline{C}^\varepsilon_{\alpha\gamma} - \overline{C}^\delta_{\varepsilon\gamma}\overline{C}^\varepsilon_{\alpha\beta}) +$$
$$+ B^i_\delta\left(\overset{0}{\delta}_\gamma\overline{C}^\delta_{\alpha\beta} - \overset{0}{\delta}_\beta\overline{C}^\delta_{\alpha\gamma}\right)\,. \tag{8.9}$$

In order to reduce this relation, we shall examine the last term on the right-hand side more closely. Differentiating (8.7) with respect to u^γ,

---

[1] BLASCHKE [4], § 47.

and using (8.8), we obtain

$$\frac{\partial \overline{C}_{\alpha\beta\delta}}{\partial u^\gamma} = \frac{\partial C_{ijh}}{\partial x^k}\, B^{ijhk}_{\alpha\beta\delta\gamma} + \overline{C}_{\beta\varepsilon\delta}(\gamma_\alpha{}^\varepsilon{}_\gamma - \overline{C}^\varepsilon_{\alpha\gamma}) + \overline{C}_{\varepsilon\alpha\delta}(\gamma_\beta{}^\varepsilon{}_\gamma - \overline{C}^\varepsilon_{\beta\gamma}) +$$
$$+ \overline{C}_{\alpha\beta\varepsilon}(\gamma_\delta{}^\varepsilon{}_\gamma - \overline{C}^\varepsilon_{\delta\gamma})\,.$$

This expression is substituted in the formal definition

$$\overset{0}{\delta}_\gamma \overline{C}_{\alpha\beta\delta} = \frac{\partial \overline{C}_{\alpha\beta\delta}}{\partial u^\gamma} - \overline{C}_{\varepsilon\beta\delta}\,\gamma_\alpha{}^\varepsilon{}_\gamma - \overline{C}_{\alpha\varepsilon\delta}\,\gamma_\beta{}^\varepsilon{}_\gamma - \overline{C}_{\alpha\beta\varepsilon}\,\gamma_\delta{}^\varepsilon{}_\gamma\,,$$

which gives, after simplification,

$$\overset{0}{\delta}_\gamma \overline{C}_{\alpha\beta\delta} = \frac{\partial C_{ijh}}{\partial x^k}\, B^{ijhk}_{\alpha\beta\delta\gamma} - \overline{C}^\varepsilon_{\alpha\gamma}\overline{C}_{\beta\varepsilon\delta} - \overline{C}^\varepsilon_{\beta\gamma}\overline{C}_{\alpha\varepsilon\delta} - \overline{C}^\varepsilon_{\delta\gamma}\overline{C}_{\alpha\varepsilon\beta}\,.$$

Inspection of this relation shows that $\overset{0}{\delta}_\gamma \overline{C}_{\alpha\delta\beta}$ is symmetric in the indices β and γ. It therefore follows that the third term on the right-hand side of (8.9) vanishes identically, so that this equation reduces to

$$B^i_\delta K^\delta_{\alpha\beta\gamma} = B^i_\delta \{k(\gamma_{\alpha\beta}\,\delta^\delta_\gamma - \gamma_{\alpha\gamma}\,\delta^\delta_\beta) + (\overline{C}^\delta_{\varepsilon\beta}\overline{C}^\varepsilon_{\alpha\gamma} - \overline{C}^\delta_{\varepsilon\gamma}\overline{C}^\varepsilon_{\alpha\beta})\}\,, \quad (8.10)$$

or,

$$K_{\delta\alpha\gamma\beta} = k(\gamma_{\alpha\beta}\,\gamma_{\delta\gamma} - \gamma_{\alpha\gamma}\,\gamma_{\beta\delta}) + \bar{S}_{\delta\alpha\gamma\beta}\,, \quad (8.10a)$$

where we have made use of the skew-symmetric properties of $K_{\alpha\delta\beta\gamma}$ and where we have put

$$\bar{S}_{\delta\alpha\gamma\beta} = \overline{C}^\varepsilon_{\alpha\gamma}\overline{C}_{\beta\delta\varepsilon} - \overline{C}^\varepsilon_{\alpha\beta}\overline{C}_{\gamma\delta\varepsilon}\,. \quad (8.11)$$

From this definition it is evident that the tensor $\bar{S}_{\delta\alpha\gamma\beta}$ is very similar in structure to the curvature tensor S_{jihk} as defined by (4.1.25), which in a Minkowskian space becomes

$$S_{jikh}(x) = F^2(C^m_{ik}(n)\, C_{jhm}(n) - C^m_{ih}(n)\, C_{jkm}(n))\,. \quad (8.12)$$

Furthermore, from this definition and (1.3.5) we have $S_{jihk}\, n^i = 0$ (with three similar equations), so that if we decompose the tensor (8.12) in terms of components with respect to the n vectors $(n_i,\, b^\alpha_i)$ it follows from (7.1), (7.7a), (7.9) and (8.11) that we will be left with the relation

$$S_{jikh} = F^2\, \bar{S}_{\delta\alpha\gamma\beta}\, b^{\delta\alpha\gamma\beta}_{jikh}\,. \quad (8.13)$$

For the indicatrix we have $F = 1 = k$, and hence in this case we have from (8.10a) and (8.13)

$$S_{jikh} = \{K_{\delta\alpha\gamma\beta} - (\gamma_{\alpha\beta}\,\gamma_{\delta\gamma} - \gamma_{\alpha\gamma}\,\gamma_{\beta\delta})\}\, b^{\delta\alpha\gamma\beta}_{jikh}\,. \quad (8.14)$$

Thus the curvature tensor S_{jihk} may be represented in terms of the curvature tensor resulting from the induced (Riemannian) metric on the indicatrix. Clearly, *this interpretation is still valid if we regard the Min-kowskian space M_n under consideration as a tangent space of a general Finsler space F_n,* for this would not affect the definition (8.12) of the tensor S_{jihk}[1].

---

[1] VARGA [8], p. 49. See also LAUGWITZ [6].

Let ξ^i, X^i be two vectors tangent to the hypersurface (8.1) at a given point P of (8.1). The Riemannian curvature R at P with respect to the orientation defined by ξ^i, X^i is defined to be [compare equation (4.4.23)]:

$$R(u, \xi, X) = \frac{K_{\delta\alpha\gamma\beta}\,\xi^\delta\,\xi^\gamma\,X^\alpha X^\beta}{(\gamma_{\alpha\beta}\gamma_{\delta\gamma} - \gamma_{\alpha\gamma}\gamma_{\beta\delta})\,\xi^\delta\,\xi^\gamma\,X^\alpha X^\beta}\,, \tag{8.15}$$

and thus we have from (8.10a),

$$R(u, \xi, X) = k + \frac{\overline{S}_{\delta\alpha\gamma\beta}\,\xi^\delta\,\xi^\gamma\,X^\alpha X^\beta}{(\gamma_{\alpha\beta}\gamma_{\delta\gamma} - \gamma_{\alpha\gamma}\gamma_{\beta\delta})\,\xi^\delta\,\xi^\gamma\,X^\alpha X^\beta}\,. \tag{8.16}$$

This suggests that we define a new scalar curvature S for the Minkowskian space by the equation[1]

$$S(x, \xi, X) = \frac{S_{jihk}\,\xi^j\,\xi^h\,X^i X^k}{(g_{jh}g_{ik} - g_{ji}g_{hk})\,\xi^j\,\xi^h\,X^i X^k}\,. \tag{8.17}$$

But since ξ^i and X^i are tangent to the indicatrix, it is easily verified with the aid of (7.8), (7.8a), (7.10a) and (8.13) that we have

$$S(x, \xi, X) = F^2\,\frac{\overline{S}_{\delta\alpha\gamma\beta}\,\xi^\delta\,\xi^\gamma\,X^\alpha X^\beta}{(\gamma_{\alpha\beta}\gamma_{\delta\gamma} - \gamma_{\alpha\gamma}\gamma_{\beta\delta})\,\xi^\delta\,\xi^\gamma\,X^\alpha X^\beta}\,. \tag{8.18}$$

Observing that $F^{-2} = k$, we finally deduce from (8.16) and (8.18) that

$$R = k(1 + S)\,. \tag{8.19}$$

The following basic theorem results from this relation: *The Riemannian metric induced by (7.8) onto the indicatrix of radius $1/\sqrt{k}$ of a Minkowskian space is of constant curvature if and only if the scalar curvature S of the space is constant. Further, the Riemannian curvature is constant and equal to k if and only if the scalar S vanishes*[2].

Let x^i, $x^i + dx^i$ be two points on the hypersurface (8.1), a distance ds apart as measured by the Riemannian metric on (8.1). Then, by (7.8),

$$ds^2 = \gamma_{\alpha\beta}\,du^\alpha\,du^\beta = g_{ij}(n)\,dx^i\,dx^j\,. \tag{8.20}$$

But according to the definition (1.7.10) the angle $d\varphi$ between two vectors x^i, $x^i + dx^i$ of equal length is given by

$$d\varphi^2 = \frac{g_{ij}(x)\,dx^i\,dx^j}{F^2(x)}\,, \tag{8.21}$$

and hence, since $n^i = \sqrt{k}\,x^i$, and $F^{-2}(x) = k$, we have from (8.20) and (8.21)

$$ds = \frac{d\varphi}{\sqrt{k}}\,. \tag{8.22}$$

Thus the angular metric defined by (1.7.10) is identical with the Riemannian metric induced on the indicatrix by (7.8).

---

[1] VARGA [8], p. 47. This definition is the direct analogue of (4.4.23).

[2] VARGA [8], p. 50. This is a generalisation of the fact that the metric induced on a sphere in a euclidean space is of constant curvature.

From this and the previous theorem we deduce the following corollary: *the angular metric of a Finsler space is of constant curvature unity* (corresponding to euclidean geometry) *provided the curvature tensor* S_{jihk} *vanishes*[1].

The indicatrix is considered from a more general point of view by VAGNER [13]. Consider the family of all tangent hyperplanes to a hypersurface M_{n-1} of a Minkowskian space. If this family defines a hypersurface in the dual space[2], M_{n-1} is called regular. On the other hand, should this family of hyperplanes depend on $n - r - 1$ parameters, M_{n-1} is said to be of the singularity class r. Similar definitions apply, of course, in the dual space. If M_{n-1} is given by equations of the type (8.2), and if the coefficients η_i of the tangent hyperplanes depend on the coordinates u^α of the point of contact according to a system of equations

$$\eta_i = \eta_i(u^\alpha) ,$$

a tensor $\overset{1}{g}_{\alpha\beta}$ (in the notation of VAGNER) may be defined by writing

$$\overset{1}{g}_{\alpha\beta} = \frac{\partial \eta_i}{\partial u^\alpha} B_\beta^i .$$

By means of (8.3) and (8.5) it is easily verified that in general the latter tensor corresponds to $\gamma_{\alpha\beta}$. A similar construction can be carried out in the dual space. It may then be shown that M_{n-1} is of the singularity class r if the matrix $\left\| \overset{1}{g}_{\alpha\beta} \right\|$ is of rank $n - r - 1$, with a corresponding theorem for the dual space[3]. Clearly the condition of LEGENDRE may be formulated in terms of these concepts if the hypersurface M_{n-1} is identified with the indicatrix. If the indicatrix is of the singularity class r, the Euler-Lagrange equations of the corresponding problem in the calculus of variations may be written in a canonical form involving additional parameters. When $r = 1$ it is possible to find a complete set of differential invariants. For a complete discussion of these results the reader is referred to VAGNER [10], where the first and second variations of the length integral corresponding to a singular metric are obtained, thus leading to the conditions of JACOBI and WEIERSTRASS for such metrics.

---

[1] CARTAN [1], p. 35, where this theorem is stated without reference to the indicatrix. A discussion similar to the one outlined above is given by KAWAGUCHI [5]. Equations (8.10a) and (8.19) are found by him for the case $k = 1$ ([5], pp. 176 to 177). In this paper the affine geometry of the indicatrix is also studied in a manner essentially different from that of VAGNER [13], § 1. See also DELENS [2].

[2] In the sense of § 4, Ch. I.

[3] VAGNER [13], p. 72.

§ 9. Comparison between the Induced and the Intrinsic Connection Parameters

In § 3 we remarked that the induced connection parameters — which we have been using throughout this chapter — of a subspace F_m of F_n do not, in general, coincide with the intrinsic parameters, i. e. with the parameters $'\Gamma^{\alpha}_{\beta\gamma}$ or $'\Gamma^{*\,\alpha}_{\beta\,\gamma}$ which are derived from the induced metric tensor $g_{\alpha\beta}$ and its derivatives in the same manner in which the Γ^{i}_{hk} and $\Gamma^{*\,i}_{hk}$ were derived from the tensor g_{ij} and its derivatives. We shall conclude this chapter by deriving formulae which exhibit the relationships between these coefficients.

We recall equation (3.7), which represents the relationship between the induced coefficients of F_m and those of F_n. By means of (3.1.30) we may write this relation in the form

$$\Gamma_{\alpha\beta\gamma} = g_{ij} B^i_\beta (B^j_{\alpha\gamma} + \gamma_h{}^j{}_k B^{h\,k}_{\alpha\gamma}) +$$
$$+ B^{jhk}_{\beta\alpha\gamma} \left(C_{hkr} \frac{\partial G^r}{\partial \dot{x}^j} - C_{jkr} \frac{\partial G^r}{\partial \dot{x}^h} \right) + C_{hjk} B^k_{\varepsilon\gamma} B^{jh}_{\beta\alpha} \dot{u}^\varepsilon . \tag{9.1}$$

From this equation we subtract (3.13a), and on simplification we then obtain

$$\Gamma_{\alpha\beta\gamma} = \gamma_{\alpha\beta\gamma} + C_{hkr} B^{hk}_{\alpha\gamma} \left(B^j_\beta \frac{\partial G^r}{\partial \dot{x}^j} + B^r_{\beta\varepsilon} \dot{u}^\varepsilon \right) -$$
$$- C_{hkr} B^{hk}_{\beta\gamma} \left(B^j_\alpha \frac{\partial G^r}{\partial \dot{x}^j} + B^r_{\alpha\varepsilon} \dot{u}^\varepsilon \right) . \tag{9.2}$$

If we write

$$\Omega^i = \tfrac{1}{2} B^i_{\alpha\gamma} \dot{u}^\alpha \dot{u}^\gamma + G^i , \tag{9.3}$$

so that

$$\Omega^r_\beta = \frac{\partial \Omega^r}{\partial \dot{u}^\beta} = B^j_\beta \frac{\partial G^r}{\partial \dot{x}^j} + B^r_{\beta\varepsilon} \dot{u}^\varepsilon , \tag{9.4}$$

equation (9.2) reduces to

$$\Gamma_{\alpha\beta\gamma} = \gamma_{\alpha\beta\gamma} + C_{hkr} (B^{hk}_{\alpha\gamma} \Omega^r_\beta - B^{hk}_{\beta\gamma} \Omega^r_\alpha) . \tag{9.5}$$

Also, using (3.1.27') and (3.14) we have in this notation

$$G^\alpha = B^\alpha_i \Omega^i , \tag{9.6}$$

and hence

$$\frac{\partial G^\alpha}{\partial \dot{u}^\gamma} = \frac{\partial B^\alpha_i}{\partial \dot{u}^\gamma} \Omega^i + B^\alpha_i \Omega^i_\gamma . \tag{9.7}$$

According to (3.1.30) the definition of the *intrinsic* connection parameters reads[1]

$$'\Gamma_{\alpha\beta\gamma} = \gamma_{\alpha\beta\gamma} - C_{\beta\gamma\delta} \frac{\partial G^\delta}{\partial \dot{u}^\alpha} + C_{\alpha\delta\gamma} \frac{\partial G^\delta}{\partial \dot{u}^\beta} . \tag{9.8}$$

---

[1] Note that due to (3.15) there is no distinction between the intrinsic $'G^\alpha$ and the induced G^α, so that no distinctive notation is required. In view of (2.22) the same applies to the $C_{\alpha\beta\gamma}$.

In virtue of (9.7) and (2.22) this becomes

$$'\Gamma_{\alpha\beta\gamma} = \gamma_{\alpha\beta\gamma} + C_{hkr}\left\{ B_{\alpha\gamma}^{hk}\left(B_{\delta}^{r}\,\frac{\partial B_{i}^{\delta}}{\partial \dot{u}^{\beta}}\,\Omega^{i} + B_{\delta}\,B_{i}^{\delta}\,\Omega_{\beta}^{i}\right) - B_{\beta\gamma}^{hk}\left(B_{\delta}^{r}\,\frac{\partial B_{i}^{\delta}}{\partial \dot{u}^{\alpha}}\,\Omega^{i} + B_{\delta}^{r}\,B_{i}^{\delta}\,\Omega_{\alpha}^{i}\right)\right\}.$$

(9.9)

But if we write

$$2\,H^{i} = N_{j}^{i}\overset{0}{H}_{\alpha\beta}^{j}\,\dot{u}^{\alpha}\,\dot{u}^{\beta}\,,$$

(9.10)

we deduce from (4.11), (4.33) and (1.3.5) that

$$H^{i} = \tfrac{1}{2}\,N_{j}^{i}(B_{\alpha\beta}^{j}\,\dot{u}^{\alpha}\,\dot{u}^{\beta} + \Gamma_{hk}^{*j}\,B_{\alpha\beta}^{hk}\,\dot{u}^{\alpha}\,\dot{u}^{\beta}) = N_{j}^{i}\,\Omega^{j}\,.$$

(9.11)

Using (2.18) and (9.4) it follows by differentiation of this result that

$$B_{\delta}^{i}\,\frac{\partial B_{j}^{\delta}}{\partial \dot{u}^{\alpha}}\,\Omega^{j} = N_{j}^{i}\,\Omega_{\alpha}^{j} - \frac{\partial H^{i}}{\partial \dot{u}^{\alpha}}\,.$$

(9.12)

If this relation is substituted in (9.9) and (2.18) applied once more, it will be noticed that a set of four terms drops out, and the final form of (9.9) is then as follows:

$$'\Gamma_{\alpha\beta\gamma} = \gamma_{\alpha\beta\gamma} + C_{hkr}(B_{\alpha\gamma}^{hk}\,\Omega_{\beta}^{r} - B_{\beta\gamma}^{hk}\,\Omega_{\alpha}^{r}) + C_{hkr}\left(B_{\beta\gamma}^{hk}\,\frac{\partial H^{r}}{\partial \dot{u}^{\alpha}} - B_{\alpha\gamma}^{hk}\,\frac{\partial H^{r}}{\partial \dot{u}^{\beta}}\right).$$

(9.13)

Comparing this result with (9.5), we obtain[1]

$$'\Gamma_{\alpha\beta\gamma} - \Gamma_{\alpha\beta\gamma} = C_{hkr}\left(B_{\beta\gamma}^{hk}\,\frac{\partial H^{r}}{\partial \dot{u}^{\alpha}} - B_{\alpha\gamma}^{hk}\,\frac{\partial H^{r}}{\partial \dot{u}^{\beta}}\right).$$

(9.14)

In particular, it follows that

$$('\Gamma_{\alpha\beta\gamma}(u,\,u') - \Gamma_{\alpha\beta\gamma}(u,\,u'))\,u'^{\gamma} = 0\,,$$

(9.14a)

and *thus the process of covariant differentiation is the same for both connections if the displacement coincides with the direction of the element of support*[2].

Equation (9.14) suffers from the defect that it contains the derivatives $\partial H^{i}/\partial \dot{u}^{\gamma}$ (which would have to be evaluated in each case), while it involves the $\Gamma_{\alpha\beta\gamma}$ instead of the $\Gamma_{\alpha\beta\gamma}^{*}$. We shall therefore derive equivalent formulae which will remedy this state of affairs, although their final form is not quite so simple. Again, for the intrinsic connection we have from (3.1.29)

$$'\Gamma_{\alpha\beta\gamma}^{*} = '\Gamma_{\alpha\beta\gamma} - C_{\alpha\beta\delta}\,\frac{\partial G^{\delta}}{\partial \dot{u}^{\gamma}}\,,$$

(9.15)

---

[1] This formula is stated in a different form (without proof) by Hombu [3], p. 71. The writer is indebted to E. T. Davies for having drawn his attention to this particular form and for providing a simple alternative proof in a personal communication.

[2] This result is already stated by Taylor [3] in a more restricted form, since the process of covariant differentiation of Taylor [1] and Synge [1] along a curve is identical with Cartan's process.

or, in view of (9.7),

$$'T^*_{\alpha\beta\gamma} = 'T_{\alpha\beta\gamma} - C_{\alpha\beta\delta}\left(\frac{\partial B_i^\delta}{\partial \dot{u}^\gamma}\,\Omega^i + B_i^\delta\,\Omega_{,\gamma}^i\right).$$
$$\tag{9.16}$$

But it follows from (3.1.29), (3.7) and (3.10a) that

$$\Gamma^*_{\alpha\beta\gamma} - \Gamma_{\alpha\beta\gamma} = -C_{hjr}\left(B_\gamma\frac{\partial G^r}{\partial \dot{x}^k} + B^r_{\varepsilon\gamma}\,\dot{u}^\varepsilon\right)B^{jh}_{\beta\alpha},$$

which, as a result of (9.4), reduces to

$$\Gamma^*_{\alpha\beta\gamma} = \Gamma_{\alpha\beta\gamma} - C_{hjr}B^{jh}_{\beta\alpha}\,\Omega^r_{,\gamma}.$$
$$\tag{9.17}$$

Thus equations (9.16), (9.17) and (2.22) yield the relation

$$'T^*_{\alpha\beta\gamma} - \Gamma^*_{\alpha\beta\gamma} = 'T_{\alpha\beta\gamma} - \Gamma_{\alpha\beta\gamma} - C_{hjr}B^{hjr}_{\alpha\beta\delta}\left(\frac{\partial B_i^\delta}{\partial \dot{u}^\gamma}\,\Omega^i + B_i^\delta\,\Omega_{,\gamma}^i\right) +$$
$$+ C_{hjr}B^{hj}_{\alpha\beta}\,\Omega^r_{,\gamma},$$

which may be simplified by means of (2.18) and (9.12) to

$$'T^*_{\alpha\beta\gamma} - \Gamma^*_{\alpha\beta\gamma} = 'T_{\alpha\beta\gamma} - \Gamma_{\alpha\beta\gamma} + C_{hjr}B^{hj}_{\alpha\beta}\frac{\partial H^r}{\partial \dot{u}^\gamma}.$$
$$\tag{9.18}$$

If we apply (9.14) to this result, we find

$$'T^*_{\alpha\beta\gamma} - \Gamma^*_{\alpha\beta\gamma} = C_{hkj}\left(B^{hk}_{\beta\gamma}\frac{\partial H^j}{\partial \dot{u}^\alpha} + B^{hk}_{\alpha\beta}\frac{\partial H^j}{\partial \dot{u}^\gamma} - B^{hk}_{\gamma\alpha}\frac{\partial H^j}{\partial \dot{u}^\beta}\right).$$
$$\tag{9.19}$$

This equation represents the required relation between the "starred" connection coefficients, and it only remains to eliminate the directional derivatives of the H^j. If we evaluate these derivatives by means of (9.12) and (2.23), the first term on the right-hand side of (9.19) becomes

$$B^{hk}_{\beta\gamma}\frac{\partial H^j}{\partial \dot{u}^\alpha}\,C_{hkj} = -B^{hk}_{\beta\gamma}C_{hkr}(2g^{\varepsilon\delta}B^{slr}_{\delta\alpha\varepsilon}C_{lsm}N^m_j\,\Omega^j - N^r_j\,\Omega^j_\alpha),$$

and in virtue of (2.22) this may be simplified to give

$$B^{hk}_{\beta\gamma}\frac{\partial H^j}{\partial \dot{u}^\alpha}\,C_{hkj} = C_{hkr}N^r_j\{B^{hk}_{\beta\gamma}\,\Omega^j_\alpha - 2C^\delta_{\beta\gamma}B^{hk}_{\delta\alpha}\,\Omega^j\}.$$
$$\tag{9.20}$$

In analogy with equations (7.69) and (7.70) we shall write

$$\widetilde{\Omega}_{\alpha\beta} = N_i I^i_{\alpha\beta} = N_i\{B^i_{\alpha\beta} + \Gamma^{*i}_{hk}B^{hk}_{\alpha\beta}\},$$
$$_\mu_\mu_\mu$$
$$\tag{9.21}$$

so that we have from (9.3) and (9.4)

$$\widetilde{\Omega}_{\alpha\beta}\,\dot{u}^\beta = N_i\left\{B^i_{\alpha\beta}\,\dot{u}^\beta + B^h_\alpha\frac{\partial G^i}{\partial \dot{x}^h}\right\} = N_i\,\Omega^i_\alpha,$$
$$_\mu_\mu_\mu$$

and

$$\widetilde{\Omega}_{\alpha\beta}\,\dot{u}^\alpha\,\dot{u}^\beta = N_i\{B^i_{\alpha\beta}\,\dot{u}^\alpha\,\dot{u}^\beta + 2G^i\} = 2N_i\,\Omega^i.$$
$$_\mu_\mu_\mu$$

Hence, by (2.19), it follows that

$$2N^r_j\,\Omega^j = \sum_\mu N^r\,\widetilde{\Omega}_{\alpha\beta}\,\dot{u}^\alpha\,\dot{u}^\beta,$$
$$_\mu_\mu_\mu$$
$$\tag{9.22}$$

together with

$$N^r_j \Omega^j_\alpha = \sum_\mu N^r_\mu \left(\widetilde{\Omega}_{\alpha\beta} \, \dot{u}^\beta \right). \tag{9.23}$$

On substituting (9.22) and (9.23) in (9.20), we find

$$B^{hk}_{\beta\gamma} \frac{\partial H^j}{\partial \dot{u}^\alpha} C_{hkj} = \sum_\mu N^j_\mu C_{hkj} \left\{ B^{hk}_{\beta\gamma} \widetilde{\Omega}_{\alpha\varepsilon} \, \dot{u}^\varepsilon - C^\delta_{\beta\gamma} B^{hk}_{\delta\alpha} \widetilde{\Omega}_{\varepsilon\lambda} \, \dot{u}^\varepsilon \, \dot{u}^\lambda \right\}. \tag{9.24}$$

Two similar equations result from this by cyclic interchange of the indices α, β, γ. Substituting these together with (9.24) in (9.19), we finally have the desired relation

$$'\Gamma^*_{\alpha\beta\gamma} - \Gamma^*_{\alpha\beta\gamma} = \sum_{\mu=m+1}^n N^j_\mu C_{hkj} \left\{ \left(B^{hk}_{\beta\gamma} \widetilde{\Omega}_{\alpha\varepsilon} + B^{hk}_{\alpha\beta} \widetilde{\Omega}_{\gamma\varepsilon} - B^{hk}_{\gamma\alpha} \widetilde{\Omega}_{\beta\varepsilon} \right) \dot{u}^\varepsilon - \right.$$
$$\left. - (C^\delta_{\beta\gamma} B^{hk}_{\delta\alpha} + C^\delta_{\alpha\beta} B^{hk}_{\delta\gamma} - C^\delta_{\gamma\alpha} B^{hk}_{\delta\beta}) \widetilde{\Omega}_{\varepsilon\lambda} \, \dot{u}^\varepsilon \, \dot{u}^\lambda \right\}. \tag{9.25}$$

The differential geometry of F_m (in relation to F_n) with respect to the intrinsic parameters $'\Gamma^{*\,\alpha}_{\beta\gamma}$ does not appear to have been studied at any time. In view of the complexity of the last relations such an investigation promises to be fairly difficult, especially since these relations do not permit us to derive a simple tensor such as $I^i_{\alpha\beta}$ whose orthogonality properties (7.22) follow directly from the relation (3.10a) between the connection parameters of F_n and the induced parameters of F_m; yet we recall that precisely these orthogonality properties lead to the second fundamental forms. It is apparent, therefore, *that there are serious obstacles to be overcome in an investigation into the relationship between the theory outlined above and the intrinsic differential geometry of the subspace F_m*[1].

---

[1] We observe that neither of the definitions of Gaussian curvature as given by BERWALD [equation (6.26)] or by FINSLER [equation (6.40)] are "intrinsic", and it would seem very doubtful to the writer whether a Gaussian "Theorema Egregium" would hold for these invariants.

Chapter VI

Miscellaneous Topics

In the first five chapters of this monograph we have endeavoured to present a general outline of the most basic aspects of Finsler geometry. There are, however, a large number of topics which we have left virtually untouched, and it is the purpose of the present chapter to fill in these gaps. The subjects to be treated are mostly unrelated to each other, and for the purpose of convenient orientation on the part of the reader the various sections of this chapter have been written in such a manner that they may be read independently, although, of course, reference will be made to earlier chapters. We shall concern ourselves briefly with the theory of groups of motions, conformal geometry, certain aspects of the equivalence problem, imbedding theories and the geometry of two-dimensional Finsler spaces. At the end of this chapter we have listed bibliographical references which are relevant to topics which could not conveniently be treated within the framework of this monograph.

§ 1. Groups of Motions

In our study of the theory of groups of motions in a Finsler space our analysis will be based chiefly on the connection coefficients G^i_{hk} of Berwald rather than on the Γ^{*i}_{hk} of CARTAN. This choice is made in view of the close affinity between the former coefficients and the general geometry of paths, but it should be remarked that it is no more difficult to develop the present theory on the basis of the latter coefficients[1].

The definition of a motion in a Finsler space F_n is similar to that prevalent in Riemannian geometry[2]. Consider a mapping of a region R of F_n onto another region $\overline{R}$ of the same space F_n, this mapping being represented analytically by the relations

$$\overline{x}^i = \psi^i\left(x^k\right) ,$$

where $\overline{x}^k$ and x^k are the coordinates of corresponding points of the regions $\overline{R}$ and R respectively. It is assumed that the functions $\psi^i\left(x^k\right)$ are at least of class C^3. Let $P\left(x^r\right)$, $Q\left(x^r + d x^r\right)$ be neighbouring points interior to R and denote by ds the length $F\left(x^r, d x^r\right)$ of the displacement $d x^r$. To the

---

[1] In fact, this has been done by Soós [1].

[2] Cf. EISENHART [1], § 27.

points P, Q there will correspond points $\overline{P}(\overline{x}^r)$, $\overline{Q}(\overline{x}^r + d\overline{x}^r)$ of $\overline{R}$, where

$$d\overline{x}^r = \frac{\partial \psi^r}{\partial x^s} \, dx^s \, .$$

The length $d\overline{s}$ of this displacement is $F(\overline{x}^r, d\overline{x}^r)$; and, in general, $d\overline{s}$ differs from ds. However, should the space F_n admit the construction of a mapping which preserves such arc-lengths, we would have

$$F(x^r, dx^r) = F\left(\psi^r(x^s), \frac{\partial \psi^r}{\partial x^s} \, dx^s\right)$$

for arbitrary displacements dx^r. Thus, writing $dx^r = \dot{x}^r \, dt$, and differentiating this condition successively with respect to $\dot{x}^h$ and $\dot{x}^k$, we obtain, by means of (1.3.1),

$$g_{hk}(x^r, \dot{x}^r) = g_{ij}\left(\psi^r(x^s), \frac{\partial \psi^r}{\partial x^s} \, \dot{x}^s\right) \frac{\partial \psi^i}{\partial x^h} \frac{\partial \psi^j}{\partial x^k} \, ,$$

where it is to be noted that the g_{hk} on the left-hand side and the g_{ij} on the right are the same functions of $(x^r, \dot{x}^r)$ and $(\psi^r, (\partial \psi^r/\partial x^s)\,\dot{x}^s)$ respectively. Thus, when the fundamental metric function of F_n is such that these conditions admit a solution of the type

$$\overline{x}^i = \psi^i(x^1, x^2, \ldots, x^n)$$

involving one or more parameters, we say that these equations define a *motion* of F_n into itself; i. e. we regard a motion as a member of a (continuous) group of point-transformations which preserves the arc-length of every curve. In particular, it follows from the definition of geodesics that a motion carries geodesics into geodesics.

Let us consider the effect of an infinitesimal transformation such as (5.5.1), namely the transformation

$$\overline{x}^i = \psi^i(x^j) = x^i + v^i(x^r) \, d\tau \, . \tag{1.1}$$

We note that in this case we have

$$\frac{\partial \psi^i}{\partial x^h} = \delta_h^i + \frac{\partial v^i}{\partial x^h} \, d\tau \, , \tag{1.1a}$$

together with

$$\dot{\overline{x}}^i = \frac{\partial \psi^i}{\partial x^h} \, \dot{x}^h = \dot{x}^i + \frac{\partial v^i}{\partial x^h} \, \dot{x}^h \, d\tau \, . \tag{1.1b}$$

The above condition which must be satisfied in order that (1.1) represents a motion may now be written in the form

$$g_{hk}(x^r, \dot{x}^r) = g_{ij}\left(x^r + v^r \, d\tau, \, \dot{x}^r + \frac{\partial v^r}{\partial x^s} \, \dot{x}^s \, d\tau\right)\left(\delta_h^i + \frac{\partial v^i}{\partial x^h} \, d\tau\right)\left(\delta_k^j + \frac{\partial v^j}{\partial x^k} \, d\tau\right).$$

Introducing the operators $\overset{v}{d}$, $\overset{m}{d}$, $\underset{L}{D}$ as defined in § 5 of Chapter V, we see that this condition reads

$$g_{hk}(x^r, \dot{x}^r) = [g_{ij}(x^r, \dot{x}^r) + \overset{v}{d}g_{ij}]\left(\delta_h^i \delta_k^j + \delta_h^i \frac{\partial v^j}{\partial x^k}d\tau + \delta_k^j \frac{\partial v^i}{\partial x^h}d\tau\right),$$

or

$$0 = \overset{v}{d}g_{hk} + \left(g_{hj}\frac{\partial v^j}{\partial x^k} + g_{ik}\frac{\partial v^i}{\partial x^h}\right)d\tau,$$

where we have neglected powers of $d\tau$ higher than the first. However, by definition,

$$\overset{m}{d}g_{hk} = \bar{g}_{hk} - g_{hk} = \bar{g}_{hk} - \bar{g}_{ij}\left(\delta_h^i + \frac{\partial v^i}{\partial x^h}d\tau\right)\left(\delta_k^j + \frac{\partial v^j}{\partial x^k}d\tau\right)$$

$$= -\left(g_{hj}\frac{\partial v^j}{\partial x^k} + g_{ik}\frac{\partial v^i}{\partial x^h}\right)d\tau,$$

so that our condition finally reduces to

$$0 = \overset{v}{d}g_{hk} - \overset{m}{d}g_{hk} = \left(\underset{L}{D}g_{hk}\right)d\tau,$$

by definition of Lie derivative. We therefore conclude that if the field $v^i(x)$ is to define an infinitesimal motion, it must satisfy the $\frac{1}{2}n(n+1)$ equations (5.5.9) of KILLING:

$$\underset{L}{D}g_{ij}(x, \dot{x}) = 0. \tag{1.2}$$

By means of the expression (5.5.8) for the Lie derivative of the metric tensor we may express these conditions in the form[1]

$$\underset{L}{D}g_{ij} = g_{ih}v^h{}_{|j} + g_{hj}v^h{}_{|i} + 2C_{ijh}v^h{}_{|r}\dot{x}^r = 0. \tag{1.2a}$$

On expanding the covariant derivatives on the right-hand side according to their definition, we have

$$g_{ih}\frac{\partial v^h}{\partial x^j} + g_{jh}\frac{\partial v^h}{\partial x^i} + v^k(\Gamma^*_{jik} + \Gamma^*_{ijk} + 2C_{ijh}\Gamma^{*h}_{kr}\dot{x}^r) +$$
$$+ 2C_{ijh}\frac{\partial v^h}{\partial x^r}\dot{x}^r = 0. \tag{1.3}$$

But from (3.1.7) and (3.1.29) it follows that

$$\Gamma^*_{jik} + \Gamma^*_{ijk} = \frac{\partial g_{ij}}{\partial x^k} - 2C_{ijh}\Gamma^{*h}_{kr}\dot{x}^r,$$

and hence (1.3) reduces to the form

$$\underset{L}{D}g_{ij} = g_{ih}\frac{\partial v^h}{\partial x^j} + g_{jh}\frac{\partial v^h}{\partial x^i} + \frac{\partial g_{ij}}{\partial x^k}v^k + 2C_{ijh}\frac{\partial v^h}{\partial x^r}\dot{x}^r = 0. \tag{1.4}$$

---

[1] The form (1.2a) of the Killing equation is given by HOKARI [1] and SOÓS [1]. See also AKBAR-ZADEH [2], where an ingenious decomposition of the Lie derivative is used.

This form of Killing's equations is due to KNEBELMAN[1], who derived these conditions by means of a direct calculation based on the definition of motion. By eliminating the derivatives $\partial v^h/\partial x^i$ between equations (1.4) and the definition of Berwald's derivative $v^h_{(i)}$, we may express (1.4) as

$$g_{ih} v^h_{(j)} + g_{jh} v^h_{(i)} + 2C_{ijh} v^h_{(r)} \dot{x}^r +$$
$$+ v^k \left(\frac{\partial g_{ij}}{\partial x^k} - \frac{\partial g_{ij}}{\partial \dot{x}^h} G^h_{kr} \dot{x}^r - g_{ih} G^h_{jk} - g_{hj} G^h_{ik} \right) = 0 ,$$

or

$$\underset{L}{D} g_{ij} = g_{ih} v^h_{(j)} + g_{jh} v^h_{(i)} + g_{ij(k)} v^k + 2C_{ijh} v^h_{(r)} \dot{x}^r = 0 . \qquad (1.5)$$

Again, as in Riemannian geometry[2] the local coordinates of the Finsler space F_n may be chosen such that the contravariant components of the field defining the infinitesimal transformation (1.1) are given by

$$v^i = \delta^i_1 , \qquad (1.6)$$

and hence the finite transformation generated by (1.1) may, in this coordinate system, be written in the form

$$\bar{x}^i = x^i + \delta^i_1 \tau . \qquad (1.6a)$$

Thus equations (1.4) now reduce to

$$\frac{\partial g_{ij}(x, \dot{x})}{\partial x^1} = 0 , \qquad (i, j = 1, \ldots, n) , \qquad (1.7)$$

which implies by (1.3.1) that

$$\frac{\partial F(x, \dot{x})}{\partial x^1} = 0 . \qquad (1.7a)$$

Conversely, if equations (1.7) are satisfied, a solution of (1.4) is given by (1.6). Thus we have the following result[3]: *The most general Finsler space admitting an infinitesimal motion may be obtained by choosing a metric function $F(x, \dot{x})$ which in some coordinate system is a function only of* $x^2, \ldots, x^n; \dot{x}^1, \dot{x}^2, \ldots, \dot{x}^n.$

Furthermore, if the g_{ij} are independent of x^1, it follows from (1.7a) that length is invariant under the finite transformation (1.6a). *Hence a Finsler space admitting an infinitesimal motion (1.1) admits the one-parameter finite continuous group G_1 of motions generated by (1.1).*

---

[1] KNEBELMAN [1], p. 557.

[2] EISENHART [1], p. 223; T. Y. THOMAS [1], p. 35.

[3] This result is due to KNEBELMAN ([1], p. 557), who stipulates in addition that $\partial F/\partial \dot{x}^1 \neq 0$. But this assumption is already contained in condition B of § 1, Ch. I; for if $\partial F/\partial \dot{x}^1$ were to vanish, we would have

$$\frac{\partial F(x, v)}{\partial \dot{x}^j} v^j = F(x, v) = 0 ,$$

as a result of (1.6). Cf. also Soós [1], p. 78.

The vector field $v^i(x)$ defines a congruence of curves in F_n, the so-called *paths* of the motion, each of which is described by a point as the latter is subjected to an infinitesimal displacement in accordance with (1.1). Let us consider two linearly independent motions, and suppose that the field defining the first motion is normalised according to (1.6). The second motion can possess the same paths if and only if its defining field is parallel to the v^i, i. e. if this field is of the form $\varphi(x)\, v^i = \varphi(x)\, \delta^i_1$. For this motion, then, the Killing equations (1.4) are of the form

$$g_{i1}\frac{\partial\varphi}{\partial x^j} + g_{j1}\frac{\partial\varphi}{\partial x^i} + 2C_{ij1}\frac{\partial\varphi}{\partial x^r}\dot{x}^r = 0$$

in view of (1.7). Multiplying by $\dot{x}^i$, $\dot{x}^j$ and taking into account (1.3.5) we deduce that

$$g_{i1}(x,\dot{x})\,\dot{x}^i\,\frac{d\varphi}{dt} = F\frac{\partial F}{\partial\dot{x}^1}\frac{d\varphi}{dt} = 0,$$

and hence, by condition B (Ch. I, § 1),

$$\frac{d\varphi}{dt} = 0.$$

But if φ is constant the two motions are not linearly independent, and hence we have the result that *two linearly independent motions cannot have the same paths.*

Following KNEBELMAN, we shall call a motion a *translation* if its paths are geodesics[1], and we shall now seek the conditions under which a Finsler space admits a one-parameter group of translations. Firstly, we shall have to discover under which circumstances the parameter curves of x^1 (i. e. the curves $x^2 = c^2$, $x^3 = c^3, \ldots, x^n = c^n$) shall be geodesics. It is easily seen that we may choose x^1 to be the arc-length s of these curves, and if the latter satisfy the differential equations (2.2.8) of the geodesics of F_n, it follows that in this special coordinate system we must have $\gamma_1{}^i{}_1 = 0$, or $\gamma_{1i1} = 0$ along such geodesics. As a result of (2.2.3) these conditions read

$$2\frac{\partial g_{1h}}{\partial x^1} - \frac{\partial g_{11}}{\partial x^h} = 0,$$

and since $g_{11} = 1$ (as a result of $x^1 = s$) along these geodesics, it follows that

$$\frac{\partial g_{1h}}{\partial x^1} = 0.$$

Hence by (1.3.1) the function $F(x,\dot{x})$ must be of the form

$$F^2(x,\dot{x}) = a(\dot{x}^1)^2 + \dot{x}^1\varphi_1(x^2,\ldots,x^n;\dot{x}^2,\ldots,\dot{x}^n) + \\ + \varphi_2(x^2,\ldots,x^n;\dot{x}^2,\ldots,\dot{x}^n), \tag{1.8}$$

---

[1] In Riemannian geometry translations are defined as motions for which each point moves through the same distance, and the definition given here is proved as a theorem. But by means of a construction similar to that of Ch. II, § 1, it is easy to prove for Finsler spaces that if the paths of a motion are geodesics, each point moves through the same distance.

where a is a constant and where φ_1 and φ_2 are homogeneous of degree 1 and 2 in $\dot{x}^2, \ldots, \dot{x}^n$ respectively. Conversely, if F is of the form (1.8) it is seen that the parameter curves of x^1 are geodesics, so that this condition on F is also sufficient. Comparing this result with the first theorem of this section we therefore have: *A Finsler space admitting a finite continuous group G_1 of translations may be obtained by choosing a metric function $F(x, \dot{x})$ for which there exists a coordinate system in which it is expressible in the form (1.8)*[1].

Also, it is possible to show that the following theorem, which is well-known in Riemannian geometry[2], is also true for Finsler spaces: *The cosine* [as defined by (1.6.8) with respect to the tangent vector of a geodesic] *between the tangent vectors of the paths of a translation and any geodesic does not vary along the geodesic.*

Returning to the Killing equations (1.2), we note that since they consist of $\frac{1}{2}n(n+1)$ independent relations we cannot solve these equations for the $v^i_{(h)}$ in terms of v^i and g_{ij}. However, we recall that a motion carries a geodesic into a geodesic. Writing the equations of the latter in the form

$$\frac{d^2 x^i}{ds^2} + 2G^i\left(x, \frac{dx}{ds}\right) = \frac{d^2 x^i}{ds^2} + G^i_{hk}\left(x, \frac{dx}{ds}\right)\frac{dx^h}{ds}\frac{dx^k}{ds} = 0,$$

it is found by means of an analysis directly analogous to that leading to (1.2) that the conditions

$$\underset{L}{D}\, G^i_{jk}(x, \dot{x}) = 0 \tag{1.9}$$

must therefore be satisfied. We remark that the evaluation of these Lie derivatives presents no difficulty: in fact we may transcribe almost word for word the calculation of the Lie derivatives of the Γ^{*i}_{hk} as presented in § 5 of Chapter V. This is possible since the G^i_{hk} satisfy the same transformation laws as the Γ^{*i}_{hk}. Hence, noting (5.5.12), we see that condition (1.9) can be written in the form

$$v^i_{(j)(k)} + H^i_{jkh}\, v^h + G^i_{jkh}\, v^h_{(r)}\, \dot{x}^r = 0. \tag{1.9a}$$

Thus the problem of finding the components of a motion reduces to the solution of (1.9a) subject to the Killing equations (1.2). The dependent variables in these equations are the $n(n+1)$ functions v^i and $v^i_{(j)}$, while the $\frac{1}{2}n(n+1)$ conditions (1.2) have to be satisfied. Thus *the greatest number of linearly independent motions which may be admitted by a Finsler space F_n is $\frac{1}{2}n(n+1)$*[3].

---

[1] Knebelman [1], p. 559.

[2] Eisenhart [1], p. 239; Knebelman, loc. cit.

[3] Knebelman [1], p. 561.

In conclusion we shall now derive the conditions of integrability of (1.9), for these have to be used in the determination of the conditions which must be satisfied in order that the space F_n admits $r \leq \frac{1}{2} n (n + 1)$ linearly independent motions. The following derivation will also furnish the reader with an example which illustrates a general method which is frequently used in the determination of integrability conditions of fairly complicated relations such as (1.9).

We note that from (4.6.10a) we have

$$v^{\,i}_{(j)\,(k)\,(l)} - v^{\,i}_{(j)\,(l)\,(k)} = - \frac{\partial v^i_{(j)}}{\partial \dot{x}^m} H^m_{kl} - v^i_{(m)} H^m_{jkl} + v^m_{(j)} H^i_{mkl} \,.$$

Taking the covariant derivative of (1.9a) with respect to x^l, and writing down a further relation resulting from the interchange of the indices k and l, we may eliminate the third covariant derivatives of v^i by means of the above commutation formula. This gives

$$\frac{\partial v^i_{(j)}}{\partial \dot{x}^m} H^m_{rkl} \dot{x}^r + v^i_{(m)} H^m_{jkl} - v^m_{(j)} H^i_{mkl}$$
$$= v^h (H^i_{jkh(l)} - H^i_{jlh(k)}) + v^h_{(l)} H^i_{jkh} - v^h_{(k)} H^i_{jlh} + \tag{1.10}$$
$$+ (G^i_{jkh(l)} - G^i_{jlh(k)}) v^h_{(r)} \dot{x}^r + G^i_{jkh} v^h_{(r)(l)} \dot{x}^r -$$
$$- G^i_{jlh} v^h_{(r)(k)} \dot{x}^r \,.$$

This relation may be simplified considerably. Firstly, by writing out the full expression for $v^i_{(j)}$, we see that

$$\frac{\partial v^i_{(j)}}{\partial \dot{x}^m} = G^i_{hjm} v^h \,. \tag{1.11}$$

Secondly, using the Bianchi identities (4.6.13c), we find that

$$v^h (H^i_{jkh(l)} - H^i_{jlh(k)} + G^i_{hjr} H^r_{lk})$$
$$= - v^h (H^i_{jlk(h)} + H^r_{kh} G^i_{rjl} + H^r_{hl} G^i_{rjk}) \,. \tag{1.12}$$

Thirdly, if we write out the full expression for the covariant derivative $G^i_{jkh(l)}$, interchange the indices k and l, and subtract the two relations thus obtained from each other, we obtain

$$G^i_{jkh(l)} - G^i_{jlh(k)} = \frac{\partial G^i_{jkh}}{\partial x^l} - \frac{\partial G^i_{jhl}}{\partial x^k} -$$
$$- \frac{\partial G^r}{\partial \dot{x}^l} G^i_{jkhr} + \frac{\partial G^r}{\partial \dot{x}^k} G^i_{jhlr} + G^i_{rl} G^r_{jkh} - G^i_{rk} G^r_{jlh} -$$
$$- G^r_{jl} G^i_{rkh} + G^r_{jk} G^i_{,lh} - G^r_{hl} G^i_{jkr} + G^r_{hk} G^i_{jlr} \,.$$

However, if we express H^i_{jkl} according to definition (4.6.7), and differentiate this expression with respect to $\dot{x}^h$, it is found that this derivative assumes precisely the form of the right-hand side above. Hence we have proved the useful identity

$$\frac{\partial H^i_{jkl}}{\partial \dot{x}^h} = G^i_{jkh(l)} - G^i_{jlh(k)} \,. \tag{1.13}$$

On substituting (1.11), (1.12), (1.13) (in this order) in (1.10), we see that (1.10) is equivalent to

$$v^i_{(h)}H^h_{jkl} - v^h_{(j)}H^i_{hkl} - H^i_{jkh}v^h_{(l)} + H^i_{jlh}v^h_{(k)} - v^h_{(r)}\dot{x}^r\frac{\partial H^i_{jkl}}{\partial \dot{x}^h} \tag{1.14}$$
$$= -v^h H^i_{jlk(h)} - G^i_{rjl}(H^r_{kh}v^h + v^r_{(s)(k)}\dot{x}^s) - G^i_{rjk}(H^r_{hl}v^h - v^r_{(s)(l)}\dot{x}^s)\ .$$

But from (1.9a) and (4.6.8a) we have

$$v^r_{(s)(k)}\dot{x}^s = -H^r_{kh}v^h\ ,$$

together with

$$v^r_{(s)(l)}\dot{x}^s = -H^r_{lh}v^h = H^r_{hl}v^h\ .$$

Thus the last four terms on the right-hand side of (1.14) drop out, and we finally have

$$v^h H^i_{jkl(h)} + v^h_{(r)}\dot{x}^r\frac{\partial H^i_{jkl}}{\partial \dot{x}^h} + v^h_{(l)}H^i_{jkh} + $$
$$+ v^h_{(k)}H^i_{jhl} - v^i_{(h)}H^h_{jkl} + v^h_{(j)}H^i_{hkl} = 0\ . \tag{1.15}$$

This equation represents the first integrability condition of (1.9a), obtained by covariant differentiation. However, we must also find a second integrability condition which results from (1.9a) by partial differentiation with respect to direction. This is done as follows: From (4.6.11a) we have

$$\left(\frac{\partial v^i_{(j)}}{\partial \dot{x}^l}\right)_{(k)} - \frac{\partial v^i_{(j)(k)}}{\partial \dot{x}^l} = v^i_{(r)}G^r_{jlk} - v^r_{(j)}G^i_{rlk}\ ,$$

and, hence, differentiating (1.9a) with respect to $\dot{x}^l$, we find

$$-\left(\frac{\partial v^i_{(j)}}{\partial \dot{x}^l}\right)_{(k)} + v^i_{(h)}G^h_{jlk} - v^h_{(j)}G^i_{hlk}$$
$$= \frac{\partial H^i_{jkh}}{\partial \dot{x}^l}v^h + \frac{\partial G^i_{jkl}}{\partial \dot{x}^h}v^h_{(r)}\dot{x}^r + G^i_{jkh}\frac{\partial v^h_{(r)}}{\partial \dot{x}^l}\dot{x}^r + G^i_{jhk}v^h_{(l)}\ . \tag{1.16}$$

By means of (1.11) and (4.6.8a) we deduce firstly that

$$\frac{\partial v^h_{(r)}}{\partial \dot{x}^l}\dot{x}^r = 0\ , \tag{1.17}$$

and secondly that

$$\left(\frac{\partial v^i_{(j)}}{\partial \dot{x}^l}\right)_{(k)} = G^i_{hjl(k)}v^h + G^i_{jhl}v^h_{(k)}\ . \tag{1.18}$$

The relations (1.17) and (1.18) together with (1.13) are substituted in (1.16). After simplification (1.16) then assumes its final form:

$$v^h G^i_{jkl(h)} + \frac{\partial G^i_{jkl}}{\partial \dot{x}^h}v^h_{(r)}\dot{x}^r - v^i_{(h)}G^h_{jlk} + v^h_{(j)}G^i_{hlk} + $$
$$+ v^h_{(l)}G^i_{jkh} + G^i_{hjl}v^h_{(k)} = 0\ . \tag{1.19}$$

Equations (1.15) and (1.19) represent the required integrability conditions of (1.9)[1]. We remark that it is not necessary to go further at this stage, for other relations resulting from (1.2) and (1.9) by partial and covariant differentiation and elimination of the $v^i_{(j)(k)}$ by means of (1.9a) are found to be linearly dependent on (1.15) and (1.19). For instance, if we differentiate the Killing equations (1.5) covariantly with respect to x^l, we obtain a relation which is derivable also from (1.19) by multiplication with $g_{is} \dot{x}^s$. This relation is therefore linearly dependent on (1.19). Similarly, partial differentiation of (1.5) with respect to the directional argument $\dot{x}^k$ yields a further relation, which in turn gives rise (by covariant differentiation) to equations derivable from (1.19). The complete analysis of this state of affairs is somewhat beyond the scope of the present section.

An alternative approach to the problem concerning the existence of motions in a Finsler space is due to WANG[2], whose method is based on the theory of linear groups.

The concept of a motion has been generalised by the introduction of *homothetic* transformations[3]. Such transformations are defined by the property that the ratios of the lengths of vectors are preserved. Infinitesimal homothetic transformations in Finsler spaces have been studied in some detail by HIRAMATU [1, 2], and by TAKANO [1], both authors obtaining generalisations of theorems of SHANKS [1] and YANO [3].

VAGNER [7] discusses the so-called *homological* transformations which affect the metric (i. e. the indicatrices) but not the geodesics, and invariants with respect to homological transformations are obtained. With regard to the treatment of Cartan's holonomy groups in Finsler spaces the reader is referred to AKBAR-ZADEH [1], NASU [2] and VAGNER [5]. The latter paper concerns itself with two-dimensional spaces, and invariant conditions that the group of holonomy, which is in general an infinite continuous group, be a finite continuous group are found.

§ 2. Conformal Geometry

Suppose that two distinct metric functions $\overline{F}(x, \dot{x})$, $F(x, \dot{x})$ are defined over a space F_n, both of which satisfy the conditions $A - C$ of § 1, Ch. I. The two metrics resulting from these functions are called *conformal* if there exists a factor of proportionality $\psi(x, \dot{x})$ between the two metric tensors [defined according to (1.3.1)]:

$$\bar{g}_{ij}(x, \dot{x}) = \psi(x, \dot{x}) \, g_{ij}(x, \dot{x}) \, . \tag{2.1}$$

This assumption implies that the magnitudes of vectors defined in the same tangent space are proportional, and furthermore, that the angle as defined by (1.7.10) between two directions in the same tangent space

---

[1] KNEBELMAN [1], p. 549 and p. 561.

[2] WANG [1], p. 5.

[3] Homothetic transformations were defined in Riemannian geometry by SHANKS [1].

is the same with respect to both metrics, for by (1.3.2) the relation (2.1) gives

$$\overline{F}(x, \dot{x}) = \psi(x, \dot{x})^{1/2} F(x, \dot{x}) . \tag{2.1a}$$

Naturally, (2.1a) does not conversely imply (2.1): in fact, it is seen that a sufficient condition for this to be the case is that ψ be independent of direction. Also, it is obvious that the alternative definition of angle as given by (1.7.2) is not invariant under (2.1) unless the same condition holds.

However, we shall see that (2.1) automatically causes this condition to be satisfied. Differentiating (2.1) with respect to $\dot{x}^k$ we have

$$\overline{C}_{ijk} = \psi C_{ijk} + \frac{1}{2} \frac{\partial \psi}{\partial \dot{x}^k} g_{ij} .$$

As a result of the symmetry properties of the C_{ijk} it follows from this relation that

$$\delta_i^h \frac{\partial \psi}{\partial \dot{x}^k} = \delta_k^h \frac{\partial \psi}{\partial \dot{x}^i} ,$$

or, putting $h = k$ and summing over k, we find that

$$\frac{\partial \psi}{\partial \dot{x}^i} = 0 , \qquad (n \neq 1) . \tag{2.2}$$

Thus the factor of proportionality is at most a point function[1] and it therefore follows from the above remarks *that both angles as defined by* (1.7.2) *and* (1.7.10) *are invariant with respect to the conformal correspondence* (2.1). A further immediate consequence of the relations (2.1) and (2.2) is the following: *If a Finsler space F_n admits a conformal correspondence* (2.1) *with a Riemannian space V_n, then F_n itself must be Riemannian*[2].

We may now write (2.1) in the form

$$\bar{g}_{ij}(x, \dot{x}) = e^{2\sigma} g_{ij}(x, \dot{x}) , \tag{2.3}$$

---

[1] This result is due to KNEBELMAN [2], p. 376.

[2] Using the form (1.7.12) of LANDSBERG's angle, GOLAB [4] shows that if a Finsler space F_n is conformal to a euclidean space, then F_n is necessarily Riemannian. It was pointed out by KNEBELMAN that this theorem is a special case of the one enunciated above, i. e. a direct consequence of (2.2), but in GOLAB [5] it is explained that while definition (2.1) implies proportionality of lengths, the equality of the (Landsberg) angles was the point of departure of GOLAB's definition of conformality. The equivalence of the two properties is established in [5]. Also, in the proof of (2.2) it has been assumed that the g_{ij}, $\bar{g}_{ij}$ have continuous directional derivatives. GOLAB [6] proves the validity of (2.2) for $n = 2$ under weaker assumptions, namely that the g_{ij} be merely continuous. This result is extended by SZMYDT [1], where it is proved that this result holds for any n, with special reference to particular metric functions. CARTAN [4] approaches the same problem in a slightly different manner and shows that for $n > 2$ the space must be Riemannian in order that the directions at every point of a Finsler space be conformal to the directions at a fixed point of a euclidean space.

where
$$\sigma = \sigma(x) = \tfrac{1}{2} \log \psi , \tag{2.4}$$

so that
$$\bar{g}^{ij}(x, \dot{x}) = e^{-2\sigma} g^{ij}(x, \dot{x}) . \tag{2.5}$$

From (2.2.3) and (2.2.7) it then follows that we have for the Christoffel symbols of the second kind

$$\bar{\gamma}_i{}^h{}_j = \gamma_i{}^h{}_j + (\sigma_i \, \delta_j^h + \sigma_j \, \delta_i^h - g^{hk} \, g_{ij} \, \sigma_k) , \tag{2.6}$$

where we have written
$$\sigma_k = \frac{\partial \sigma}{\partial x^k} . \tag{2.7}$$

Noting that in view of (3.3.6) we may write Berwald's connection coefficients in the form

$$G_i{}^h{}_j = \gamma_i{}^h{}_j + \frac{\partial \gamma_r{}^h{}_j}{\partial \dot{x}^i} \dot{x}^r + \frac{\partial \gamma_i{}^h{}_s}{\partial \dot{x}^j} \dot{x}^s + \frac{1}{2} \frac{\partial^2 \gamma_r{}^h{}_s}{\partial \dot{x}^i \partial \dot{x}^j} \dot{x}^r \dot{x}^s ,$$

it is easily verified by means of (2.6) that we have

$$\bar{G}_{ij}^h = G_{ij}^h + (\sigma_i \, \delta_j^h + \sigma_j \, \delta_i^h - g^{hk} \, g_{ij} \, \sigma_k) -$$
$$- \left(\frac{\partial g^{hk}}{\partial \dot{x}^j} g_{ir} \dot{x}^r + \frac{\partial g^{hk}}{\partial \dot{x}^i} g_{jr} \dot{x}^r + \frac{1}{2} F^2 \frac{\partial^2 g^{hk}}{\partial \dot{x}^i \partial \dot{x}^j} \right) \sigma_k .$$

Hence, if we put
$$B^{hk} = \tfrac{1}{2} F^2 g^{hk} - \dot{x}^h \dot{x}^k , \tag{2.8}$$

it follows that
$$\bar{G}_{ij}^h = G_{ij}^h - \frac{\partial^2 B^{hk}}{\partial \dot{x}^i \partial \dot{x}^j} \sigma_k . \tag{2.9}$$

A few simple corollaries result from this formula. Since, by definition (2.8), the B^{hk} are homogeneous of the second degree in their directional arguments, we have from (2.9)

$$\bar{G}^h = G^h - B^{hk} \sigma_k , \tag{2.9a}$$

and hence the differential equation (2.2.8) of a geodesic Γ of F_n with respect to the metric function $F(x, \dot{x})$ may be expressed in the form

$$0 = \frac{d^2 x^i}{ds^2} + 2G^i \left(x, \frac{dx}{ds} \right) = \frac{d^2 x^i}{ds^2} + 2\bar{G}^i \left(x, \frac{dx}{ds} \right) + 2B^{ik} \left(x, \frac{dx}{ds} \right) \sigma_k . \tag{2.10}$$

Let $\bar{s}$ denote the arc-length of Γ with respect to the second metric function $\bar{F}(x, \dot{x})$. Then by (2.3) we have

$$\frac{d\bar{s}}{ds} = e^\sigma , \tag{2.11}$$

and on transforming (2.10) according to (2.11) we find that the differential equations of the geodesic Γ in terms of the parameter $\bar{s}$ now become, after some simplification,

$$e^{2\sigma} \left\{ \frac{d^2 x^i}{d\bar{s}^2} + 2\bar{G}^i \left(x, \frac{dx}{d\bar{s}} \right) \right\} + e^{2\sigma} \left\{ g^{ik} F^2 \left(x, \frac{dx}{d\bar{s}} \right) - \frac{dx^i}{d\bar{s}} \frac{dx^k}{d\bar{s}} \right\} \sigma_k = 0 . \tag{2.12}$$

Thus the geodesics of F_n with respect to the metric function $F(x, \dot{x})$ are not, in general, geodesics with respect to the conformal metric $\overline{F}(x, \dot{x})$. In fact, we may deduce from (2.12) that the geodesics of the two metrics can coincide if and only if

$$\left(g^{ik} - \frac{\dot{x}^i \dot{x}^k}{F^2(x, \dot{x})} \right) \sigma_k = 0 , \tag{2.13}$$

for all line elements $(x, \dot{x})$. But this is possible only if $\sigma_k = 0$, provided $n > 1$, i. e. if σ is constant [equation (2.7)]. The truth of this statement is easily verified as follows: Multiplying (2.13) by g_{ij} we find

$$\sigma^k \left(g_{kj} - \frac{y_j y_k}{F^2} \right) = 0 , \qquad \sigma^k = g^{hk}(x, y) \sigma_h ,$$

and by (1.3.1) and (1.5.6) it follows that the coefficients of σ^k are simply the $F F_{\dot{x}^k \dot{x}^j}(x, \dot{x})$. But the matrix of these coefficients is of rank $(n - 1)$ by condition C (§ 1, Ch. I), and hence by (1.1.12) condition (2.13) implies $\sigma^k = \lambda \dot{x}^k$, where λ is homogeneous of degree -1 in the $\dot{x}^j$. If we differentiate this result with respect to $\dot{x}^h$, noting that σ_k is independent of $\dot{x}^h$, we find on contracting over h and k that $\lambda(n - 1) = 0$, which proves our assertion. *Thus the conformal transformation* (2.1) *leaves geodesics invariant if and only if the factor of proportionality is constant*[1]. Also, it is clear that a transformation of the type (2.1) with a constant ψ will leave unchanged all metric relationships existing in the Finsler space F_n. In the terminology of WEYL we can therefore say that *the projective and conformal properties of a Finsler space determine its metric relationships uniquely.*

Recalling the definition of the curvature tensor (4.6.7) constructed by means of the G^i_{hk}, we find by substitution from (2.9) in the corresponding expression for the curvature tensor $\overline{H}^i_{jkh}$, constructed by means of the $\overline{G}^i_{hk}$, the following relation[2]:

$$\overline{H}^i_{jkh} = H^i_{jkh} - \left(\frac{\partial^2 B^{im}}{\partial \dot{x}^j \partial \dot{x}^k} \sigma_{m(h)} - \frac{\partial^2 B^{im}}{\partial \dot{x}^j \partial \dot{x}^h} \sigma_{m(k)} \right) -$$
$$- \sigma_r \sigma_m \frac{\partial}{\partial \dot{x}^j} \left(\frac{\partial^2 B^{ir}}{\partial \dot{x}^k \partial \dot{x}^l} \frac{\partial B^{lm}}{\partial \dot{x}^h} - \frac{\partial^2 B^{ir}}{\partial \dot{x}^h \partial \dot{x}^l} \frac{\partial B^{lm}}{\partial \dot{x}^k} \right) - \tag{2.14}$$
$$- \sigma_m \frac{\partial}{\partial \dot{x}^j} \left\{ \left(\frac{\partial B^{im}}{\partial \dot{x}^k} \right)_{(h)} - \left(\frac{\partial B^{im}}{\partial \dot{x}^h} \right)_{(k)} \right\} .$$

Comparing this result with the corresponding relation in Riemannian geometry, we see that the latter contains no term involving the σ_m

---

[1] This is the generalisation of a theorem of Riemannian geometry due to WEYL [1] which is of great importance in the theory of relativity. A different approach to the same problem is given by LAUGWITZ [4]. The projective relationships between two conformal metrics are studied also by HOSOKAWA [1].

[2] KNEBELMAN [2], p. 377. The corresponding relation in Riemannian geometry is given by EISENHART [1], p. 89, eqn. (28.5).

separately, since the last expression on the right-hand side vanishes for this case, and the algebraic elimination of the $\sigma_{m(h)}$ carries with it the elimination of the σ_r. The result of this process is the conformal curvature tensor of WEYL[1]. In the general case it may be shown that the elimination of the $\sigma_{m(h)}$ yields a conformal invariant which is not a tensor.

In order to indicate how a complete set of conformal invariants may be found, we follow a method which is a direct generalisation of a theory developed for Riemannian geometry[2]. Let us write

$$\Phi = F^2 |g|^{-1/n}, \tag{2.15}$$

where $|g| = \det(g_{ij})$, so that Φ is a relative scalar of weight $-2/n$. In addition to the assumptions of Ch. I concerning the metric function F, we now have to assume that the Hessian of Φ with respect to the $\dot{x}^i$ does not vanish identically. If we put

$$G_{ij} = \frac{1}{2} \frac{\partial^2 \Phi}{\partial \dot{x}^i \, \partial \dot{x}^j}, \tag{2.16}$$

we obtain a tensor which is invariant under the conformal change (2.1). For since

$$|\bar{g}|^{-1/n} = \psi^{-1} |g|^{-1/n},$$

it follows from (2.15) and (2.1a) that Φ and hence G_{ij} is invariant under (2.1). Hence G_{ij} is often referred to as the *fundamental conformal tensor*. Its explicit expression reads

$$G_{ij} = g_{ij} |g|^{-1/n} + g_{ih} \dot{x}^h \frac{\partial}{\partial \dot{x}^j} |g|^{-1/n} + g_{hj} \dot{x}^h \frac{\partial}{\partial \dot{x}^i} |g|^{-1/n} +$$
$$+ \frac{1}{2} F^2 \frac{\partial^2}{\partial \dot{x}^i \, \partial \dot{x}^j} |g|^{-1/n}. \tag{2.16a}$$

Under the transformation (1.1.1) with non-vanishing Jacobian $|A_i^{i'}| \equiv \det(A_i^{i'}) \equiv \det\left(\frac{\partial x^{i'}}{\partial x^i}\right)$, the function Φ transforms according to the law

$$\Phi' = \Phi |A_{i'}^i|^{-2/n},$$

and hence it follows by differentiation that the law of transformation of the G_{ij} is

$$G_{i'j'} = |A_{i'}^i|^{-2/n} G_{ij} A_{i'}^i A_{j'}^j, \tag{2.17}$$

while for the normalised cofactors G^{ij} of G_{ij} in $\det(G_{ij})$ we have

$$G^{i'j'} = |A_{i'}^i|^{2/n} G^{ij} A_i^{i'} A_j^{j'}. \tag{2.17a}$$

With the aid of the fundamental tensor G_{ij} we are now in a position to generalise directly the theory which has been developed for Riemannian geometry. However, since the corresponding calculations are fairly

---

[1] EISENHART [1], p. 90 et seq.

[2] T. Y. THOMAS [1], p. 66 et seq.

involved, we shall content ourselves with a brief outline of a few initial steps.

We first form the Christoffel symbols with respect to the G_{ij}:

$$\lambda^i_{jk} = G^{ih}\,\lambda_{jhk} = \frac{1}{2}\,G^{ih}\left(\frac{\partial G_{hk}}{\partial x^j} + \frac{\partial G_{jh}}{\partial x^k} - \frac{\partial G_{jk}}{\partial x^h}\right). \tag{2.18}$$

These quantities permit us to form the analogue of the connection coefficients G^i_{jk}, namely the *conformal connection parameters*:

$$\Lambda^i_{jk} = \lambda^i_{jk} + \frac{\partial \lambda^i_{rk}}{\partial \dot{x}^j}\,\dot{x}^r + \frac{\partial \lambda^i_{rj}}{\partial \dot{x}^k}\,\dot{x}^r + \frac{1}{2}\,\frac{\partial^2 \lambda^i_{rs}}{\partial \dot{x}^j\,\partial \dot{x}^k}\,\dot{x}^r\dot{x}^s. \tag{2.19}$$

The transformation law satisfied by these coefficients is not the same as that satisfied by the G^i_{hk}. In fact, if we write

$$\psi_{k'} = \frac{1}{n}\,\frac{\partial}{\partial x^{k'}}\{\log\det(A^i_{i'})\}\,, \tag{2.20}$$

we find after a straight-forward calculation that under (1.1.1)

$$\Lambda^{i'}_{j'k'}A^i_{i'} = \Lambda^i_{jk}A^j_{j'}A^k_{k'} + \partial_{j'}A^i_{k'} - A^i_{j'}\,\psi_{k'} - A^i_{k'}\,\psi_{j'} + \tag{2.21}$$
$$+\,\psi_{h'}A^i_{i'}\left[G_{j'k'}G^{i'h'} + \frac{1}{2}\,\frac{\partial\Phi'}{\partial \dot{x}^{j'}}\,\frac{\partial G^{i'h'}}{\partial \dot{x}^{k'}} + \frac{1}{2}\,\frac{\partial\Phi'}{\partial \dot{x}^{k'}}\,\frac{\partial G^{i'h'}}{\partial \dot{x}^{j'}} + \frac{1}{2}\,\Phi'\,\frac{\partial^2 G^{i'h'}}{\partial \dot{x}^{j'}\,\partial \dot{x}^{k'}}\right].$$

Putting

$$D^{i'h'} = \tfrac{1}{2}\,\Phi'G^{i'h'} - \dot{x}^{i'}\,\dot{x}^{h'}\,, \tag{2.22}$$

we find that (2.21) assumes the simpler form[1]

$$\Lambda^{i'}_{j'k'}A^i_{i'} = \Lambda^i_{jk}A^j_{j'}A^k_{k'} + \partial_{j'}A^i_{k'} + \frac{\partial^2 D^{i'h'}}{\partial \dot{x}^{j'}\,\partial \dot{x}^{k'}}\,\psi_{h'}A^i_{i'}\,. \tag{2.23}$$

Again, the connection parameters (2.19) may be used to define a process of covariant differentiation, while in analogy with (4.6.7) a curvature tensor is defined in terms of the Λ^i_{jk} and their derivatives. By means of this curvature tensor a complete set of conformal invariants may be constructed as in Riemannian geometry[2]. These conformal invariants are not in general tensors.

In conclusion we may remark that it is possible to develop a similar theory in which the Γ^{*i}_{jk} instead of the G^i_{jk} play a central role. Again, in virtue of the close affinity between the G^i_{jk} and the general geometry of paths, the latter coefficients are preferable. Also, a simple calculation

---

[1] KNEBELMAN [2], p. 378.

[2] A process of this kind is sketched very briefly by KNEBELMAN [2], who also indicates how conformal tensors may be obtained if further assumptions are made. HOMBU [1] derives conformal scalar invariants for two-dimensional Finsler spaces. A further discussion, indicating the construction of a complete set of conformal invariants, is given by HOMBU [3].

based on (2.4.4) and (2.9) shows that under the conformal transformation (2.1) the Γ_{jk}^{*i} transform as follows:

$$\overline{\Gamma}_{ij}^{*h} = \Gamma_{ij}^{*h} + (\sigma_i\,\delta_j^h + \sigma_j\,\delta_i^h - g^{hk}\,g_{ij}\,\sigma_k) -$$
$$- \frac{F^2}{2}\,g^{mh}\left(C_{ijr}\,\frac{\partial g^{rk}}{\partial \dot{x}^m} - C_{imr}\,\frac{\partial g^{rk}}{\partial \dot{x}^j} - C_{jmr}\,\frac{\partial g^{rk}}{\partial \dot{x}^i}\right)\sigma_k - \qquad (2.24)$$
$$- g^{rk}(C_{irj}\,\dot{x}^h - C_{ir}^h\,y_j - C_{jr}^h\,y_i)\,\sigma_k - C_i{}^h{}_j\,\dot{x}^k\,\sigma_k\,.$$

These relations are obviously more complicated than the corresponding equations (2.9) for the $\overline{G}_{ij}^h$, and in consequence it is to be expected that the subsequent analysis would become still more cumbersome.

§ 3. The Equivalence Problem

The problem of the local equivalence of two Finsler spaces may be approached from two distinct points of view, which are exemplified by the work of CHERN[1] on the one hand and of VARGA on the other. Although the approach of the latter is devoted primarily to the study of the equivalence of spaces of line elements endowed with a set of connection parameters, the method is applicable also to Finsler spaces. For by this method the solution of the equivalence problem is formulated in terms of the torsion and curvature tensors and their successive covariant derivatives, which, furthermore, leads to results which are direct generalisations of well-known theorems of Riemannian geometry[2]. On the other hand, the method of CHERN is based on the formulation of the equivalence of two Finsler geometries in terms of the equivalence of two systems of Pfaffian forms.

In the present section we shall follow the method of CHERN, which enjoys a considerable degree of generality (although this is true also for the method of VARGA) in the sense that it may also be used to derive a whole class of euclidean connections in the Finsler space, of which Cartan's connection is a special case. We shall use the method of exterior differential forms[3]; again, a reader who is not familiar with this method may omit this discussion, since no further developments are based on the results of this section.

Two Finsler spaces will be called *locally equivalent* if there exists an analytic point-wise correspondence between two neighbourhoods such that corresponding arcs have the same length. We shall thus base our initial considerations on the metric function $F(x^k, \dot{x}^k)$, rather than on the connection coefficients Γ_{hk}^{*i}, since, as we have remarked above, the theory ultimately involves the problem of equivalence of a family of euclidean connections in the space.

---

[1] CHERN [1, 2]; VARGA [11].
[2] Compare, for instance, T. Y. THOMAS [1], p. 206.
[3] References to this method are listed under 3°, § 1 (Ch. IV).

The element of arc-length $ds = F(x^k, dx^k)$ may be represented by the Pfaffian form

$$\omega = F_{\dot{x}^i}(x, \dot{x})\, dx^i .\tag{3.1}$$

By introducing $n(n-1)$ new variables v_k^α satisfying the relations[1]

$$v_k^\alpha\, \dot{x}^k = 0 ,\tag{3.2}$$

and by writing

$$v_k^n = F_{\dot{x}^k} ,\tag{3.3}$$

we may construct n linearly independent forms

$$\omega^i = v_k^i\, dx^k\tag{3.4}$$

for which we have, by (3.1) and (3.3),

$$\omega = \omega^n .\tag{3.4a}$$

If the analytic correspondence between two equivalent Finsler spaces is represented in the form

$$\bar{x}^i = \bar{x}^i(x^j) ,\tag{3.5}$$

it is adjoined by the transformation

$$\dot{\bar{x}}^i = \dot{\bar{x}}^i(x^j, \dot{x}^j) , \qquad \bar{v}_k^\alpha = \bar{v}_k^\alpha(x^j, \dot{x}^j, v_j^\alpha) ,\tag{3.5a}$$

such that under (3.5) and (3.5a) we have

$$\overline{\omega}^i = \omega^i .\tag{3.6}$$

Conversely, if a transformation of the form

$$\bar{x}^i = \bar{x}^i(x^j, \dot{x}^j, v_j^\alpha) , \qquad \dot{\bar{x}}^i = \dot{\bar{x}}^i(x^j, \dot{x}^j, v_j^\alpha) , \qquad \bar{v}_k^\alpha = \bar{v}_k^\alpha(x^j, \dot{x}^j, v_j^\alpha) ,\tag{3.7}$$

exists for which (3.6) is satisfied, it follows from (3.4) that the functions $\bar{x}^i$ will be independent of $\dot{x}^j$ and v_j^α and thus define a local point-wise correspondence which establishes the local equivalence of the two Finsler geometries. Hence the problem of equivalence is reduced to the problem of the equivalence of linearly independent Pfaffian forms in the two sets of $n^2 + n$ variables x^i, $\dot{x}^i$, v_i^α and $\bar{x}^i$, $\dot{\bar{x}}^i$, $\bar{v}_i^\alpha$ respectively[2].

In order to evaluate the exterior derivatives of the forms ω^i it is necessary to introduce the elements u_k^i of the matrix inverse to v_k^i:

$$u_k^i\, v_j^k = \delta_j^i = v_k^i\, u_j^k .\tag{3.8}$$

Also, from (3.3) and the homogeneity properties of F we have

$$u_n^k = \frac{\dot{x}^k}{F} .\tag{3.9}$$

By means of (3.4) and (3.8) we may now write the exterior derivative of (3.4a) in the form

$$(\omega^n)' = F_{\dot{x}^i \dot{x}^j}\, u_h^i\, u_k^j\, [\omega^h\, \omega^k] + F_{\dot{x}^i \dot{x}^j}\, u_h^j\, [d\dot{x}^i\, \omega^h] .$$

---

[1] Throughout this section Latin indices run from 1 to n, Greek indices from 1 to $n-1$.

[2] CHERN [2], p. 98.

But it follows from (3.9) and the fact that $F_{\dot{x}^i \dot{x}^j}$ is homogeneous of degree zero in the $\dot{x}^k$ that the coefficient of $[d\,\dot{x}^i\,\omega^n]$ vanishes identically. Thus we may choose $(n-1)$ Pfaffian forms $\omega_\alpha{}^n$ such that

$$(\omega^n)' = [\omega^\alpha\,\omega_\alpha{}^n] ,\tag{3.10}$$

where, in view of (3.9),

$$\omega_\alpha{}^n = -\,u_\alpha^i\,F_{\dot{x}^i \dot{x}^j}\,d\,\dot{x}^j + \frac{1}{F}\,u_\alpha^i\,(F_{x^i} - F_{x^j \dot{x}^i}\,\dot{x}^j)\,\omega^n +$$
$$+\,u_\alpha^i\,u_\beta^j\,F_{x^i \dot{x}^j}\,\omega^\beta + \lambda_{\alpha\beta}\,\omega^\beta ,\tag{3.10a}$$

the $\frac{1}{2}n(n-1)$ quantities $\lambda_{\alpha\beta}$ being arbitrary except for the symmetry condition

$$\lambda_{\alpha\beta} = \lambda_{\beta\alpha} .\tag{3.10b}$$

It is easily established that *the Pfaffian forms ω^i, $\omega_\alpha{}^n$ are linearly independent*. For if there were to exist a relation of the form

$$\mu^\alpha\,\omega_\alpha{}^n + v_i\,\omega^i = 0 ,$$

it would follow that

$$\mu^\alpha\,u_\alpha^i F_{\dot{x}^i \dot{x}^j}\,d\,\dot{x}^j = 0 ,$$

and hence by (1.1.12) and condition C of § 1, Ch. I (which, as we have seen, implies that the matrix of the $F_{\dot{x}^i \dot{x}^j}$ is of rank $n-1$) we would have

$$\mu^\alpha\,u_\alpha^i = \lambda\,\dot{x}^i ,$$

from which, by multiplication with v_i^n, one may deduce $\lambda F = 0$ in virtue of (3.3) and (3.8), or $\lambda = 0$ (since $F > 0$). Since the matrix $\|u_\alpha^i\|$ is of rank $n-1$, this would imply $\mu^\alpha = 0$, and hence $v_i\,\omega^i = 0$, which is possible if and only if $v_i = 0$. This proves our assertion.

Furthermore, differentiating (3.8) and solving for dv_j^k, we find

$$dv_j^i = -\,v_h^i\,v_j^k\,du_k^h ,\tag{3.11}$$

and hence, on forming the exterior derivative of (3.4), it follows that

$$(\omega^\alpha)' = -\,v_h^\alpha\,[d\,u_k^h\,\omega^k] .$$

In the expression on the right-hand side we split up the summation over k into a summation over β, which gives us an additional term. Using (3.2) and (3.9) this leads to the relation

$$(\omega^\alpha)' = -\,v_h^\alpha\,[d\,u_\beta^h\,\omega^\beta] - \frac{1}{F}\,v_j^\alpha\,[d\,\dot{x}^j\,\omega^n] .\tag{3.12}$$

We may therefore write[1]

$$(\omega^\alpha)' \equiv [\omega^n\,\omega_n{}^\alpha] \pmod{\omega^\beta} ,\tag{3.13}$$

---

[1] The notation

$$\Omega \equiv 0 \pmod{\omega_1,\,\omega_2,\,\ldots,\,\omega_r}$$

indicates that the form Ω is a member of the ideal generated by the homogeneous forms $\omega_1,\,\omega_2,\,\ldots,\,\omega_r$.

where

$$\omega_n{}^\alpha \equiv \frac{1}{F}\, v_j^\alpha\, d\,\dot x^j \quad (\mathrm{mod}\,\omega^i)\,. \tag{3.13a}$$

Now, using (3.10a) we see that the coefficient of $d\,\dot x^j$ in the expression $\omega_\alpha{}^n + \delta_{\alpha\beta}\,\omega_n{}^\beta$ will vanish if

$$u_\alpha^i F F_{\dot x^i \dot x^j} = \delta_{\alpha\beta}\, v_j^\beta\,. \tag{3.14}$$

We may choose the u_j^i, v_j^i such that this condition is satisfied, provided this does not violate the condition $\det(v_j^i) \neq 0$. It may be verified immediately as follows that (3.14) cannot have this undesirable effect: If we write (3.14) in the form

$$v_k^\alpha\, u_\alpha^i F F_{\dot x^i \dot x^j} = \delta_{\alpha\beta}\, v_k^\alpha\, v_j^\beta\,,$$

and note that in view of (3.8), (3.9) and (1.1.12) the left-hand side is equivalent to the expressions

$$(v_k^h\, u_h^i - v_k^n\, u_n^i)\, F F_{\dot x^i \dot x^j} = F F_{\dot x^k \dot x^j}\,,$$

it follows from (3.3) that condition (3.14) implies

$$F F_{\dot x^i \dot x^j} = \delta_{\alpha\beta}\, v_i^\alpha\, v_j^\beta = \delta_{hk}\, v_i^h\, v_j^k - F_{\dot x^i} F_{\dot x^j}\,, \tag{3.14a}$$

which, by means of (1.1.19) and (1.3.1) reduces to

$$g_{ij} = \delta_{hk}\, v_i^h\, v_j^k\,, \tag{3.14b}$$

together with

$$g_{ij}\, u_\alpha^i\, u_\beta^j = \delta_{\alpha\beta}\,. \tag{3.14c}$$

But from condition C of § 1, Ch. I, we have $\det(g_{ij}) > 0$ [equation (1.1.25)], and hence $\det(v_i^h) \neq 0$, so that the required condition is satisfied.

Having thus eliminated the terms involving $d\,\dot x^j$ in the expression $\omega_\alpha{}^n + \delta_{\alpha\beta}\,\omega_n{}^\beta$, we are now at liberty to choose the forms $\omega_n{}^\beta$ such that

$$\omega_\alpha{}^n + \delta_{\alpha\beta}\,\omega_n{}^\beta = 0\,. \tag{3.15}$$

Using (3.10a) and (3.14) we then find that (3.12) may be written as

$$(\omega^\alpha)' = [\omega^n\, \omega_n{}^\alpha] + [\omega^\beta\, d\,u_\beta^h]\, v_h^\alpha - [\omega^\beta\, \omega^n]\, \delta^{\alpha\gamma}(F_{\dot x^i \dot x^j}\, u_\gamma^i\, u_\beta^j + \lambda_{\gamma\beta})\,, \tag{3.16}$$

or,

$$(\omega^\alpha)' = [\omega^n\, \omega_n{}^\alpha] + [\omega^\beta\, \omega_\beta{}^\alpha]\,, \tag{3.16a}$$

where the most general expression for the $(n-1)^2$ Pfaffian forms $\omega_\beta{}^\alpha$ is[1]

$$\omega_\beta{}^\alpha = v_h^\alpha\, d\,u_\beta^h - \delta^{\alpha\gamma}(F_{\dot x^i \dot x^j}\, u_\gamma^i\, u_\beta^j + \lambda_{\gamma\beta})\, \omega^n + \mu_{\beta\gamma}^\alpha\, \omega^\gamma\,, \tag{3.17}$$

the $\mu_{\beta\gamma}^\alpha$ being arbitrary except for the symmetry condition

$$\mu_{\beta\gamma}^\alpha = \mu_{\gamma\beta}^\alpha\,. \tag{3.17a}$$

---

[1] CHERN [1], p. 34.

The $\lambda_{\alpha\beta}$, $\mu^{\alpha}_{\gamma\beta}$ may now be specified as follows: By differentiation of (3.14a) and repeated application of (3.8) we find

$$- \delta_{\beta\delta}(v^{\beta}_j\, du^j_\varepsilon) - \delta_{\alpha\varepsilon}(v^{\alpha}_i\, du^i_\delta) = d(FF_{\dot{x}^i \dot{x}^j})\, u^i_\delta\, u^j_\varepsilon \,.$$

In these equations we substitute for the two terms on the left-hand side from (3.17), and after some simplification we obtain

$$\delta_{\alpha\varepsilon}\, \omega^{\alpha}_\delta + \delta_{\alpha\delta}\, \omega^{\alpha}_\varepsilon = - d(FF_{\dot{x}^i \dot{x}^j})\, u^i_\delta\, u^j_\varepsilon - 2\lambda_{\varepsilon\delta}\, \omega^n - $$
$$- (F_{x^i \dot{x}^j} + F_{x^j \dot{x}^i})\, u^i_\delta\, u^j_\varepsilon\, \omega^n + (\delta_{\alpha\delta}\, \mu^{\alpha}_{\varepsilon\gamma} + \delta_{\alpha\varepsilon}\, \mu^{\alpha}_{\delta\gamma})\, \omega^\gamma \,. \qquad (3.18)$$

However, the term $d(FF_{\dot{x}^i \dot{x}^j})$ may be decomposed in the following manner: From (3.10a) we see that if v_k is the coefficient of $d\dot{x}^j$ in any linear combination of the ω^n_α, we have $v_j\, \dot{x}^j = 0$. Since the ω^n_α are $(n-1)$ linearly independent forms with respect to $d\dot{x}^j$, it follows conversely that any form $v_j\, d\dot{x}^j$, satisfying $v_j\, \dot{x}^j = 0$, is congruent to a linear combination of the ω^n_α, mod ω^i. But in the differential form $d(FF_{\dot{x}^i \dot{x}^h})$ the coefficient of $d\dot{x}^j$ is

$$v_{ihj} = (F_{\dot{x}^j} F_{\dot{x}^i \dot{x}^h} + FF_{\dot{x}^i \dot{x}^h \dot{x}^j}) \,,$$

and from the homogeneity properties of F it follows that

$$v_{ihj}\, \dot{x}^j = 0 \,,$$

so that $d(FF_{\dot{x}^i \dot{x}^h})$ has the required property. We may therefore write

$$d(FF_{\dot{x}^i \dot{x}^j}) = F^{\alpha}_{ij}\, \omega^n_\alpha + F_{ijh}\, \omega^h \,. \qquad (3.19)$$

On substituting this result in (3.18) we shall find that we may specify the auxiliary variables $\mu^{\alpha}_{\varepsilon\gamma}$, $\lambda_{\varepsilon\delta}$ completely by demanding that the final result of this substitution be of the form

$$\delta_{\alpha\varepsilon}\, \omega^{\alpha}_\delta + \delta_{\alpha\delta}\, \omega^{\alpha}_\varepsilon = H^{\alpha}_{\varepsilon\delta}\, \omega^n_\alpha \,. \qquad (3.20)$$

For from (3.18) and (3.19) we deduce that this is possible if we choose

$$-2\lambda_{\varepsilon\delta} = (F_{x^i \dot{x}^j} + F_{x^j \dot{x}^i} + F_{ijn})\, u^i_\varepsilon\, u^j_\delta \,, \qquad (3.21)$$

together with

$$\delta_{\alpha\varepsilon}\, \mu^{\alpha}_{\delta\gamma} + \delta_{\alpha\delta}\, \mu^{\alpha}_{\varepsilon\gamma} = F_{ij\gamma}\, u^i_\varepsilon\, u^j_\delta \,, \qquad (3.22)$$

and, since it is easily verified that the last equation is identical with the relation

$$\mu^{\alpha}_{\varepsilon\delta} = \tfrac{1}{2}\, \delta^{\alpha\beta}(F_{ij\delta}\, u^i_\varepsilon\, u^j_\beta + F_{ij\varepsilon}\, u^i_\delta\, u^j_\beta - F_{ij\beta}\, u^i_\varepsilon\, u^j_\delta) \,, \qquad (3.22a)$$

the variables $\mu^{\alpha}_{\varepsilon\delta}$, $\lambda_{\varepsilon\delta}$, and hence the Pfaffians ω^n_α, ω^{α}_β are completely determined. Conversely, the fact that by a suitable choice of these quantities the validity of (3.20) may be ensured is of fundamental importance[1].

Of the $n + n + n(n-1)$ variables x^i, $\dot{x}^i$, v^{α}_i only the ratios of the $\dot{x}^i$ are essential, while the $\dot{x}^i$ and v^{α}_i are linked by the $(n-1)$ equations

---

[1] CHERN [1], p. 35.

(3.2) and the $\frac{1}{2} n (n - 1)$ equations (3.14c). Thus in effect there are only

$$n + (n - 1) + \tfrac{1}{2} (n - 1) (n - 2)$$

essential variables. We now have the *same* number of completely determined Pfaffian forms

$$\omega^i , \qquad \omega_\alpha{}^n , \qquad \omega_\beta{}^\alpha \tag{3.23}$$

[of which the latter are linked by equations (3.20)], and the process by means of which we arrived at these forms is intrinsic, i. e. the same for locally equivalent spaces.

The problem of equivalence is thus solved in the sense that it has been reduced to the problem of the equivalence of systems of Pfaffian forms involving a number of forms equal to the number of variables, and, according to CARTAN, the problem of the equivalence of two such sets of Pfaffians may be solved by a finite algorithm and there exists a process for its solution[1]. From the general theory it also follows that further differential invariants of the Finsler space are obtained by forming exterior derivatives of the Pfaffian forms (3.23). Thus we shall not pursue the subject beyond this stage and we shall confine ourselves to a few remarks concerning the geometrical background of the theory outlined above.

It is not difficult to show[2] that the coefficients $H_{\varepsilon\delta}{}^\alpha$ on the right-hand side of (3.20) are given by

$$H_{\beta\varepsilon\delta} = 2F \, C_{ijk} \, u_\beta^i \, u_\varepsilon^j \, u_\delta^k , \tag{3.24}$$

so that Riemannian spaces are characterised by the condition $H_{\varepsilon\delta}{}^\alpha = 0$.

From the point of view of the geometrical interpretation it is convenient to define a new set of forms by putting

$$\pi^i = \omega^i , \qquad \pi_j{}^i = \omega_j{}^i + \gamma_j{}^{i\alpha} \, \omega_\alpha{}^n , \tag{3.25}$$

where we suppose that

$$\gamma_{ji}{}^\alpha + \gamma_{ij}{}^\alpha = -H_{ij}{}^\alpha , \tag{3.26}$$

with the understanding that $H_{ik\alpha}$ is zero if any one of its indices is equal to n, and where the Kronecker δ is used to raise and lower suffixes. Clearly the relations (3.26) are not sufficient to determine the $\gamma_{ij}{}^\alpha$. Putting $\gamma_{ni}{}^\alpha = \gamma_{jn}{}^\alpha = 0$, we see that the most general expression for these quantities is given by

$$2\gamma_{\alpha\beta\gamma} = \lambda_{\alpha\beta\gamma} - \lambda_{\gamma\alpha\beta} - \lambda_{\beta\alpha\gamma} - H_{\alpha\beta\gamma} , \tag{3.27}$$

where the $\lambda_{\alpha\beta\gamma}$ are a set of quantities skew-symmetric in their last pair of indices.

The forms (3.25) may now be used to define a euclidean connection in the Finsler space by Cartan's method of the "repère mobile", i. e. by attaching to each set of variables $(x^i, \dot{x}^i, v_i^\alpha)$ an n-dimensional euclidean space with an orthonormal frame $(e_1, \ldots, e_n)$, such that a displacement PP_1 is represented in the form $\pi^i e_i$, while the increments of e_i are subject to the relation $de_i = \pi_i{}^j e_j$[3]. These equations

---

[1] CHERN [2], p. 105; CARTAN [5, 6].

[2] CHERN [2], p. 107.

[3] See Ch. IV, § 1, 3°; also Ch. III, § 1.

determine the infinitesimal displacement between two neighbouring euclidean spaces or a euclidean connection. The properties of the latter depend on the expressions for the differential forms

$$\Pi^i = (\pi^i)' - [\pi^j \, \pi_j{}^i] \,, \tag{3.28}$$

and

$$\Pi_j{}^i = (\pi_j{}^i)' - [\pi_j{}^k \, \pi_k{}^i] \,, \tag{3.28a}$$

where it is necessary to impose the condition that the forms (3.28) and (3.28a) are exterior quadratic differential forms in ω^i and $\omega_\alpha{}^n$. It can then be shown that there are as many euclidean connections as there are invariants satisfying the above conditions. In particular, if we put $\lambda_{\alpha\beta\gamma} = 0$, equation (3.27) becomes

$$\gamma_{\alpha\beta\gamma} = - \tfrac{1}{2} H_{\alpha\beta\gamma} \,, \tag{3.29}$$

and we obtain the euclidean connection of CARTAN [1].

In conclusion we remark that of all the euclidean connections belonging to the class of connections derived above, each one has the property that the equivalence of the euclidean connection is a necessary and sufficient condition for the equivalence of the Finsler spaces.

§ 4. The Theory of Non-linear Connections

In Chapter III, § 4, we briefly mentioned the possibility of introducing non-linear connections in a Finsler space. The most general treatment of such connections is due to VAGNER, while more specialised types of such connections have been studied by KAWAGUCHI and BARTHEL[2]. The original motivation for the introduction of non-linear connections is due to the fact that one can dispense — to some extent, at least — with the notion of element of support. In fact, in the work of VAGNER the set of local indicatrices is regarded as the fundamental entity of Finsler geometry, and it was from this point of view that VAGNER developed the so-called "theory of the composite manifold"[3].

Let us consider a differentiable vector field X^i in the Finsler space F_n and suppose that we are given a set of functions $\overset{1}{\Gamma}{}_k^i(x, X)$, depending on this field, whose transformation properties are the same as those of the combination $\overset{1}{\Gamma}{}_{hk}^i(x, X) \, X^h$. The differentials defined by the relations

$$\delta X^i = d X^i + \overset{1}{\Gamma}{}_k^i(x, X) \, d x^k \,, \tag{4.1}$$

are then the components of a vector, namely a type of covariant differential of X^i. The coefficients $\overset{1}{\Gamma}{}_k^i(x, X)$ thus give rise to what we may call a non-linear connection, since the second term on the right-hand

---

[1] For the somewhat complicated analysis involving the equations of structure on which these statements are based, the reader is referred to CHERN [2], pp. 107 to 120.

[2] VAGNER [13], especially §§ 3—4; KAWAGUCHI [5]; BARTHEL [3, 4, 6]. From the point of view of the general geometry of paths FRIESECKE [1], BORTOLOTTI [3], KAWAGUCHI [3] should be mentioned.

[3] VAGNER [11, 15], also [13], pp. 65—67.

side of (4.1) will, in general, be non-linear in the X^h. Similarly, given a covariant vector field Y_i, a set of coefficients $\overset{2}{\Gamma}_{ik}(x, Y)$ with suitable transformation properties define a covariant differential of Y_i:

$$\overset{2}{\delta} Y_i = d Y_i - \overset{2}{\Gamma}_{ik}(x, Y)\, d x^k \, . \tag{4.1a}$$

Again, parallel displacement may be defined by the conditions $\overset{1}{\delta} X^i = 0$ (or $\overset{2}{\delta} Y_i = 0$), and, as before, this would represent a mapping of two neighbouring tangent spaces $T_n(x^i)$ and $T_n(x^i + d x^i)$ (or $T'_n(x^i)$ and $T'_n(x^i + d x^i)$) onto each other.

It is easily shown that a necessary and sufficient condition that a pair of parallel contravariant vectors of $T_n(x^i)$ remain parallel after undergoing a parallel displacement (in this sense) to $T_n(x^i + d x^i)$ is that the functions $\overset{1}{\Gamma}{}^i_k(x, X)$ be positively homogeneous of the first degree in their directional arguments. Assuming this condition to be satisfied, and also that the $\overset{1}{\Gamma}{}^i_k(x, X)$ are at least of class C^2, we may then write

$$\overset{1}{\Gamma}{}^i_{jk}(x, X)\, X^j = \overset{1}{\Gamma}{}^i_k(x, X) \, , \tag{4.2}$$

where we have put

$$\overset{1}{\Gamma}{}^i_{jk}(x, X) = \frac{\partial \overset{1}{\Gamma}{}^i_k(x, X)}{\partial X^j} \, . \tag{4.3}$$

On introducing similar assumptions regarding the functions $\overset{2}{\Gamma}_{ik}(x, Y)$ we have correspondingly[1]

$$\overset{2}{\Gamma}{}^i_{jk}(x, Y)\, Y_i = \overset{2}{\Gamma}_{jk}(x, Y) \, , \tag{4.2a}$$

where

$$\overset{2}{\Gamma}{}^i_{jk}(x, Y) = \frac{\partial \overset{2}{\Gamma}_{jk}}{\partial Y_i} \, . \tag{4.3a}$$

Naturally the coefficients of the non-linear connection give rise to the following curvature tensor:

$$\overset{1}{R}{}^i_{hk} = \frac{\partial \overset{1}{\Gamma}{}^i_h}{\partial x^k} - \frac{\partial \overset{1}{\Gamma}{}^i_k}{\partial x^h} + \Gamma^i_{jk}\overset{1}{\Gamma}{}^j_h - \Gamma^i_{jh}\overset{1}{\Gamma}{}^j_k \, , \tag{4.4}$$

---

[1] Essentially this represents the point of view taken ab initio by MIKAMI [1] and VAGNER ([13], p. 108). KAWAGUCHI [5], starting with a less general set of connection coefficients $\overset{1}{\Gamma}{}^i_k$ uses equations corresponding to (4.2) to define the $\overset{1}{\Gamma}{}^i_{hk}$ from the $\overset{1}{\Gamma}{}^i_k$, noting, of course, that these equations do not uniquely define the $\overset{1}{\Gamma}{}^i_{hk}$ in terms of the $\overset{1}{\Gamma}{}^i_k$. In fact, in order to be able to derive the equivalent of Ricci's lemma, KAWAGUCHI used connection coefficients which do not correspond to equations (4.3) directly ([5], p. 197).

and similarly[1],

$$\overset{2}{R}_{ihk} = \frac{\partial \overset{2}{\Gamma}_{ih}}{\partial x^k} - \frac{\partial \overset{2}{\Gamma}_{ik}}{\partial x^h} + \overset{2}{\Gamma}^j_{ik}\overset{2}{\Gamma}_{jh} - \overset{2}{\Gamma}^j_{ih}\overset{2}{\Gamma}_{jk} . \tag{4.4a}$$

The theory resulting from this point of departure can be extended significantly by the introduction of the so-called "ex-central" connection. For in view of the transformation properties of the connection parameters it is clear that the difference between any pair of distinct coefficients represents the components of a tensor. In particular, we may thus define new "ex-central" connection coefficients by putting

$$\widetilde{\Gamma}^i_j(x, X) = \overset{1}{\Gamma}^i_j(x, X) + \delta^i_j , \tag{4.5}$$

and

$$\widetilde{\Gamma}_{ij}(x, Y) = \overset{2}{\Gamma}_{ij}(x, Y) - Y_i Y_j . \tag{4.5a}$$

As a result of (4.1) and (4.1a) the corresponding covariant differentials may be written as

$$\overset{1}{\widetilde{\delta}} X^i = \overset{1}{\delta} X^i + d x^i , \tag{4.6}$$

and

$$\overset{2}{\widetilde{\delta}} Y_i = \overset{2}{\delta} Y_i - Y_j d x^j Y_i . \tag{4.6a}$$

The parallel displacement defined by the conditions $\overset{1}{\widetilde{\delta}} X^i = 0$, $\overset{2}{\widetilde{\delta}} Y_i = 0$, may thus be interpreted as the result of two mappings, the first being the above-mentioned mapping of the two respective neighbouring tangent spaces onto each other, and the second by a translation

$$'X^i = X^i - d x^i , \tag{4.7}$$

and

$$'Y_i = Y_i + Y_j d x^j Y_i , \tag{4.7a}$$

in T_n and T'_n respectively.

Again, one may define curvature tensors $\overset{1}{\widetilde{R}}^i_{hk}$ and $\overset{2}{\widetilde{R}}_{ihk}$ for the connections (4.5) and (4.5a) in a manner exactly analogous to that suggested by equations (4.4) and (4.4a). By doing so, and substituting from (4.5) and (4.5a) we find

$$\overset{1}{\widetilde{R}}^i_{hk} = \overset{1}{R}^i_{hk} + \overset{1}{\Gamma}^i_{hk} - \overset{1}{\Gamma}^i_{kh} , \tag{4.8}$$

and

$$\overset{2}{\widetilde{R}}_{ihk} = \overset{2}{R}_{ihk} - Y_i (\overset{2}{\Gamma}_{kh} - \overset{2}{\Gamma}_{hk}) . \tag{4.8a}$$

---

[1] Note for instance, that we have from (4.1.7)

$$\dot{x}^j K^i_{jhk} = \frac{\partial \Gamma^{*i}_h}{\partial x^k} - \frac{\partial \Gamma^{*i}_k}{\partial x^h} + \frac{\partial \Gamma^{*i}_k}{\partial \dot{x}^j} \Gamma^{*j}_h - \frac{\partial \Gamma^{*i}_h}{\partial \dot{x}^j} \Gamma^{*j}_k ,$$

where we have put $\Gamma^{*i}_j = \Gamma^{*i}_{jh} \dot{x}^h$.

This naturally leads to the following "torsion" tensors:

$$\overset{1}{S}{}^{i}_{hk}= \tfrac{1}{2}\left(\overset{1}{\tilde{R}}{}^{i}_{hk}-\overset{1}{R}{}^{i}_{hk}\right)= \tfrac{1}{2}\left(\overset{1}{\Gamma}{}^{i}_{hk}-\overset{1}{\Gamma}{}^{i}_{kh}\right),\qquad(4.9)$$

together with

$$\overset{2}{S}{}_{ihk}= \tfrac{1}{2}\left(\overset{2}{\tilde{R}}{}_{ihk}-\overset{2}{R}{}_{ihk}\right)= \tfrac{1}{2}\,Y_i\left(\overset{2}{\Gamma}{}_{hk}-\overset{2}{\Gamma}{}_{kh}\right).\qquad(4.9a)$$

These formulae represent the bare outline of the analytical apparatus corresponding to the theory of non-linear connections in its most general form.

Specialisations may be obtained as follows: Firstly, if two vectors Y_i and X^i are connected by the relation

$$Y_i = g_{ij}(x, X)\,X^j = F\,\frac{\partial F}{\partial X^i},\qquad(4.10)$$

and if we require that this relationship should hold also after parallel displacement as defined by $\overset{1}{\delta}X^i = 0,\ \overset{2}{\delta}Y_i = 0$, it is easily seen that we must stipulate:

$$\overset{2}{\Gamma}{}_{ik}(x, Y) = \frac{\partial g_{ij}(x, X)}{\partial x^k}\,X^j - g_{ij}(x, X)\,\overset{1}{\Gamma}{}^{j}_{k}(x, X),\qquad(4.11)$$

where we note that this relation completely specifies the $\overset{2}{\Gamma}{}_{ik}$ in terms of the $\overset{1}{\Gamma}{}^{i}_{k}$ [1].

Secondly, one may require that the connection be *metric* (in the sense of VAGNER[2]): in view of (4.1) this implies that the coefficients $\overset{1}{\Gamma}{}^{i}_{k}$ satisfy the relationship

$$0 = dF = \left(\frac{\partial F}{\partial x^k} - \frac{\partial F}{\partial X^i}\,\overset{1}{\Gamma}{}^{i}_{k}\right)d x^k,$$

---

[1] The relation (4.11) is given by KAWAGUCHI ([5], p. 188) for his special class of metric connections, but this relation is derived from the condition that $X^i\,\delta g_{ij}(x, X) = 0$, which is not generally satisfied in the present case, at least not until further assumptions are made.

[2] VAGNER [13], p. 113. In this context the term "metric" implies that the length of a (contravariant) vector remains unchanged under parallel displacement. In the theory of non-linear connections this does not necessarily imply that the covariant derivative of the metric tensor vanishes identically (which state of affairs we had previously described by means of the term "metric connection"). It should be noted also that in view of (4.11) condition (4.12) implies that the length of covariant vectors remains unchanged under parallel displacement. This result follows immediately from the fact that $\partial F/\partial x^k = -\,\partial H/\partial x^k$. Condition (4.12) may also be derived from a consideration of the covariant derivatives of scalars (RUND [15]). Furthermore, VAGNER (loc. cit.) also introduces the more general notion of a "semi-metric" connection, which implies that under parallel displacement of two vectors the ratio of their lengths remains constant. In this respect the theory of non-linear connections is closely connected with the conformal geometry discussed in § 2.

or, by (4.10)

$$Y_i \overset{1}{\Gamma^i_k}(x, X) \equiv g_{ij}(x, X) X^j \overset{1}{\Gamma^i_k}(x, X) = F \frac{\partial F}{\partial x^k} . \qquad (4.12)$$

Using (1.3.2) we see that this condition can be expressed in the form

$$g_{ij}(x, X) X^j \overset{1}{\Gamma^i_k}(x, X) = \frac{1}{2} \frac{\partial g_{ij}(x, X)}{\partial x^k} X^i X^j . \qquad (4.13)$$

Differentiating the latter equation with respect to X^h, noting the homogeneity relation (1.3.5) and (4.3), we find

$$g_{ih}(x, X) \overset{1}{\Gamma^i_k} + Y_i \overset{1}{\Gamma^i_{hk}}(x, X) = \frac{\partial g_{ih}(x, X)}{\partial x^k} X^i , \qquad (4.13a)$$

and hence equation (4.11) becomes

$$\overset{2}{\Gamma_{ik}}(x, Y) = Y_j \overset{1}{\Gamma^j_{ik}}(x, X) . \qquad (4.14)$$

Further differentiation of this result with respect to Y_h yields in virtue of (4.3a)

$$\overset{2}{\Gamma^h_{ik}}(x, Y) = \overset{1}{\Gamma^h_{ik}}(x, X) + g^{hl}(x, Y) Y_j \frac{\partial \overset{1}{\Gamma^j_{ik}}}{\partial X^l} . \qquad (4.15)$$

If we define the covariant derivative of the metric tensor g_{ij} by the equations

$$\overset{2}{\delta} g_{ij} = \frac{\partial g_{ij}}{\partial x^k} d x^k + \frac{\partial g_{ij}}{\partial X^h} d X^h - g_{ih} \overset{2}{\Gamma^h_{jk}} d x^k - g_{hj} \overset{2}{\Gamma^h_{ik}} d x^k , \qquad (4.16)$$

a simple calculation based on (4.11) and (4.15) shows that

$$\overset{2}{\delta} g_{ij} = \frac{\partial g_{ij}}{\partial X^h} \overset{1}{\delta} X^h - Y_h \frac{\partial \overset{1}{\Gamma^h_{ik}}}{\partial X^j} d x^k . \qquad (4.17)$$

From this result we deduce that if the covariant derivative of the g_{ij} is defined according to (4.16), it is impossible to find further restrictions which, when imposed upon the connection coefficients, cause the covariant derivatives of the g_{ij} to vanish[1].

Let us now consider the geodesics of F_n in the light of this theory. Suppose that the vectors X^i are the tangent vectors dx^i/ds of a geodesic, so that if we use the notation (3.1.26) we may write the differential

---

[1] This conclusion does not apply to the covariant derivative as defined by KAWAGUCHI [5], p. 196, for the method of the latter is formally similar to that of CARTAN (Ch. III, §§ 1—2). For instance, KAWAGUCHI (loc. cit.) defines

$$D V^i = d V^i + \Gamma^i_{jk} V^j d x^k + F^{-1} A^i_{jk} V^j D X^k ,$$

for a vector $V^i = V^i(x, X)$. Thus even if the vector V^i happens to be *independent* of X^k, its covariant derivative still depends on the derivatives of X^i, which, in this context, assumes the role of the element of support. Hence the Γ^i_{hk} may be adjusted such that $D g_{ij} = 0$.

equations (2.2.8) of the latter in the form

$$\frac{dX^i}{ds} + 2G^i(x, X) = 0 .\tag{4.18}$$

In order to discover under which conditions the geodesics are the auto-parallels with respect to the parallel displacement defined by the $\overset{1}{\Gamma}{}^i_k(x, X)$, we have to re-write the conditions (4.13) in a different form. Applying (2.2.10) to the right-hand side of (4.13) and noting (3.1.28), (1.3.5) and (3.1.27′), we find that (4.13) is equivalent to

$$Y_i\overset{1}{\Gamma}{}^i_k = Y_i\frac{\partial G^i}{\partial X^k} .\tag{4.19}$$

Differentiating this relation with respect to X^h and multiplying the result by g^{hj}, we obtain

$$\overset{1}{\Gamma}{}^j_k + g^{hj}Y_i\overset{1}{\Gamma}{}^i_{hk} = \frac{\partial G^j}{\partial X^k} + Y_i g^{hj}G^i_{hk} ,$$

and hence, using (4.19) once more, together with the fact that the G^i are homogeneous of the second degree, we deduce that

$$2G^i = \overset{1}{\Gamma}{}^i_k X^k + g^{ih}Y_j\left(\overset{1}{\Gamma}{}^j_{hk}X^k - \overset{1}{\Gamma}{}^j_h\right) .\tag{4.20}$$

In view of (4.1) the differential equations (4.18) of the geodesics thus become

$$\frac{\delta X^i}{\delta s} + g^{ih}Y_j\left(\overset{1}{\Gamma}{}^j_{hk}X^k - \overset{1}{\Gamma}{}^j_h\right) = 0 .\tag{4.21}$$

Thus the geodesics will coincide with the autoparallel curves of F_n if and only if[1]

$$Y_j\left(\overset{1}{\Gamma}{}^j_{hk}X^k - \overset{1}{\Gamma}{}^j_h\right) = 0 .\tag{4.22}$$

This condition is satisfied (although in a rather trivial sense) if the $\overset{1}{\Gamma}{}^j_{hk}$ are symmetric in h and k. For then we have, as a result of (4.2),

$$\overset{1}{\Gamma}{}^j_{hk}X^k = \overset{1}{\Gamma}{}^j_{kh}X^k = \overset{1}{\Gamma}{}^j_h.$$

But in this case the $\overset{1}{\Gamma}{}^j_{hk}$ are completely determined, for equation (4.20) then becomes

$$2G^i = \overset{1}{\Gamma}{}^i_k X^k ,$$

and repeated differentiation with respect to X^h and X^k [during which process (4.2) and the symmetry of $\overset{1}{\Gamma}{}^i_{hk}$ are taken into account] will yield

$$\overset{1}{\Gamma}{}^i_{hk} = G^i_{hk} .$$

---

[1] Rund [15].

But the coefficients on the right-hand side are the connection parameters as defined by BERWALD[1], and these are completely determined by the metric of F_n. Hence we have the theorem[2]:

If the coefficients $\overset{1}{\Gamma}{}^i_{hk}$ of a metric non-linear connection are symmetric in their lower indices h and k, they are uniquely determined and equivalent to the connection parameters of BERWALD.

We observe that under these circumstances the $\overset{2}{\Gamma}{}^i_{hk}$ are determined by (4.15), but they will coincide with the $\overset{1}{\Gamma}{}^i_{hk}$ if and only if $C^i_{hk|r}X^r = 0$. This result is easily established by a brief calculation.

Further developments in the theory of non-linear connections due to VAGNER are based on methods involving the centro-affine geometry of the local indicatrices, for which connection parameters are defined, which are related to the $\overset{1}{\Gamma}{}^i_h$, $\overset{2}{\Gamma}_{ih}$ by what might be called a process of projection. It is possible to characterise "affinely-connected" spaces (i. e. spaces for which the connection coefficients are functions of position only) by conditions which are essentially equivalent to the conditions discussed at the end of § 3 of Chapter III. In particular, for two-dimensional Finsler spaces of non-vanishing curvature it may be shown that a necessary and sufficient condition for the space to be affinely-connected is that the indicatrices all possess the same constant centrally affine curvature[3]. By the same methods the necessary and sufficient conditions that the space be flat may be derived[4]. Using the theory of the composite manifold it is possible to treat the theory of CARTAN from this alternative point of view, but owing to the complexity of this method the reader must be referred to the original publications of VAGNER[5].

In conclusion we remark that the class of non-linear metric connections studied by KAWAGUCHI are of particular geometrical interest since they arise naturally in the course of an investigation into the properties of motions in Minkowskian spaces. Consider the correspondence between two neighbouring tangent spaces $T_n(x^i)$ and $T_n(x^i + dx^i)$ defined by the equations[6]

$$X^{*i} = \frac{F(x, X)}{F(x + dx, \overline{X})}\, \overline{X}^i\,, \tag{4.23}$$

where

$$\overline{X}^i = (\delta^i_j - \xi^i_{jk}(x, X)\, dx^k)\, X^j\,, \tag{4.24}$$

---

[1] Ch. III, § 3.

[2] This result seems to be the essence of the fundamental uniqueness theorem due to VAGNER ([13], p. 116). A detailed treatment of the geodesics, including such aspects as the second variation and the Jacobi equation, from the point of view of the theory of non-linear connections, is given by VAGNER [13], § 4.

[3] VAGNER [13], p. 124.

[4] See Ch. IV; VAGNER [13], p. 126.

[5] VAGNER [11, 15]; [13], §§ 5—6.

[6] KAWAGUCHI [5], p. 186.

the functions $\xi^i_{jk}(x, X)$ being homogeneous of degree zero in the X^i (in fact, it is suggested that one might use functions, which are "generalised" homogeneous functions[1]). It is easily verified that equations (4.23) are equivalent to a "parallel displacement" which may be expressed in the form

$$dX^i = -\left\{B_j\,\xi^i_{hk}X^h + X^i\,\frac{1}{F}\,\frac{\partial F}{\partial x^k}\right\}dx^k,\qquad(4.25)$$

where

$$B^i_j(x, X) = \delta^i_j - \frac{X^i}{F}\,\frac{\partial F}{\partial X^j}.\qquad(4.26)$$

Hence one may write

$$\overset{1}{\Gamma}{}^i_k = B^i_j\,\xi^j_{hk}\,X^h + X^i\,\frac{1}{F}\,\frac{\partial F}{\partial x^k},\qquad(4.27)$$

and direct substitution in (4.12) shows that the parallel displacement thus defined is metric. It may be shown[2] that the connection parameters (4.27) are such that the geodesics are autoparallel, provided the functions ξ^i_{hk} satisfy the conditions

$$B^i_j\,\xi^j_{hk}\,X^h X^k = (g^{ij} - F^{-2}\,X^i\,X^j)\,X^k\,X^h\left\{\frac{\partial g_{jh}(x, X)}{\partial x^k} - \frac{1}{2}\,\frac{\partial g_{kh}(x, X)}{\partial x^j}\right\}.\qquad(4.28)$$

As we have already observed above, further conditions may be imposed on the coefficients (4.27) which will ensure the vanishing of the covariant derivative of the metric tensor.

§ 5. The Local Imbedding Theories

In this section we shall briefly treat the question of the imbedding of an n-dimensional Finsler space F_n in a space of higher dimension and of a different type. In other words, we shall inquire under what conditions a F_n can be regarded as a subspace (holonomic or non-holonomic) of another space. This problem has not yet been dealt with as fully as the corresponding problem in Riemannian geometry, and we shall thus merely outline a few of the principal known results.

An unusual and very fruitful approach is due to EISENHART [3], who showed that *a Finsler space can be derived from a Riemannian space by the application of a homogeneous contact transformation*[3]. This raises the problem of developing a tensor calculus with respect to the group of homogeneous contact transformations. To a certain extent such a development is contained already in the work of HOSOKAWA [1], YANO [2, 3], and EISENHART and KNEBELMAN [1]. The latter authors introduced the so-called first contact frame (see below), to which a second frame is adjoined in the subsequent work of MUTO [1] and DOYLE [1]. That these methods are applicable to the study of Finsler geometry,

---

[1] The idea of a "generalised" homogeneous function is also due to KAWAGUCHI and is described in detail in [4].

[2] KAWAGUCHI [5], p. 191.

[3] It should be mentioned, however, that RACHEVSKI [1] had previously shown how a two-dimensional Finsler metric can be obtained by a contact transformation from a metric involving the line element.

provided certain special restrictions are imposed, is shown by MUTO and YANO [1]. The work of DOYLE [1] suggests the introduction of non-holonomic subspaces, an idea that was taken up once more by DAVIES [5, 6] in connection with the theory of general metric spaces. Again, the resulting theory contains certain restrictions, which give rise to the difficulty that the work of MUTO and YANO [1] cannot be included, nor is the relation to the theory of EISENHART [3] clearly exhibited.

Recently, in a long and detailed paper, YANO and DAVIES [1] developed the so-called contact tensor calculus in its most general form, this development being based on the theory of non-holonomic subspaces (of a Riemannian space), and the gaps mentioned above have thus been filled. It is shown that Cartan's theory of Finsler spaces can be incorporated as a special case when the contact transformations reduce to the extended point transformations which had hitherto been used in the study of Finsler geometry. This suggests that *possibly Finsler spaces may be regarded as non-holonomic subspaces of certain Riemannian spaces*, and a further paper by YANO and DAVIES [2] treats this question.

It is this latter approach which we shall discuss very briefly in the present section, for it is manifestly impossible within the bounds of the present monograph to give a reasonably comprehensible account of the developments outlined above, especially as we would be compelled to give a full treatment of the contact tensor calculus as well as of the — somewhat complicated and extensive — theory of non-holonomic subspaces. For a complete discussion of these topics and their applications to Finsler spaces the reader must be referred to YANO and DAVIES [1][1].

We shall consider the $2n$-dimensional space V_{2n} of the line elements $(x^i, \dot{x}^i)$ of the Finsler space F_n, and we introduce the notation $x^i = x^i$, $\dot{x}^i = x^{n+i}$, $x^A = (x^i, x^{n+i})$ [2]. The coordinate transformations

$$x^{A'} = x^{A'}(x^A) \tag{5.1}$$

in V_{2n} are assumed to be of the form (1.1.1):

$$x^{i'} = x^{i'}(x^j), \quad \dot{x}^{i'} \equiv x^{(n+i)'} = \frac{\partial x^{i'}}{\partial x^i}\,\dot{x}^i = \frac{\partial x^{i'}}{\partial x^i}\,x^{n+i}. \tag{5.1a}$$

Suppose that we are given $2n$ linearly independent contravariant vector fields $B_i{}^A$, $N_{n+j}{}^A$ in V_{2n}. We may then define the inverse $(B^i{}_A, N^{n+j}{}_A)$ of

<hr>

[1] Results similar to those given by YANO and DAVIES [1] were found independently and almost simultaneously by TAKANO [2], while a slightly more general discussion is given by SUGURI and NAKAYAMA [1]. For the literature concerning non-holonomic subspaces of Riemannian or affinely connected spaces see YANO and DAVIES [1].

[2] Throughout this section it is understood that capital Latin letters, A, B, ... run from 1 to $2n$, while small Latin letters run from 1 to n as before.

the matrix $(B_i{}^A, N_{n+j}{}^A)$, such that

$$B_j{}^A B^i{}_A = \delta_j^i\,, \qquad B_j{}^A N^{n+i}{}_A = 0\,, \tag{5.2}$$

$$N_{n+j}{}^A B^i{}_A = 0\,, \qquad N_{n+j}{}^A N^{n+i}{}_A = \delta_{n+j}^{n+i}\,, \tag{5.2a}$$

and

$$B_i{}^A B^i{}_B + N_{n+i}{}^A N^{n+i}{}_B = \delta_B^A\,. \tag{5.3}$$

The $2n$ vectors $B_i{}^A$, $N_{n+i}{}^A$ define the so-called *non-holonomic frame*.

This terminology is justified by the following definition. In each tangent space of V_{2n} the n vectors $B_i{}^A$ span an n-dimensional "plane". The set of all these planes (taken over the various tangent spaces) define a *non-holonomic* subspace of V_{2n}, the word non-holonomic indicating that in general there does not exist a family of n-dimensional surfaces in V_{2n} whose tangent planes (in the tangent spaces of V_{2n}) coincide with the planes spanned by the vectors $B_i{}^A$. Should, however, such a family of surfaces exist, the non-holonomic subspace reduces to an ordinary (holonomic) subspace of the type considered in Chapter V.

An arbitrary displacement dx^A in V_{2n} may, by (5.3), be expressed in the form

$$dx^A = \delta_B^A\, dx^B = B_i{}^A (dx)^i + N_{n+i}{}^A (dx)^{n+i}\,, \tag{5.4}$$

where we have written

$$(dx)^i = B^i{}_B\, dx^B\,, \qquad (dx)^{n+i} = N^{n+i}{}_B\, dx^B\,, \tag{5.5}$$

the brackets indicating that the $(dx)^i$ and $(dx)^{n+i}$ are not, in general, exact differentials. In particular, we note that if, for instance, $(dx)^{n+i} = 0$, so that $dx^A = B_i{}^A (dx)^i$, then dx^A is tangent to *the non-holonomic subspace V_{2n}^n of V_{2n} spanned by the n vectors $B_i{}^A$.*

We shall now show that the extended point-transformations (5.1a) lead to a specialisation of the non-holonomic frame which gives rise to a non-holonomic affine connection in V_{2n}. This defines an induced connection on the non-holonomic subspace, and by a restriction on the connection in V_{2n} we shall thus obtain the coefficients of the connection in the Finsler space F_n.

Let $f = f(x^A)$ be an arbitrary function of the coordinates x^A of V_{2n}. Using (5.4) we may write

$$df = \frac{\partial f}{\partial x^A}\, dx^A = (dx)^i X_i\, f + (dx)^{n+i} X_{n+i}\, f\,, \tag{5.6}$$

where we have defined the so-called non-holonomic partial derivatives by

$$X_i\, f = B_i{}^A \frac{\partial f}{\partial x^A}\,, \qquad X_{n+i}\, f = N_{n+i}{}^A \frac{\partial f}{\partial x^A}\,. \tag{5.7}$$

For instance, for a displacement along V_{2n}^n we have

$$df = (dx)^i X_i f .$$

Using equations (5.2) it is easily seen that the commutation formulae for the operators (5.7) may be written in the form[1]

$$(X_h X_k - X_k X_h) f = \Omega_{hk}{}^i X_i f + \Omega_{hk}{}^{n+i} X_{n+i} f , \tag{5.8}$$

$$(X_h X_{n+k} - X_{n+k} X_h) f = \Omega_{h,n+k}{}^i X_i f + \Omega_{h,n+k}{}^{n+i} X_{n+i} f , \tag{5.8a}$$

$$\begin{aligned}(X_{n+h} X_{n+k} - X_{n+k} X_{n+h}) f \\ = \Omega_{n+h,n+k}{}^i X_i f + \Omega_{n+h,n+k}{}^{n+i} X_{n+i} f ,\end{aligned} \tag{5.8b}$$

where we have written

$$\Omega_{hk}{}^i = (X_h B_k{}^A - X_k B_h{}^A) B^i{}_A , \tag{5.9}$$

$$\Omega_{hk}{}^{n+i} = (X_h B_k{}^A - X_k B_h{}^A) N^{n+i}{}_A , \tag{5.9a}$$

$$\Omega_{h,n+k}{}^i = (X_h N_{n+k}{}^A - X_{n+k} B_h{}^A) B^i{}_A , \tag{5.9b}$$

$$\Omega_{h,n+k}{}^{n+i} = (X_h N_{n+k}{}^A - X_{n+k} B_h{}^A) N^{n+i}{}_A , \tag{5.9c}$$

$$\Omega_{n+h,n+k}{}^i = (X_{n+h} N_{n+k}{}^A - X_{n+k} N_{n+h}{}^A) B^i{}_A , \tag{5.9d}$$

$$\Omega_{n+h,n+k}{}^{n+i} = (X_{n+h} N_{n+k}{}^A - X_{n+k} N_{n+h}{}^A) N^{n+i}{}_A . \tag{5.9e}$$

Let us now suppose that we are given an affine connection Γ^A_{BC} in V_{2n}, which gives rise to a covariant differential of a vector field v^A of V_{2n}:

$$\delta v^A = dv^A + \Gamma^A_{BC} v^C dx^B .$$

In particular, if v^A is tangent to V_{2n}^n, i. e. if $v^A = B_i{}^A v^i$, we have

$$\delta v^A = B_i{}^A dv^i + \Gamma^A_{BC} v^C dx^B + (dB_i{}^A) v^i ,$$

and if we apply (5.4) and (5.6) this may be written in the form

$$\begin{aligned}\delta v^A = B_i{}^A dv^i + (dx)^j [\Gamma^A_{BC} B_k{}^C B_j{}^B + X_j B_k{}^A] v^k + \\ + (dx)^{n+j} [\Gamma^A_{BC} B_k{}^C N_{n+j}{}^B + X_{n+j} B_k{}^A] v^k .\end{aligned} \tag{5.10}$$

The projection of this differential on V_{2n}^n may then be expressed as

$$B_i{}^A [dv^i + (dx)^j \Gamma^i_{jk} v^k + (dx)^{n+j} \Gamma^i_{n+j,k} v^k] ,$$

where we have put[2]

$$\Gamma^i_{jk} = B^i{}_A (B_j{}^B B_k{}^C \Gamma^A_{BC} + X_j B_k{}^A) , \tag{5.11}$$

together with

$$\Gamma^i_{n+j,k} = B^i{}_A (N_{n+j}{}^B B_k{}^C \Gamma^A_{BC} + X_{n+j} B_k{}^A) . \tag{5.11a}$$

Interchanging j and k in the first of these equations and subtracting, we obtain in view of (5.9),

$$\Gamma^i_{jk} - \Gamma^i_{kj} = T_{jk}{}^i + \Omega_{jk}{}^i \tag{5.12}$$

---

[1] DAVIES and YANO [1], p. 5.

[2] These coefficients must not be confused with the parameters (3.1.30).

where we have written

$$T_{jk}{}^{i} = B^{i}{}_{A} B_{j}{}^{B} B_{k}{}^{C} (\Gamma^{A}_{BC} - \Gamma^{A}_{CB}) .$$

Similar relations are found for quantities involving indices $> n$.

The covariant differential of v^i may thus be defined as

$$\delta v^{i} = d v^{i} + (d x)^{j} \Gamma^{i}_{jk} v^{k} + (d x)^{n+j} \Gamma^{i}_{n+j,k} v^{k} , \tag{5.13}$$

and the covariant derivative of v^i along V^n_{2n} is given by

$$\nabla_{j} v^{i} = X_{j} v^{i} + \Gamma^{i}_{jk} v^{k} , \tag{5.13a}$$

while along the non-holonomic space defined by the $N_{n+j}{}^{A}$ we have

$$\nabla_{n+j} v^{i} = X_{n+j} v^{i} + \Gamma^{i}_{n+j,k} v^{k} , \tag{5.13b}$$

so that

$$\delta v^{i} = (d x)^{j} \nabla_{j} v^{i} + (dx)^{n+j} \nabla_{n+j} v^{i} . \tag{5.13c}$$

Naturally, this process of covariant differentiation may also be applied to tensors of arbitrary rank.

Furthermore, let us now suppose that a Riemannian metric with metric tensor g_{AB} has been imposed on V_{2n} (we shall specify g_{AB} presently). Assuming that the vectors $B_i{}^A$, $N_{n+i}{}^A$ are mutually orthogonal with respect to this metric:

$$g_{AB} B_{i}{}^{A} N_{n+j}{}^{B} = 0 ,$$

it follows from (5.4) that we have

$$ds^{2} = g_{AB} d x^{A} d x^{B} = g_{jk} (d x)^{j} (d x)^{k} + g_{n+j,n+k} (d x)^{n+j} (d x)^{n+k} ,$$

where

$$g_{jk} = g_{AB} B_{j}{}^{A} B_{k}{}^{B} , \qquad g_{n+j,n+k} = g_{AB} N_{n+j}{}^{A} N_{n+k}{}^{B} . \tag{5.14}$$

These tensors define a metric in the non-holonomic subspaces.

We now stipulate that the coefficients Γ^{A}_{BC} of the affine connection be so chosen that $\nabla_{i} g_{jk} = 0$ and $\nabla_{n+i} (g_{n+j,n+k}) = 0$; in other words, it is assumed that the connection is metric[1]. This entails that the Γ^{i}_{jk} have to satisfy certain conditions which we shall derive as follows. For indices $i, j, k \leq n$ we have, by definition,

$$0 = \nabla_{i} g_{jk} = X_{i} g_{jk} - g_{hk} \Gamma^{h}_{ij} - g_{jh} \Gamma^{h}_{ik} .$$

Cyclic permutation of the indices i, j, k yields two further equations, from the sum of which the original equation is subtracted. In this manner we find that

$$2 g_{ih} \Gamma^{h}_{jk} = (X_{k} g_{ij} + X_{j} g_{ki} - X_{i} g_{jk}) +$$

$$+ g_{jh} (\Gamma^{h}_{ik} - \Gamma^{h}_{ki}) + g_{hk} (\Gamma^{h}_{ij} - \Gamma^{h}_{ji}) + g_{ih} (\Gamma^{h}_{jk} - \Gamma^{h}_{kj}) .$$

---

[1] In fact, the reader may verify that these conditions result from the requirement that $\delta g_{AB} = 0$. See YANO and DAVIES [1], p. 13.

Introducing the notation

$$\Omega^{i}{}_{jk} = g^{im} g_{hk}\, \Omega_{mj}{}^{h},$$

and

$$T^{i}{}_{jk} = g^{im}\, g_{hk}\, T_{mj}{}^{h},$$

we see that with the aid of (5.12) the above condition on the connection coefficients may be expressed in the form

$$2\Gamma^{i}_{jk} = g^{ih}(X_k\, g_{hj} + X_j\, g_{kh} - X_h\, g_{jk}) + \\ + (\Omega_{jk}{}^{i} + \Omega^{i}{}_{jk} + \Omega^{i}{}_{kj}) + (T_{jk}{}^{i} + T^{i}{}_{jk} + T^{i}{}_{kj}),\qquad (5.15)$$

together with similar conditions for indices greater than n. The $T_{jk}{}^{i}$ may be regarded as the non-holonomic components of the torsion of V_{2n}.

Thus far we have described a few essential notions of the theory of non-holonomic subspaces of a Riemannian space from a fairly general point of view. We shall now adapt this theory to our special needs.

As we have noted above, the non-holonomic frame may be specialised in various ways. In order to do so now, we notice that linear combinations of the type $C^{i}_{j}B_{i}{}^{A}$ of the $B_{i}{}^{A}$ span a non-holonomic subspace identical with V^{n}_{2n}, provided that $\det(C^{i}_{j}) \neq 0$. In view of the linear independence of the $B_{i}{}^{A}$ we may find values C^{i}_{j} such that $C^{i}_{j}B_{k}{}^{j} = \delta^{i}_{k}$. Having carried out such a transformation, and furthermore specialising the $N_{n+j}{}^{A}$ such that

$$N_{n+j}{}^{A} = \frac{\partial x^{A}}{\partial x^{n+j}} = (0,\ \delta^{n+i}_{n+j})$$

in accordance with equations (5.1a), our non-holonomic frame may be represented by

$$B_{j}{}^{A} = (\delta^{i}_{j},\ B_{j}{}^{n+i}),\qquad N_{n+j}{}^{A} = (0,\ \delta^{n+i}_{n+j}),\qquad (5.16)$$

while in virtue of (5.2) we must have for the inverse matrix a representation of the type

$$B^{i}{}_{A} = (\delta^{i}_{j},\ 0),\qquad N^{n+i}{}_{A} = (-B_{j}{}^{n+i},\ \delta^{n+i}_{n+j}).\qquad (5.17)$$

However, it will be seen that these special values are not invariant under a point transformation such as (5.1a). Since the $B_{i}{}^{A}$ are components of contravariant vectors in V_{2n} it follows that

$$B_{j}{}^{A'} = \frac{\partial x^{A'}}{\partial x^{A}}\, B_{j}{}^{A} = \left(\frac{\partial x^{i'}}{\partial x^{j}},\ \frac{\partial x^{(n+i)'}}{\partial x^{j}} + \frac{\partial x^{(n+i)'}}{\partial x^{n+i}}\, B_{j}{}^{n+i}\right).$$

But by (5.1a) we have

$$\frac{\partial x^{(n+i)'}}{\partial x^{j}} = \frac{\partial \dot{x}^{i'}}{\partial x^{j}} = \frac{\partial^{2} x^{i'}}{\partial x^{j} \partial x^{h}}\, \dot{x}^{h},\qquad \frac{\partial x^{(n+i)'}}{\partial x^{n+i}} = \frac{\partial x^{i'}}{\partial x^{i}},$$

so that

$$B_{j}{}^{A'} = \left(\frac{\partial x^{i'}}{\partial x^{j}},\ \frac{\partial^{2} x^{i'}}{\partial x^{j} \partial x^{h}}\, \dot{x}^{h} + \frac{\partial x^{i'}}{\partial x^{i}}\, B_{j}{}^{n+i}\right).\qquad (5.18)$$

If we adjoin to the transformation (5.1a) a change in the non-holonomic frame given by

$$B_{j'}{}^{A'} = B_j{}^{A'} \frac{\partial x^j}{\partial x^{j'}} \, ,$$

equation (5.18) becomes

$$B_{j'}{}^{A'} = \left(\delta_{j'}^{i'} \, , \, \frac{\partial^2 x^{i'}}{\partial x^j \, \partial x^h} \frac{\partial x^j}{\partial x^{j'}} \dot{x}^h + \frac{\partial x^{i'}}{\partial x^i} \frac{\partial x^j}{\partial x^{j'}} B_j{}^{n+i} \right) ,$$

so that we may write in analogy with (5.16)

$$B_{j'}{}^{A'} = (\delta_{j'}^{i'} \, , \, B_{j'}{}^{(n+i)'}) , \tag{5.19}$$

where we have put

$$B_{j'}{}^{(n+i)'} = \frac{\partial x^{i'}}{\partial x^i} \left(\frac{\partial x^j}{\partial x^{j'}} B_j{}^{n+i} - \frac{\partial^2 x^i}{\partial x^{j'} \, \partial x^{k'}} \dot{x}^{k'} \right) . \tag{5.19a}$$

Similarly, the $N_{n+i}{}^A$ lose their special form under the transformation (5.1a): but again, if we adjoin a change of frame of the type

$$N_{(n+i)'}{}^{A'} = \frac{\partial x^i}{\partial x^{i'}} N_{n+i}{}^{A'} \, ,$$

the special form of these quantities may be recovered. The same applies to the inverse quantities (5.17), and consequently we have to stipulate that in this special theory all point transformations (5.1a) be followed by a suitable change of the non-holonomic frame[1].

Equation (5.19a) represents the transformation of the $B_j{}^{n+i}$ under these combined transformations, and if we put

$$B_j{}^{n+i} = - \Gamma_j^i , \tag{5.20}$$

we see that (5.19a) becomes

$$\Gamma_{j'}^{i'} = \frac{\partial x^{i'}}{\partial x^i} \left(\Gamma_j^i \frac{\partial x^j}{\partial x^{j'}} + \frac{\partial^2 x^i}{\partial x^{j'} \, \partial x^{k'}} \dot{x}^{k'} \right) . \tag{5.20a}$$

We are now in a position to fully exploit the special forms (5.16) and (5.17) of the non-holonomic frame. For from these equations we may now deduce that (5.5) and (5.7) become

$$(d x)^i = d x^i , \qquad (d x)^{n+i} = - B_j{}^{n+i} d x^j + d x^{n+i} ,$$

---

[1] Equations (5.16) and (5.17) exhibit the special form of the frame as used by Yano and Davies [2], p. 412. In Yano and Davies [1] (p. 14) the special form is taken to be

$$B_j{}^A = (\delta_j^i, B_j{}^{n+i}) , \qquad B^i{}_A = (\delta_j^i - B^i{}_{n+h} B_j{}^{n+h}, B^i{}_{n+j}) ,$$

$$N_{n+j}{}^A = (- B^i{}_{n+j}, \delta_{n+j}^{n+i} - B_k{}^{n+i} B^k{}_{n+j}) , \qquad N^{n+i}{}_A = - (B_j{}^{n+i}, \delta_{n+j}^{n+i}) .$$

The case considered by Davies [5] is more special, namely:

$$B^i{}_{n+k} B_j{}^{n+k} = B_k{}^{n+i} B^k{}_{n+j} = 0 ,$$

while Doyle [1] stipulates

$$B_j{}^A = (\delta_j^i, B_j{}^{n+i}) , \qquad N_{n+j}{}^A = (N^i{}_{n+j}, \delta_{n+j}^{n+i}) .$$

and

$$X_i f = \frac{\partial f}{\partial x^i} + B_i{}^{n+j} \frac{\partial f}{\partial x^{n+j}}, \quad X_{n+i} f = \frac{\partial f}{\partial x^{n+i}}.$$

In particular, if we write $X_{n+i} f = \dot{X}_i f$, $(dx)^{n+i} = (d\dot{x})^i$, and take heed of (5.20), the latter equations may be written in the form

$$(dx)^i = dx^i, \quad (d\dot{x})^i = d\dot{x}^i + \Gamma^i_j dx^j, \tag{5.21}$$

and

$$X_i f = \frac{\partial f}{\partial x^i} - \Gamma^j_i \frac{\partial f}{\partial \dot{x}^j}, \quad \dot{X}_i f = \frac{\partial f}{\partial \dot{x}^i}. \tag{5.21a}$$

Incidentally, we also find that the commutation formulae (5.8) assume a much simpler form as a result of (5.16) and (5.17):

$$(X_h X_k - X_k X_h) f = \Omega_{hk}{}^{n+i} \dot{X}_i f = R^i{}_{hk} \dot{X}_i f, \tag{5.22}$$

$$(X_h \dot{X}_k - X_k \dot{X}_h) f = \Omega_{h,n+k}{}^{n+i} \dot{X}_i f = S^i{}_{hk} \dot{X}_i f, \tag{5.22a}$$

$$(\dot{X}_h \dot{X}_k - \dot{X}_k \dot{X}_h) f = 0, \tag{5.22b}$$

where we have written

$$R^i{}_{hk} = \Omega_{hk}{}^{n+i}, \quad S^i{}_{hk} = \Omega_{h,n+k}{}^{n+i}. \tag{5.22c}$$

The latter tensors may be expressed more directly as follows. From (5.9a) and (5.9b) we find on application of (5.16) and (5.17) that

$$\Omega_{hk}{}^{n+i} = (X_h B_k{}^{n+i} - X_k B_h{}^{n+i}),$$

and

$$\Omega_{h,n+k}{}^{n+i} = - X_{n+k} B_h{}^{n+i},$$

and hence, using (5.20) and (5.22c), we obtain

$$R^i{}_{hk} = X_k \Gamma^i_h - X_h \Gamma^i_k, \tag{5.23}$$

together with[1]

$$S^i{}_{hk} = \dot{X}_k \Gamma^i_h. \tag{5.23a}$$

With the aid of (5.21a) we may now also express the covariant derivatives (5.13a) and (5.13b) in a more specialised form, namely as

$$\nabla_j v^i = \frac{\partial v^i}{\partial x^j} - \Gamma^h_j \frac{\partial v^i}{\partial \dot{x}^h} + \Gamma^i_{jk} v^k, \tag{5.24}$$

and

$$\nabla_{n+j} v^i = \frac{\partial v^i}{\partial \dot{x}^j} + \Gamma^i_{n+j,k} v^k. \tag{5.24a}$$

At this stage we introduce the metric tensor of the given Finsler space F_n. If $F(x^k, \dot{x}^k)$ is the metric function of F_n from which the metric tensor g_{ij} has been derived according to (1.3.1), we now identify the Riemannian metric of V_{2n} as exemplified by (5.14) with this tensor, stipulating in particular that:

$$g_{n+i,n+j} = g_{ij}. \tag{5.25}$$

---

[1] YANO and DAVIES [2], p. 414.

The conditions (5.15) imposed upon the connection coefficients Γ^i_{jk} now involve the Finsler metric. Naturally, they are not completely determined by these conditions. We therefore impose further conditions, namely that $V_j \dot{x}^i = 0$, $V_{n+j} \dot{x}^i = \delta^i_j$, [which, by (5.13c), is equivalent to the requirement that $\delta \dot{x}^i = (dx)^{n+i}$]. Substituting these conditions in (5.24) and (5.24a), we find that we must have

$$-\Gamma^i_j + \Gamma^i_{jk}\, \dot{x}^k = 0 , \tag{5.26}$$

together with

$$\Gamma^i_{n+j,k}\, \dot{x}^k = 0 . \tag{5.27}$$

These relations suggest that we should put

$$\Gamma^i_j = \frac{\partial G^i}{\partial \dot{x}^j} , \qquad \Gamma^i_{jk} = \Gamma^{*i}_{jk} , \qquad \Gamma^i_{n+j,k} = C^i_{jk} , \tag{5.28}$$

the quantities on the right-hand side being defined with respect to the Finsler metric $F(x^k, \dot{x}^k)$ by equations (3.1.26), (3.1.28), (1.3.4). In fact, it follows from (3.1.27') and (1.3.5) that under these circumstances (5.26) and (5.27) would be satisfied.

Let us therefore take

$$\Gamma^i_j = \Gamma^i_{jk}\, \dot{x}^k = \frac{\partial G^i}{\partial \dot{x}^j} . \tag{5.29}$$

In view of (5.21a) this will affect (5.15); in fact, noting also (5.9), (5.16) and (5.17) we see that (5.15) now reduces to

$$2\Gamma^i_{jk} = g^{ih} \left(\frac{\partial g_{hj}}{\partial x^k} - \frac{\partial G^l}{\partial \dot{x}^k} \frac{\partial g_{hj}}{\partial \dot{x}^l} + \frac{\partial g_{hk}}{\partial x^j} - \frac{\partial G^l}{\partial \dot{x}^j} \frac{\partial g_{hk}}{\partial \dot{x}^l} - \frac{\partial g_{jk}}{\partial x^h} + \frac{\partial G^l}{\partial \dot{x}^h} \frac{\partial g_{jk}}{\partial \dot{x}^l} \right) + (T_{jk}{}^i + T^i{}_{jk} + T^i{}_{kj}) . \tag{5.30}$$

Comparing this relation with (3.1.28), we see that the coefficients Γ^i_{jk} become the Γ^{*i}_{jk} provided we impose the condition

$$T_{jk}{}^i + T^i{}_{jk} + T^i{}_{kj} = 0 \tag{5.31}$$

on the components of the torsion of V_{2n}.

Similarly, by writing down the relation corresponding to (5.15) involving indices greater than n, we find by means of (5.21a) and (5.25):

$$2\Gamma^i_{n+j,k} = g^{ih} \frac{\partial g_{hk}}{\partial \dot{x}^j} + \Omega^i{}_{k,n+j} + T_{n+j,k}{}^i + T^i{}_{n+j,k} + T^i{}_{k,n+j} . \tag{5.32}$$

Again, a comparison with (1.3.4) shows that we may identify the $\Gamma^i_{n+j,k}$ with the C^i_{jk} provided that

$$\Omega^i{}_{k,n+j} + T_{n+j,k}{}^i + T^i{}_{n+j,k} + T^i{}_{k,n+j} = 0 . \tag{5.33}$$

We have therefore proved the following theorem:

A Finsler space F_n may always be regarded as a non-holonomic subspace of a Riemannian space V_{2n} whose metric is given by (5.25) and whose torsion satisfies (5.31) and (5.33)[1].

Another ingenious approach to the imbedding problem — which is in some respects similar to the one outlined above — is due to GALVANI[2]. Again, owing to the wide range and the complexity of the concepts involved in this method, we shall merely give a very cursory sketch of the fundamental ideas and of the most important results obtained by GALVANI, omitting all proofs. While in the classical treatment dealing with the problem of imbedding of Riemannian spaces in higher dimensional euclidean spaces the generating element of the imbedding space is the point, an entirely different point of view is adopted in the present theory: each point of the Finsler space F_n is to correspond to a configuration S in a euclidean space E_N of N dimensions where $N > n$, and where S is defined by the triplet

$$S = (M, \varDelta, \varPi) , \tag{5.34}$$

M being a point, $\varDelta$ a line through M, and $\varPi$ an n-dimensional plane containing $\varDelta$. Clearly

$$L = L(M, \varDelta) \tag{5.35}$$

is simply a line-element. Also, we recall that according to the general theory of Pfaffian forms a euclidean connection $\mathfrak{L}_{2n-1}$ is defined in the space of line elements by a set of linearly independent forms ω_i, ω_{ij} $(i, j = 1, \ldots, n; \omega_{ij} = -\omega_{ji})$. In general, the system

$$\omega_i = 0 , \tag{5.36}$$

is not completely integrable, but should the contrary be true we shall call the corresponding connection $\mathfrak{L}_{2n-1}$ a *semi-punctual* connection.

Let us consider a space M_n of n dimensions on which local coordinates x^i are defined. Again, when determining directions $\dot{x}^i$ in the tangent spaces of M_n only the ratios of the $\dot{x}^i$ are of importance, so that the points x^i of M_n together with the directions $\dot{x}^i$ define a $(2n - 1)$-dimensional space W_{2n-1}. By imposing the connection $\mathfrak{L}_{2n-1}$ on W_{2n-1} such that the ω_i do not depend on the $d\dot{x}^i$, a semi-punctual space L_{2n-1} is obtained. Furthermore, we may introduce a Finsler metric in W_{2n-1}, with respect to which the ω_i, ω_{ij} define a covariant differential (in the sense of Ch. III), and conversely, this metric and the ω_{ij} characterise the space L_{2n-1}.

Returning to the configuration (5.34) of E_N $(N > n)$, we now assume that the multilinear elements generate a manifold M_{2n-1}, and by means of a mapping of M_{2n-1} onto E_n an $\mathfrak{L}_{2n-1}$ is induced on M_{2n-1}. This process represents a local realisation by means of M_{2n-1} of the connection $\mathfrak{L}_{2n-1}$.

---

[1] This theorem is due to YANO and DAVIES [2], p. 416. Closely related to this is the work of DEICKE [2, 3], who proved that it is in general not possible to imbed a Finsler space in a Riemannian space without torsion, but that it is always possible to determine the metric and torsion tensors of a $(2n - 1)$-dimensional space V_{2n-1} such that a given F_n may be regarded as a non-holonomic subspace of V_{2n-1}. In this process no conditions such as (5.33) are imposed; instead it is stipulated with respect to the surrounding space that the autoparallel curves shall be geodesics.

[2] GALVANI [1—6]. A similar method for two-dimensional Finsler spaces is suggested by RACHEVSKY [2].

Having thus defined this concept, we are now in a position to quote the principal theorems:

(1) Every analytic $\mathfrak{L}_{2n-1}$ is locally realisable in a euclidean space E_N of $N = 2n^2 - n$ dimensions, the general solution depending on $n(n^2 - 1)$ arbitrary functions of $(2n - 1)$ arguments[1].

(2) Every analytic semi-punctual $\mathfrak{L}_{2n-1}$ is locally realisable by a semi-punctual M_{2n-1} in an E_N of $N = 2(n^2 - n + 1)$ dimensions, the general solution depending on $n(n - 1)^2$ arbitrary functions of $(2n - 1)$ arguments[2].

(3) Every space L_{2n-1} is locally realisable by M_{2n-1} in an E_N, where $N = 2n^2 - n$.

(4) Every space L_{2n-1} is locally realisable in an E_N with $N = 2(n^2 - n + 1)$ by a semi-punctual M_{2n-1}[3].

(5) Each analytic Finsler space F_n is locally realisable under the conditions stated in theorems (3) and (4)[4].

A special study may also be made of the case $n = 2$, for which, however, the manifold of the elements (5.34) may be imbedded either in an E_3 or in a 3-dimensional Riemannian space V_3. Again, the following theorems may be proved:

(6) In order that a two-dimensional Finsler space F_2 be locally realisable in an E_3 it is necessary that it be endowed with an absolute parallelism of its line elements; if this condition is not satisfied, it is realisable in a V_3[5].

(7) Under such a realisation the images of points of F_2 are the orthogonal trajectories of a one-parameter family of ruled developables of V_3 (or of E_3), and the images of geodesics of F_2 are the generators of those developables[6].

In conclusion mention should be made of another approach to the imbedding problem which is due to VAGNER [2], this approach being based on VAGNER's concept of a Finsler space in terms of fields of local hypersurfaces. A field of M-dimensional surfaces in an affine E_N is called a *constant* field if under a translation in which a point P_1 is mapped into a point P_2 the M-dimensional surface associated with P_1 is mapped into the M-dimensional surface associated with P_2. It may be shown that a field of local m-dimensional surfaces in an X_n may be imbedded in a constant field of $(m + k)$-dimensional surfaces of E_{n+k}, where $k \leq n$. Explicit formulae for such an imbedding are given by INGARDEN [1].

§ 6. Two-dimensional Finsler Spaces

The theory of two-dimensional Finsler spaces has been the subject of a great many investigations, and it is almost impossible to present a complete account of the many and varied results within the bounds of the present monograph. We shall thus content ourselves with a brief description of the principal aspects of the theory. In particular, it should be emphasised that the methods and techniques which are available for the two-dimensional case differ significantly from the methods applicable to the general case, especially as a set of three useful types of derivatives may be introduced. We shall thus begin our con-

---

[1] GALVANI [1]; [4], p. 133.

[2] GALVANI [4], p. 137.

[3] GALVANI [4], p. 141.

[4] GALVANI [4], p. 143

[5] GALVANI [2, 3]; [6], p. 431.

[6] GALVANI [6], p. 444 et seq.

siderations with a brief description of this formalism and its immediate consequences, which are chiefly due to BERWALD and CARTAN[1].

1°. Formal Aspects

In a two-dimensional Finsler space F_2 a unique unit vector m^i normal to the unit vector l^i in the direction of the element of support $\dot{x}^i$ may be defined as follows. Introducing the matrices

$$\varepsilon_{ik} = \begin{pmatrix} 0 & \sqrt{g} \\ -\sqrt{g} & 0 \end{pmatrix}, \qquad \varepsilon^{ik} = \begin{pmatrix} 0 & \dfrac{1}{\sqrt{g}} \\ -\dfrac{1}{\sqrt{g}} & 0 \end{pmatrix}, \tag{6.1}$$

where $g = \det(g_{ij})$ and $\sqrt{}$ indicates the positive square root, the vector m^i is defined by[2]

$$l^i = \varepsilon^{ik} m_k, \qquad l_i = \varepsilon_{ik} m^k, \tag{6.2a}$$

so that the following relations hold identically:

$$m^i = -\varepsilon^{ik} l_k, \qquad m_i = -\varepsilon_{ik} l^k; \tag{6.2b}$$

$$l^i m^k - l^k m^i = \varepsilon^{ik}, \qquad l_i m_k - l_k m_i = \varepsilon_{ik}; \tag{6.2c}$$

and, since $l^i_{|k} = 0$, $g_{ij|k} = 0$, we also have

$$m^i_{|k} = m_{i|k} = 0. \tag{6.2d}$$

Also, the metric tensor may be decomposed as follows:

$$g_{ij} = l_i l_j + m_i m_j, \qquad \delta^k_i = l_i l^k + m_i m^k, \tag{6.3}$$

and for future reference we note that

$$\varepsilon_{ik} + \varepsilon_{ki} = 0, \qquad \varepsilon^{ik} \varepsilon_{jk} = \delta^i_j, \qquad \varepsilon^{ik} \varepsilon_{ik} = 2. \tag{6.4}$$

As in equation (5.6.9) we introduce the principal (or main) scalar[3] J, and a decomposition in consequence of (1.3.5) leads to the relations

$$A_{ijk} = J m_i m_j m_k = F C_{ijk} = \tfrac{1}{2} F \frac{\partial g_{ij}}{\partial \dot{x}^k}. \tag{6.5}$$

An immediate consequence of this is that

$$F \frac{\partial \sqrt{g}}{\partial \dot{x}^k} = J \sqrt{g}\, m_k. \tag{6.6}$$

---

[1] BERWALD [5], CARTAN [5]. A report on earlier results is given by BERWALD [6]. We should also draw the reader's attention to the fact that a number of results concerning the curvature of two-dimensional Finsler spaces has been given in §§ 4—5 of Ch. IV. For the sake of completeness, we should mention that GRÜSS [1] made a detailed study of systems of meshes in an F_2. The concept of "geodesic conics" in an F_2 is discussed by GRÜSS [2] and NAKAJIMA [1]. A treatment of Cartan's parallelism in two-dimensional Finsler spaces is given by GALVANI [6].

[2] Note that this is essentially the construction used in § 6 of Ch. V.

[3] This scalar appears for the first time in BERWALD [5], p. 204, where it differs by the factor $\tfrac{1}{2}$ from the scalar used in the present text.

We shall now introduce three types of derivatives peculiar to two-dimensional spaces as follows. Firstly, for the sake of brevity, let us write

$$\Phi_{[i]} = \frac{\partial \Phi}{\partial x^i} - \frac{\partial \Phi}{\partial \dot{x}^l}\frac{\partial G^l}{\partial \dot{x}^i} , \qquad X^k_{[i]} = \frac{\partial X^k}{\partial x^i} - \frac{\partial X^k}{\partial \dot{x}^l}\frac{\partial G^l}{\partial \dot{x}^i} , \ldots, \text{etc.}, \qquad (6.7)$$

where Φ is an arbitrary homogeneous scalar function, X^k an arbitrary vector field, and similarly for tensors of any rank. If we now define

$$\Phi_s = \Phi_{[i]}\, l^i , \qquad X^k_s = X^k_{[i]}\, l^i , \ldots, \text{etc.} , \qquad (6.7\,\text{a})$$

it is clear that the subscript s (which is *not* an index assuming the values $1, 2$) in effect denotes differentiation with respect to the arc-length s of the geodesic of F_2 tangent to the line-element $(x, \dot{x})$. By means of (6.7) the ordinary covariant derivatives may be expressed in the form

$$X^i_{|j} = X^i_{[j]} + X^h \Gamma^{*i}_{hj} , \qquad (6.7\,\text{b})$$

$$T_{ik|j} = T_{ik[j]} - T_{hk}\Gamma^{*h}_{ij} - T_{ih}\Gamma^{*h}_{kj} , \ldots, \text{etc.} , \qquad (6.7\,\text{c})$$

while, in particular,

$$F_{|i} = F_{[i]} = 0 . \qquad (6.7\,\text{d})$$

The following commutation formulae are easily verified with the help of (3.3.8) and (4.6.10):

$$\Phi_{[h][k]} - \Phi_{[k][h]} = -\frac{\partial \Phi}{\partial \dot{x}^j} K^j_{rhk}\, \dot{x}^r , \qquad (6.8)$$

$$\frac{\partial \Phi_{[h]}}{\partial \dot{x}^k} - \left(\frac{\partial \Phi}{\partial \dot{x}^k}\right)_{[h]} = -\frac{\partial \Phi}{\partial \dot{x}^j} (A^i_{hk|r}\, l^r + \Gamma^{*i}_{hk}) . \qquad (6.9)$$

A derivative similar to (6.7 a) may be defined as follows:

$$\Phi_b = \Phi_{[i]}\, m^i , \qquad X^j_b = X^j_{[i]}\, m^i , \ldots, \text{etc.} , \qquad (6.10)$$

so that, for instance,

$$\Phi_{[i]} = \Phi_s\, l_i + \Phi_b\, m_i . \qquad (6.10\,\text{a})$$

As a result of these definitions, we have the following formulae, whose verification is left to the reader:

$$l^i_{[j]} = -\frac{1}{F}\frac{\partial G^i}{\partial \dot{x}^j} , \qquad m^i_{[j]} = -m^h\, \Gamma^{*i}_{jh} ; \qquad (6.11)$$

$$l_{i[j]} = l_h G^h_{ij} = l_{j[i]} , \qquad m_{i[j]} = m_h \Gamma^{*h}_{ij} = m_{j[i]} ; \qquad (6.12)$$

$$l^i_s = -\frac{2G^i}{F^2} , \qquad m^i_s = -\frac{1}{F}\frac{\partial G^i}{\partial \dot{x}^j}\, m^j ; \qquad (6.13)$$

$$l^i_b = -\frac{1}{F}\frac{\partial G^i}{\partial \dot{x}^j}\, m^j , \qquad m^i_b = -\Gamma^{*i}_{jh}\, m^j m^h ; \qquad (6.13\,\text{a})$$

$$(l_i)_s = \frac{1}{F}\frac{\partial G^h}{\partial \dot{x}^i}\, l_h , \qquad (m_i)_s = \frac{1}{F}\frac{\partial G^h}{\partial \dot{x}^i}\, m_h ; \qquad (6.14)$$

$$(l_i)_b = G^h_{ij}\, l_h\, m^j , \qquad (m_i)_b = \Gamma^{*h}_{ij}\, m_h\, m^j . \qquad (6.14\,\text{a})$$

The third derivative may be approached by recalling the expression (5.6.8) for the angle φ. In terms of our notation (6.1) we see that we may write

$$\varphi = \int \varepsilon_{ik} \frac{\dot{x}^i d\,\dot{x}^k}{F^2} = \int m_k\, dl^k \,. \tag{6.15}$$

Hence

$$F \frac{\partial \varphi}{\partial \dot{x}^k} = m_k \,, \tag{6.16}$$

and thus, if Φ is homogeneous of degree zero in the $\dot{x}^k$, it follows that

$$F \frac{\partial \Phi}{\partial \dot{x}^i} = m_i \frac{\partial \Phi}{\partial \varphi} \,, \qquad \frac{\partial \Phi}{\partial \varphi} = F \frac{\partial \Phi}{\partial \dot{x}^i} m^i \,. \tag{6.17}$$

Denoting derivatives with respect to φ by the subscript φ, we have from (6.5) and (6.6)

$$(g_{ij})_\varphi = 2J\, m_i\, m_j \,, \qquad (\sqrt{g})_\varphi = J \sqrt{g} \,, \tag{6.18}$$

while

$$(\varepsilon_{ik})_\varphi = J\, \varepsilon_{ik} \,, \qquad (\varepsilon^{ik})_\varphi = -\, J\, \varepsilon^{ik} \,. \tag{6.19}$$

If we recall that

$$\frac{\partial l^i}{\partial \dot{x}^j} = \frac{1}{F} \left(\delta^i_j - l^i\, l_j \right) \,, \qquad \frac{\partial l_i}{\partial \dot{x}^j} = \frac{1}{F} \left(g_{ij} - l_i\, l_j \right) \,, \tag{6.20}$$

the following formulae result directly from (6.19):

$$l^i_\varphi = m^i \,, \qquad (l_k)_\varphi = m_k \,, \tag{6.21}$$

and

$$m^i_\varphi = -\, l^i - J\, m^i \,, \qquad (m_i)_\varphi = -\, l_i + J\, m_i \,. \tag{6.22}$$

As regards the curvature tensors we observe that in view of the skew-symmetry (4.2.16) we have

$$R^j_{ihk}\, l^i\, l_j = 0 = K^j_{ihk}\, l^i\, l_j \,,$$

and thus in consequence of (6.4) we may write

$$R^j_{ihk}\, l^i = K^j_{ihk}\, l^i = K\, m^j\, \varepsilon_{hk} \,, \tag{6.23}$$

where

$$K = \tfrac{1}{2} K^j_{ihk}\, \varepsilon^{hk}\, l^i\, m_j \,. \tag{6.23a}$$

On the basis of these analytical preliminaries it is possible to derive the fundamental properties of F_2 with great ease. Firstly, we shall obtain commutation formulae for the three types of derivatives defined above, from which the Bianchi identity follows almost as a direct corollary. Firstly, we note that as a result of (6.2d), (6.5) and (6.7a) we have

$$A^i_{hk|r}\, l^r = J_s\, m_h\, m^j\, m_k \,, \tag{6.24}$$

and secondly, it is seen that in virtue of (6.17) and (6.23) equation (6.8) becomes

$$\Phi_{[h][k]} - \Phi_{[k][h]} = -F \frac{\partial \Phi}{\partial \dot{x}^j} K^j_{rhk}\, l^r = -\frac{\partial \Phi}{\partial \varphi} K\, \varepsilon_{hk} \,, \tag{6.25}$$

while similarly equation (6.9) may be written as

$$\frac{\partial \Phi_{[h]}}{\partial \dot{x}^k} - \left(\frac{\partial \Phi}{\partial \dot{x}^k}\right)_{[h]} = -\frac{1}{F}\frac{\partial \Phi}{\partial \varphi}\left(J_s\, m_h\, m_k + m_j \Gamma^{*\,j}_{h\,k}\right). \tag{6.26}$$

Hence a simple calculation based on (6.11), (6.17), (6.22) and (6.26) yields the formula

$$\Phi_{b\varphi} - \Phi_{\varphi b} = -J_s\,\Phi_\varphi - \Phi_s - J\,\Phi_b, \tag{6.27}$$

and similarly

$$\Phi_{\varphi s} - \Phi_{s\varphi} = -\Phi_b, \tag{6.27a}$$

together with

$$\Phi_{sb} - \Phi_{bs} = -K\,\Phi_\varphi. \tag{6.27b}$$

These relations are the commutation formulae of CARTAN[1]. Let us consider now the following obvious Jacobi identity:

$$(\Phi_{sb\varphi} - \Phi_{bs\varphi}) + (\Phi_{\varphi bs} - \Phi_{\varphi sb}) + (\Phi_{\varphi sb} - \Phi_{s\varphi b}) +$$
$$+ (\Phi_{b\varphi s} - \Phi_{\varphi bs}) + (\Phi_{bs\varphi} - \Phi_{b\varphi s}) + (\Phi_{s\varphi b} - \Phi_{sb\varphi}) = 0.$$

On substituting from (6.27) in each of these brackets we obtain

$$-K_\varphi\,\Phi_\varphi - J_{ss}\,\Phi_\varphi + J_s(\Phi_{s\varphi} - \Phi_{\varphi s} - \Phi_b) + J(\Phi_{sb} - \Phi_{bs}) = 0,$$

and if we apply (6.27) once more and drop the term Φ_φ, we finally find that

$$K_\varphi + JK + J_{ss} = 0. \tag{6.28}$$

This is the *Bianchi identity* for an F_2 as given by BERWALD[2].

Furthermore, if we differentiate (3.3.8) with respect to $\dot{x}^h$ and make use of (6.24) and (6.17) we see that

$$F\,G^i_{jkh} = F\,\frac{\partial \Gamma^{*\,i}_{jk}}{\partial \dot{x}^h} + (J_s\, m_j\, m^i\, m_k)_\varphi\, m_h, \tag{6.29}$$

---

[1] CARTAN [5], p. 121.

[2] BERWALD [5], p. 206. In this connection the work of MOÓR [2] concerning two-dimensional Finsler spaces of constant curvature should be mentioned. By means of (6.23) the commutation formulae for a scalar function Φ may be written as

$$\Phi_{|hk} - \Phi_{|kh} = -F\,\frac{\partial \Phi}{\partial \dot{x}^r}\,K\,\varepsilon_{hk}\,m^r.$$

In particular, it follows from (6.2a) and (6.4) that we have (BERWALD [5], p. 203)

$$\Phi_{|12} - \Phi_{|21} = FK\left(\frac{\partial \Phi}{\partial \dot{x}^1}\frac{\partial F}{\partial \dot{x}^2} - \frac{\partial \Phi}{\partial \dot{x}^2}\frac{\partial F}{\partial \dot{x}^1}\right).$$

If we assume that the space is such that $K_{|1} = K_{|2} = 0$, $K \neq 0$, and apply this formula to $\Phi = K$, we find $K_{\dot{x}^1} F_{\dot{x}^2} - K_{\dot{x}^2} F_{\dot{x}^1} = 0$. But by homogeneity, $K_{\dot{x}^1}\dot{x}^1 + K_{\dot{x}^2}\dot{x}^2 = 0$, so that $K_{\dot{x}^1} = K_{\dot{x}^2} = 0$, and hence, from the definition of $K_{|1}$, $K_{|2}$ it also follows that $K_{x^1} = K_{x^2} = 0$. Hence the conditions $K_{|i} = 0$ and $K = \text{const.}$ are equivalent (MOÓR [2], p. 4). From (6.23) we deduce that these conditions are also equivalent to $(K^i_{r\,hk}\,l^r)_{|j} = 0$. The Bianchi identities become

$$JK + J_{ss} = 0.$$

A detailed discussion of this equation is given by MOÓR [2].

while in view of (6.22) we have

$$(J_s\, m_j\, m^i\, m_k)_\varphi = (J_{s\varphi} + J\,J_s)\, m_j\, m^i\, m_k -$$
$$- J_s(l_j\, m^i\, m_k + m_j\, l^i\, m_k + m_j\, m^i\, l_k)\,. \tag{6.30}$$

But from (6.2d), (6.5) and (6.10a) it follows that

$$A^h_{jr|i} = (J_s\, l_i + J_b\, m_i)\, m_j\, m^h\, m_r\,, \tag{6.31}$$

and if we substitute these values in (3.3.14), we find after some simplification

$$F\,\frac{\partial\Gamma^{*i}_{jk}}{\partial\dot x^h} = m_j\, m_k\, m_h(J_b\, m^i - J\,J_s\, m^i - J_s\, l^i) +$$
$$+ m^i\, m_h(J_s\, m_k\, l_j + J_s\, l_k\, m_j)\,. \tag{6.32}$$

Substitution of (6.30) and (6.32) in (6.29) yields the following fundamental decomposition:

$$F\,G^i_{jkh} = m_j\, m_h\, m_k\,\{m^i\,(J_{s\varphi} + J_b) - 2l^i\, J_s\}\,. \tag{6.33}$$

In particular,

$$F\,G^i_{jki} = m_j\, m_k\,(J_{s\varphi} + J_b)\,. \tag{6.33a}$$

If we put

$$G^i_{jkhr} = \frac{\partial}{\partial\dot x^r}\,(G^i_{jkh})\,,$$

and differentiate (6.33a) with respect to $\dot x^r$, noting (6.17) and (6.22), this gives

$$F\,l_r\,G^i_{jki} + F^2 G^i_{jkir} = m_r\, m_j\, m_k\,\{(J_{s\varphi} + J_b)_\varphi +$$
$$+ 2J\,(J_{s\varphi} + J_b)\} - (J_{s\varphi} + J_b)\,\{m_r\, m_k\, l_j + m_j\, m_r\, l_k\}\,,$$

and hence we obtain from (6.33a)

$$F^2 G^i_{jkri} = m_r\, m_j\, m_k\,\{(J_{s\varphi} + J_b)_\varphi + 2J\,(J_{s\varphi} + J_b)\} -$$
$$- (J_{s\varphi} + J_b)\,\{l_j\, m_r\, m_k + m_j\, l_r\, m_k + m_j\, m_r\, l_k\}\,. \tag{6.34}$$

From (6.33) it is evident that *the two-dimensional "affinely connected spaces"* (i. e. those Finsler spaces for which the G^i_{hk} are independent of direction) *are characterised by the conditions*

$$J_b = J_s = 0\,. \tag{6.35}$$

Applying (6.27b) to the principal scalar J we see that *these conditions will imply that either* (1) $K = 0$, *or* (2) $J = $ const. From (6.23) we deduce that if $K = 0$, then $l^i\, K^j_{ihk} = 0$, which, as we have seen in Ch. IV, § 7, is a sufficient condition for the existence of a coordinate system in which $\Gamma^{*i}_{hk} = 0$, and hence by (3.3.8a), (6.31) and (6.35) we find that $G^i_{hk} = 0$ for the affinely connected space. From (6.7d) it then follows that $F_{\dot x^i}(x^k, \dot x^k) = 0$ for all directions $\dot x^k$, which implies that the space is Minkowskian (in agreement with the last theorem of § 7 of Ch. IV). Thus we may dismiss the first alternative $K = 0$. We shall return to the second alternative in 3^0.

2°. Certain Projective Changes Applied to F_2.
Spaces with Rectilinear Geodesics

We shall now prove the following theorem[1]:

A necessary and sufficient condition in order that in a two-dimensional Finsler space the functions G^i_{hk} can be changed projectively in such a way that the transformed functions $\overline{G}^i_{hk}$ be independent of direction is that the differential equation

$$(J_{s\varphi} + J_b)_\varphi + 2J(J_{s\varphi} + J_b) + 6J_s = 0 \tag{6.36}$$

holds.

Proof: Consider a projective change as exemplified by (4.8.3):

$$\overline{G}^i = G^i + P(x, \dot{x})\, \dot{x}^i, \tag{6.37}$$

where the functions $P(x, \dot{x})$ are positively homogeneous of the first degree in the $\dot{x}^k$. We then have

$$\overline{G}^i_{hk} = G^i_{hk} + \frac{\partial P}{\partial \dot{x}^k}\,\delta^i_h + \frac{\partial P}{\partial \dot{x}^h}\,\delta^i_k + \dot{x}^i\,\frac{\partial^2 P}{\partial \dot{x}^h\,\partial \dot{x}^k}, \tag{6.37a}$$

together with

$$\overline{G}^i_{hkj} = G^i_{hkj} + \frac{\partial^2 P}{\partial \dot{x}^k\,\partial \dot{x}^j}\,\delta^i_h + \frac{\partial^2 P}{\partial \dot{x}^j\,\partial \dot{x}^h}\,\delta^i_k + \frac{\partial^2 P}{\partial \dot{x}^h\,\partial \dot{x}^k}\,\delta^i_j + \dot{x}^i\,\frac{\partial^3 P}{\partial \dot{x}^j\,\partial \dot{x}^h\,\partial \dot{x}^k}. \tag{6.37b}$$

Also, by contracting the first derivative with respect to direction of (6.37), we obtain

$$P = \frac{1}{3}\left(\frac{\partial \overline{G}^r}{\partial \dot{x}^r} - \frac{\partial G^r}{\partial \dot{x}^r}\right). \tag{6.38}$$

Let us now suppose that $\overline{G}^i_{hk}$ is independent of direction, so that the right-hand side of (6.37b) vanishes. Substitution of (6.38) in (6.37b) then gives

$$G^i_{hkj} - \tfrac{1}{3}\left(\delta^i_h G^r_{rkj} + \delta^i_k G^r_{rjh} + \delta^i_j G^r_{rhk}\right) - \tfrac{1}{3}\,\dot{x}^i G^r_{rhkj} = 0. \tag{6.39}$$

Conversely, if this condition is satisfied, we define a function P for the projective change (6.37) by putting

$$P = \frac{1}{3}\left(\overline{G}^r_{rk}\,\dot{x}^k - \frac{\partial G^r}{\partial \dot{x}^r}\right),$$

where the $\overline{G}^i_{rk} = \overline{G}^i_{kr}$ are arbitrary functions of position only, so that

$$\frac{\partial^2 P}{\partial \dot{x}^j\,\partial \dot{x}^k} = -\frac{1}{3}\,G^r_{rkj}.$$

In view of condition (6.39) substitution of the latter relation in (6.37b) will then yield $\overline{G}^i_{hjk} = 0$. Hence condition (6.39) is also sufficient. Finally, by means of equations (6.33), (6.33a) and (6.34) we transform

---

[1] BERWALD [10, III], p. 97.

equations (6.39) into the given form (6.36) [in the course of the latter calculation a number of simplifications will result from (6.3)].

It is seen that the theorem just proved may be stated in the form: *A necessary and sufficient condition that the geodesics of a Finsler space form a quasi-geodesic system is that* (6.36) *be satisfied*[1].

The following theorem should be adjoined to the one enunciated above:

In order that there may exist a projective change of the G^i_{hk} which transforms the curvature tensor (4.6.6) *of a two-dimensional Finsler space to zero, it is necessary and sufficient that the equations*

$$K_{\varphi s} - 3K_b = 0 \tag{6.40}$$

be satisfied.

Proof: By means of (4.8.7b) it is easy to show that under the projective change (6.37) together with (6.37a) the curvature tensor H^i_{jk} transforms as follows:

$$\bar{H}^i_{jk} = H^i_{jk} - \delta^i_k \left(P_{(j)} - P\frac{\partial P}{\partial \dot{x}^j} \right) + \delta^i_j \left(P_{(k)} - P\frac{\partial P}{\partial \dot{x}^k} \right) - \\ - \left(\frac{\partial P_{(j)}}{\partial \dot{x}^k} - \frac{\partial P_{(k)}}{\partial \dot{x}^j} \right) \dot{x}^i . \tag{6.41}$$

Recalling the notation (4.6.16a) and (4.6.16b):

$$H_j = H^i_{ji} , \qquad \frac{\partial H_j}{\partial \dot{x}^h} = H_{hj} ,$$

we see that if we contract the indices i and k in (6.41), the condition $\bar{H}^i_{jk} = 0$ becomes

$$H_j = 2 P_{(j)} - P\frac{\partial P}{\partial \dot{x}^j} - \frac{\partial P_{(k)}}{\partial \dot{x}^j} \dot{x}^k . \tag{6.42}$$

Differentiating this equation with respect to $\dot{x}^h$, we find, since P is homogeneous of the first degree,

$$H_{hj} \dot{x}^j = 2 \frac{\partial P_{(j)}}{\partial \dot{x}^h} \dot{x}^j - P\frac{\partial P}{\partial \dot{x}^h} - P_{(h)} .$$

With the aid of the latter equations, (6.42) may be reduced to the form

$$\frac{2}{3} H_j + \frac{1}{3} H_{jk}\dot{x}^k = \left(P_{(j)} - P\frac{\partial P}{\partial \dot{x}^j} \right) . \tag{6.43}$$

These are the differential equations which must be satisfied by P in order that $\bar{H}^i_{jk} = 0$. The conditions of integrability of (6.43) read

$$P_{(j)(k)} - P_{(k)(j)} - \frac{1}{2}\left(\frac{\partial P^2}{\partial \dot{x}^j} \right)_{(k)} + \frac{1}{2}\left(\frac{\partial P^2}{\partial \dot{x}^k} \right)_{(j)} \\ = \frac{2}{3}(H_{j(k)} - H_{k(j)}) + \frac{1}{3}\{(H_{jh} \dot{x}^h)_{(k)} - (H_{kh} \dot{x}^h)_{(j)}\} . \tag{6.44}$$

---

[1] "Quasi-geodesic" systems are systems of differential equations in which $\dfrac{d^2 x^i}{dt^2}$ is a homogeneous second order polynomial in $\dfrac{d x^k}{dt}$. Cf. BLASCHKE and BOL [1], § 29; BERWALD [10, III], p. 98.

In view of the commutation formulae (4.6.10) and (4.6.11) the left-hand side of (6.44) may be written in the form

$$-\frac{\partial P}{\partial \dot{x}^i}\, H^i_{jk} + P_{(j)}\,\frac{\partial P}{\partial \dot{x}^k} - P_{(k)}\,\frac{\partial P}{\partial \dot{x}^j} + P\left(\frac{\partial P_{(j)}}{\partial \dot{x}^k} - \frac{\partial P_{(k)}}{\partial \dot{x}^j}\right),$$

and, if we add and subtract the same term in this expression, it becomes

$$-\frac{\partial P}{\partial \dot{x}^i}\left\{ H^i_{jk} - \delta^i_k\left(P_{(j)} - \frac{1}{2}\frac{\partial P^2}{\partial \dot{x}^j}\right) + \delta^i_j\left(P_{(k)} - \frac{1}{2}\frac{\partial P^2}{\partial \dot{x}^k}\right) - \left(\frac{\partial P_{(j)}}{\partial \dot{x}^k} - \frac{\partial P_{(k)}}{\partial \dot{x}^j}\right)\dot{x}^i \right\}.$$

But this is the right-hand side of (6.41) multiplied by $\partial P/\partial \dot{x}^i$, which vanishes in view of the hypothesis $\overline{H}^i_{jk}= 0$. Hence, by (6.44), the integrability conditions of (6.43) are simply

$$2\,(H_{j\,(k)} - H_{k\,(j)}) + (H_{jh}\,\dot{x}^h)_{(k)} - (H_{kh}\,\dot{x}^h)_{(j)} = 0\,. \tag{6.45}$$

Conversely, if (6.45) is satisfied, it follows that the conditions of integrability of (6.43) give rise to $\overline{H}^i_{jk} = 0$ in virtue of (6.41). Thus conditions (6.45) are necessary and sufficient for the existence of a function P which is such that $\overline{H}^i_{jk}= 0$[1].

We shall now express the relations (6.45) in terms of the s-, b- and φ-derivatives. In view of (4.6.6), (4.6.9d) and (6.23) we have

$$H^j_{hk}= F K\, m^j\, \varepsilon_{hk}\,, \tag{6.46}$$

and hence by (6.2a), (6.17) and (6.20) we obtain

$$H_h= F K\, l_h\,, \tag{6.47}$$

together with

$$H_{jh}= K\, g_{jh} + K_\varphi\, m_j\, l_h\,, \tag{6.47a}$$

and

$$H_{jh}\,\dot{x}^h = F\,(K\, l_j + K_\varphi\, m_j)\,. \tag{6.47b}$$

Since the G^i_{kj} are symmetric it follows that

$$H_{j\,(k)} - H_{k\,(j)} = H_{j[k]} - H_{k[j]}\,,$$

and if we expand the terms on the right-hand side by means of (6.10a), simplifying the result with the help of (6.12) and (6.2c) we find

$$H_{j\,(k)} - H_{k\,(j)} = F K_b\, \varepsilon_{jk}\,. \tag{6.48}$$

Similarly, a simple reduction based on (6.47b) gives

$$(H_{jh}\,\dot{x}^h)_{(k)} - (H_{kh}\,\dot{x}^h)_{(j)} = F\,(K_b - K_{\varphi s})\, \varepsilon_{jk}\,. \tag{6.48a}$$

---

[1] This, incidentally, is an alternative proof of the latter part of the second theorem stated on p. 144.

On substituting (6.48) and (6.48a) in the integrability conditions (6.45) the latter reduce to the form

$$F\{3 K_b \, \varepsilon_{jk} - K_{\varphi s} \, \varepsilon_{jk}\} = 0 \,,$$

from which the desired condition (6.40) follows directly.

A further interpretation of the conditions (6.36) and (6.40) is contained in the following theorem:

A necessary and sufficient condition that the geodesics of a two-dimensional Finsler space be rectilinear, is that conditions (6.36) and (6.40) be satisfied[1].

For suppose that there exists a coordinate system in which the differential equations of the geodesics are of the form

$$\dot{x}^i \, \ddot{x}^k - \dot{x}^k \, \ddot{x}^i = 0 \,. \tag{6.49}$$

We must then have by (4.8.2):

$$G^i = F q \, \dot{x}^i \,, \tag{6.50}$$

where $q = q(x, \dot{x})$ is some function homogeneous of degree zero in the $\dot{x}^k$. In the projective change defined by (6.37) put $P = -Fq$. Then $\overline{G}^i = 0$ as a result of (6.50). Thus (6.36) and (6.40) are satisfied. The converse is established similarly.

3°. Two-dimensional Finsler Spaces whose Principal Scalar is a Function of Position only. Landsberg Spaces

These spaces have been investigated in great detail, as a result of which a large number of fairly complicated theorems have been obtained. Due to their intrinsic interest we shall quote some of these results, but as regards the proofs we must refer the reader to the original literature.

All two-dimensional Finsler spaces for which J is a non-vanishing constant may be listed under the following table[2], in which $\alpha_i \, \dot{x}^i$ and $\beta_i \, \dot{x}^i$ represent two linearly independent Pfaffian forms:

$$J^2 < 4 : F = [(\alpha_i \, \dot{x}^i)^2 + (\beta_i \, \dot{x}^i)^2]^{1/2} \exp\left\{\frac{J}{\sqrt{4 - J^2}} \text{ arc tg } \frac{\beta_i \, \dot{x}^i}{\alpha_i \, \dot{x}^i}\right\} ; \tag{6.51}$$

$$J^2 = 4 : F = (\alpha_i \, \dot{x}^i) \exp\left\{\frac{1}{2} J \frac{\beta_i \, \dot{x}^i}{\alpha_i \, \dot{x}^i}\right\} ; \tag{6.51a}$$

$$J^2 > 4 : \log F = \frac{1}{2}\left(1 - \frac{J}{\sqrt{J^2 - 4}}\right) \log(\alpha_i \, \dot{x}^i) + \frac{1}{2}\left(1 + \frac{J}{\sqrt{J^2 - 4}}\right) \log(\beta_i \, \dot{x}^i) \,. \tag{6.51b}$$

---

[1] This may also be proved directly from considerations concerning the integrability of the conditions that the geodesics be rectilinear (BERWALD [10, III], p. 96). See also FUNK [2].

[2] BERWALD [5], p. 215 et seq.

Furthermore, if in a two-dimensional Finsler space for which J is a non-vanishing constant the curvature K is a function of position only, the space is Minkowskian. [Indeed, this follows directly from (6.35) and the Bianchi identity (6.28).]

Also, if in a two-dimensional Finsler space with rectilinear geodesics J is a function of position only, $J \neq 0$, $J^2 \neq {}^9/_2$, the space is Minkowskian[1].

The metric function of a two-dimensional Finsler space with rectilinear geodesics which is not a Minkowskian space, and for which $J^2 = {}^9/_2$, may be transformed to the form

$$F = \frac{(\dot{x}^1 + z\,\dot{x}^2)^2}{\dot{x}^2}\,, \tag{6.52}$$

where z is a non-constant solution of the partial differential equation

$$z\,\frac{\partial z}{\partial x_1} - \frac{\partial z}{\partial x_2} = 0\,. \tag{6.53}$$

The solutions of this equation are given by[2]

$$x^1 + z\,x^2 = \psi(z)\,, \tag{6.54}$$

where ψ is an arbitrary analytic function of z.

The space is Minkowskian if and only if z assumes either of the forms

$$z = c^3\,, \qquad z = -\frac{x^1 - c^1}{x^2 - c^2}\,, \tag{6.55}$$

where c^1, c^2, c^3 are arbitrary constants.

A *Landsberg space* is a two-dimensional Finsler space with $J_{\bullet} = 0$[3].

All Landsberg spaces with rectilinear geodesics may be listed as follows:

(I) $K = 0$: Minkowskian spaces.

(II) $K = \text{const.} \neq 0$: Non-euclidean spaces.

$$(1)\quad K = \frac{1}{k^2} : F = k\,\frac{[(a^2 + (x^2)^2)\,(\dot{x}^1)^2 - 2\,x^1 x^2\,\dot{x}^1 \dot{x}^2 + (a^2 + (x^1)^2)\,(\dot{x}^2)^2]^{1/2}}{a^2 + (x^1)^2 + (x^2)^2}\,,$$

$$(2)\quad K = -\frac{1}{k^2} : F = k\,\frac{[(a^2 - (x^2)^2)\,(\dot{x}^1)^2 + 2\,x^1 x^2\,\dot{x}^1 \dot{x}^2 + (a^2 - (x^1)^2)\,(\dot{x}^2)^2]^{1/2}}{a^2 - (x^1)^2 - (x^2)^2}\,,$$

where a and k are positive constants.

(III) K variable:

$$F = \frac{(\dot{x}^1 + z\,\dot{x}^2)^2}{\dot{x}^2}\,, \qquad x^1 + z\,x^2 = \psi(z)\,,$$

where $\psi(z)$ is an arbitrary function with $\psi''(z) \neq 0$.

Appendix

Bibliographical References to Related Topics

Unfortunately it was not feasible to include accounts of various important aspects of Finsler geometry within the bounds of the present monograph. We shall therefore merely list these topics under separate headings, to which are appended the relevant references to the

---

[1] BERWALD [10, III], p. 104.

[2] BERWALD [10, III], p. 107 et seq.

[3] Such spaces were first considered by LANDSBERG [2], p. 334 et seq.; CARTAN [5], p. 33 et seq.; BERWALD [5], p. 208; BERWALD [10, III], p. 109 et seq.

bibliography. This list does not include all the possible generalisations of Finsler geometry which have been studied during the past thirty years.

1. *Recent survey articles:*

Busemann [7]; Golab [12]; Hölder [1]; of which the first is an exceptionally lucid account of Busemann's approach to Finsler geometry. A more recent and very comprehensive treatment of this approach is presented in Busemann's book [10].

2. *Finsler spaces whose metric functions are assumed to be of a special form:*

It should be pointed out that the metric functions studied by the authors listed below do not always satisfy conditions B and C of § 1, Ch. I, so that the spaces concerned are, strictly speaking, not always Finsler or Minkowskian spaces as defined in Ch. I.

Delens and Devisme [1], Devisme [1, 2], are concerned with a metric function of the form

$$F(x', y', z') = [(x')^3 + (y')^3 + (z')^3 - 3x'y'z']^{1/3}. \tag{1}$$

A series of papers by Humbert [1—5] is devoted to the properties of the metric function

$$F(x', y') = [(x')^3 + (y')^3]^{1/3}, \tag{2}$$

from which certain theorems concerning plane algebraic curves may be derived. More, generally, a two-dimensional metric of the type

$$ds = [a_{\alpha_1 \alpha_2 \cdots \alpha_p}(x) \, dx^{\alpha_1} \, dx^{\alpha_2} \cdots \cdots dx^{\alpha_p}]^{1/p} \tag{3}$$

with $p \geqq 3$, $(\alpha_1, \ldots, \alpha_p = 1, 2)$ is studied in detail by Liber [1], who shows that it is possible to define a linear affine connection whose parameters depend solely on position and not on direction.

Moór [3] considers a 2-dimensional metric function of the type $F = f/g$, where

$$f = \sum_{k=0}^{n} a_k(x, y) \, \dot{x}^{n-k} \, \dot{y}^k, \tag{4}$$

and

$$g = \sum_{k=0}^{n-1} b_k(x, y) \, \dot{x}^{n-k-1} \, \dot{y}^k, \tag{4a}$$

while Moór [7] is devoted to the special case $p = 4$ of the form (3), although there is little similarity between the methods and results of the latter paper and that of Liber [1]. Wegener [1, 3] treats the case $p = 3$ in the form (3), but extends his results from two to three dimensions.

Finsler spaces whose invariants possess special pre-assigned properties (apart from constant curvature, etc.) are discussed by Moór [6]; Nobuhara and Nagai [1], Tachibana [1]; Wrona [1].

3. *Non-holonomic spaces:*

Under this heading we shall list papers concerning non-holonomic subspaces of a Finsler space, as well as non-holonomic spaces proper. For the corresponding literature with regard to Riemannian and Non-Riemannian spaces the reader is referred to SCHOUTEN [1], while the literature concerning the problem of LAGRANGE in the calculus of variations (which is intimately related to the theory of non-holonomic spaces) is given by CARATHÉODORY [1]. We shall therefore not mention the latter literature, but shall confine ourselves to the following: HOMBU [3]; HOSOKAWA [2]; KATSURADA [1, 3]; McFARLAN [1]; RUND [12, 13, 14]; VAGNER [3, 6, 8, 9, 11, 12, 14]; WUNDHEILER [1].

4. *Integral geometry:*

Fundamental aspects of integral geometry against the background of Finsler geometry are described by BLASCHKE [5] and OWENS [1].

5. *The "generalised variation" spaces:*

A direct and very ingenious generalisation of the idea of a Finsler space is due to LICHNEROWICZ [2, 3, 4, 5]. These spaces are based on a non-holonomic problem in the calculus of variations with respect to an integral of the type

$$J = \int\limits_{u_0}^{u_1} H\left[F_{u_0}^u,\ x^i(u),\ \dot{x}^i(u)\right] du ,$$

where

$$F_{u_0}^u = \int\limits_{u_0}^{u} \omega\left[x^i(v),\ \dot{x}^i(v)\right] dv ,$$

the functions H and ω being homogeneous of the first degree in the $\dot{x}^i$. In such spaces connection parameters may be defined, giving rise to a parallel displacement as well as to a theory of curvature. The definition of generalised variation spaces is a natural consequence of the study of non-conservative dynamical systems.

6. *Infinitely-dimensional Finsler spaces:*

From the point of view of the calculus of variations infinitely dimensional spaces are studied by GOLDSTINE [1]. A more geometrical approach is due to LAUGWITZ [5], which seems to be related to the results of ARONSZAJN [1]. It should be possible to establish some connection between these methods and those of A. D. MICHAL (not listed in the bibliography). The fact that a Minkowskian space is a finitely-dimensional Banach space clearly exhibits the plausibility of infinitely dimensional Finsler spaces.

7. *Further generalisations of Finsler spaces: Cartan spaces, Kawaguchi spaces:*

A complete bibliography listing papers concerned with Cartan spaces, Kawaguchi spaces and their generalisations up to and including 1950 has been compiled by H. SCHUBERT and is published as an appendix to the reprint [1] of Finsler's thesis.

8. *Applications of the methods of metric differential geometry to theoretical physics:*

Some attempts to apply the methods described in this monograph to theoretical physics seem to suggest promising openings for future avenues of research. A recent collection of individual papers, entitled "Memoirs of the unifying study of basic problems in engineering sciences by means of geometry", Vol. 1, Tokyo (1955), seems to confirm this view. We list the following papers, whose titles indicate to which branch of theoretical physics an application is intended: ELIOPOULOS [1]; FUJINAKA [1]; HORVÁTH and MOÓR [1, 2]; KONDO [1]; HOSOKAWA [4]; LEVASOV [1]; LICHNEROWICZ [1, 4, 5]; LICHNEROWICZ and THIRY [1]; PIHL [1]; RUND [2, 12, 13, 14]; SEGRE [1] (this paper, together with BLASCHKE [6], shows interesting implications with regard to geometrical optics); SYNGE [4]; THÉODORESCO [1, 2, 3]; WUNDHEILER [1].

Again, it should be remarked that very frequently the metric function which is given by a homogeneous Lagrangian function of a dynamical system does not always satisfy conditions B and C of § 1, Ch. I. The singularities which may occur as a result of the relaxation of condition C are usually ignored, but it is well possible that an investigation of these singularities in connection with physical applications cannot be avoided and might furthermore prove to be fruitful.

Bibliography

AKBAR-ZADEH, H.: [1] Sur la réductibilité d'une variété finslérienne. C. R. Acad. Sci. (Paris) **239**, 945—947 (1954). — [2] Sur les isométries infinitésimales d'une variété finslérienne. C. R. Acad. Sci. (Paris) **242**, 608—610 (1956).

ALT, F.: [1] Dreiecksungleichung und Eichkörper in verallgemeinerten Minkowskischen Räumen. Ergebn. math. Kolloqu. H. **8**, 32—33 (1937).

ANCOCHEA, G.: [1] Kovariante Differentiation und die Riccischen Identitäten in Finslerschen Räumen. Rev. mat. hisp.-amer. **2**, s. 8, 261—264 (1933).

ARONSZAJN, N.: [1] Sur quelques problèmes concernant les espaces de MINKOWSKI et les espaces vectoriels généraux. Atti Accad. naz. Lincei, Rend. (6), **26**, 374—376 (1937).

AUSLANDER, L.: [1] The use of forms in variational calculations. Pacific J. Math. **5**, 853—859 (1955). — [2] Remark on the use of forms in variational calculations. Pacific J. Math. **6**, 209—210 (1956). — [3] On curvature in Finsler geometry. Trans. Amer. math. Soc. **79**, 378—388 (1955).

BARTHEL, W.: [1] Zum Inhaltsbegriff der Minkowskischen Geometrie. Math. Z. **58**, 358—375 (1953). — [2] Über die Minimalflächen in gefaserten Finsler-Räumen. Ann. Mat. Pura Appl. (4) **36**, 159—190 (1954). — [3] Über eine Parallelverschiebung mit Längeninvarianz in lokal-Minkowskischen Räumen. I, II. Arch. Math. **4**, 346—365 (1953). — [4] Über Minkowskische und Finslersche Geometrie. Convegno Internazionale di Geometria Differenziale, Italia 71—76, Roma (1954). — [5] Variationsprobleme der Oberflächenfunktion in der Finslerschen Geometrie. Math. Z. **62**, 23—36 (1955). — [6] Extremalprobleme in der Finslerschen Inhaltsgeometrie. Ann. Univ. Sarav. **4**, 171—183 (1956).

BEKE, M.: [1] Transversalität und Orthogonalität. Mat. fiz. Lap. **45**, 157—161 (1938) (Hungarian).

BERWALD, L.: [1] Über Parallelübertragung in Räumen mit allgemeiner Maßbestimmung. Jber. dtsch. Math. Ver. **34**, 213—220 (1926). — [2] Untersuchung der Krümmung allgemeiner metrischer Räume auf Grund des in ihnen herrschenden Parallelismus. Math. Z. **25**, 40—73 (1926). Correction: Math. Z. **26**, 176 (1927). — [3] Über die erste Krümmung der Kurven bei allgemeiner Maßbestimmung. Lotos (Prag) **67/68**, 52—56 (1919/1920). — [4] Parallelübertragung in allgemeinen Räumen. Atti Congr. Bologna **4**, 263—270 (1931). — [5] Über zwei-dimensionale allgemeine metrische Räume. I, II. J. reine angew. Math. **156**, 191—210 and 211—222 (1927). — [6] Zur Geometrie ebener Variationsprobleme. Lotos **74**, 43—52 (1926). — [7] Über Finslersche und verwandte Räume. Cas. mat. fys. **64**, 1—16 (1935). — [8] Über die Hauptkrümmungen einer Fläche im dreidimensionalen Finslerschen Raum. Mh. Math. Phys. **43**, 1—14 (1936). — [9] Über Beziehungen zwischen den Theorien der Parallelübertragung in Finslerschen Räumen. Nederl. Akad. Wetensch. Proc. Ser. A **49**, 642—647 (1946). — [10] Über Finslersche und Cartansche Geometrie. I. Geometrische Erklärungen der Krümmung und des Hauptskalars im zweidimensionalen Finslerschen Raum. Mathematica, Timişoara **17**, 34—55 (1941). II. Invarianten bei der Variation vielfacher Integrale und Parallelhyperflächen in Cartanschen Räumen. Compositio math. **7**, 141—176 (1939). III. Two-dimensional Finsler spaces with rectilinear extremals. Ann. Math. (2) **42**,

84—112 (1941). IV. Projektivkrümmung allgemeiner affiner Räume und Finslersche Räume skalarer Krümmung. Ann. Math. (2) **48**, 755—781 (1947). — [11] Sui differenziali secondi covarianti. Atti Accad. Lincei Rend. (6) **5**, 763—768 (1927). — [12] Una forma normale invariante della seconda variazione. Atti Accad. Lincei Rend. (6) **7**, 301—306 (1928). — [13] Differentialinvarianten in der Geometrie. Riemannsche Mannigfaltigkeiten und ihre Verallgemeinerungen. Enzykl. math. Wiss. 3 D 11, Leipzig (1927). — [14] Über die n-dimensionalen Geometrien konstanter Krümmung, in denen die Geraden die Kürzesten sind. Math. Z. **30**, 449—469 (1929). — [15] Über Systeme von gewöhnlichen Differentialgleichungen zweiter Ordnung, deren Integralkurven mit dem System der geraden Linien topologisch äquivalent sind. Ann. Math. (2) **48**, 193—215 (1947).

BIANCHI, L.: [1] Lezioni di Geometria Differenziale. Bologna 1924.

BIELECKI, A., et S. GOLAB: [1] Sur un problème de la métrique angulaire dans les espaces de Finsler. Ann. Soc. polon. Math. **18**, 134—144 (1945).

BLASCHKE, W.: [1] Über die Figuratrix in der Variationsrechnung. Arch. Math. Phys. **20**, 28—44 (1912). — [2] Geometrische Untersuchungen zur Variationsrechnung. I. Über Symmetralen. Math. Z. **6**, 281—285 (1920). — [3] Räumliche Variationsprobleme mit symmetrischer Transversalitätsbedingung. Ber. kgl. Sächs. Ges. Wiss., Math. Phys. Kl. **68**, 50—55 (1916). — [4] Vorlesungen über Differentialgeometrie, Vol. I., 3rd ed., Berlin 1924. Repr. Dover, New York 1945. — [5] Integralgeometrie. XI. Zur Variationsrechnung. Abh. math. Sem. Hamburg Univ. **11**, 359—366 (1936). — [6] Zur Variationsrechnung. Rev. Fac. Sci. Univ. Istanbul. Sér. A. **19**, 106—107 (1954).

—, u. G. BOL: [1] Geometrie der Gewebe. Berlin 1938.

BLISS, G. A.: [1] A generalisation of the notion of angle. Trans. Amer. math. Soc. **27**, 61—67 (1925). — [2] Generalisations of geodesic curvature and a theorem of Gauss concerning geodesic triangles. Amer. J. Math. **37**, 1—18 (1915). — [3] Lectures on the calculus of variations. Chicago 1946.

BLUMENTHAL, L. M.: [1] Theory and applications of distance geometry. Oxford 1953.

BOHNENBLUST, F.: [1] Convex regions and projections in Minkowski spaces. Ann. Math. (2) **39**, 301—308 (1938).

BOL, G: See BLASCHKE and BOL.

BOLZA, O.: [1] Vorlesungen über Variationsrechnung. Leipzig 1909. Unveränderter Neudruck, Leipzig (1949).

BOMPIANI, E.: [1] Sulle connessioni affini non posizionali. Arch. Math. **3**, 183—186 (1952). — [2] Studi sugli spazi curvi. Atti Ist. Veneto (9), **5** (80), 355—386, 839—859 (1921). — [3] Proprietà d'immersione di una varietà in uno spazio di RIEMANN. Rend. Sem. Mat. Milano **22**, 3—26 (1951). — [4] Studi sugli spazi curvi; la seconda forma fondamentale di una V_m in V_n. Atti Ist. Veneto (9), **5** (80), 1113—1145 (1921).

BORTOLOTTI, E.: [1] Trasporti non lineari. I. Atti Accad. naz. Lincei, Rend. (6), **23**, 16—21 (1936). II. Atti Accad. naz. Lincei, Rend. (6), **23**, 104—110 (1936). III. Atti Accad. naz. Lincei, Rend. (6), **23**, 175—180 (1936). — [2] Spazi subordinati: equazioni di GAUSS e CODAZZI. Boll. Un. Mat. Ital. **6**, 134—137 (1927). — [3] Differential invariants of direction and point displacements. Ann. Math. (2) **32**, 361—377 (1931).

BOSQUET, J. P.: [1] Quelques formules fondamentales de la théorie invariantive du calcul des variations. Bull. Acad. Bruxelles (5) **15**, 270—277 (1929). — [2] Contribution à la théorie invariantive du calcul des variations. Bull. Acad. Bruxelles (5) **15**, 1002—1017 (1929).

BOULIGAND, G., et G. CHOQUET: Problèmes liés à des métriques variationnelles. C. R. Acad. Sci. (Paris) **218**, 696—698 (1944).

BUSEMANN, H.: [1] Über die Geometrien, in denen die „Kreise mit unendlichem Radius" die kürzesten Linien sind. Math. Ann. 106, 140—160 (1932). — [2] Über die Räume mit konvexen Kugeln und Parallelenaxiom. Nachr. Ges. Wiss. Göttingen 1933, 116—140. — [3] Metric methods in Finsler spaces and the foundations of geometry. Ann. Math. Studies No. 8, Princeton 1942. — [4] The foundations of Minkowskian geometry. Comm. Math. Helv. 24, 156—187 (1950). — [5] Intrinsic area. Ann. Math. 48, 234—267 (1947). — [6] Angular measure and integral curvature. Canad. J. Math. 1, 279—296 (1949). — [7] The geometry of Finsler spaces. Bull. Amer. math. Soc. 56, 5—15 (1950). — [8] On geodesic curvature in two-dimensional Finsler spaces. Ann. Mat. Pura Appl. (4) 31, 281—295 (1950). — [9] Metric conditions for symmetric Finsler spaces. Proc. nat. Acad. Sci. USA 27, 533—539 (1941). — [10] The geometry of geodesics. New York 1955. — [11] On normal coordinates in Finsler spaces. Math. Ann. 129, 417—423 (1955). — [12] Quasihyperbolic geometry. Rend. Circ. Mat. Palermo (2) 4, 256—269 (1955). — [13] The isoperimetric problem for Minkowski area. Amer. J. Math. 71, 743—762 (1949).

—, and W. MAYER: [1] On the foundations of the calculus of variations. Trans. Amer. math. Soc. 49, 173—198 (1941).

CAIRNS, S. S.: [1] Normal coordinates for extremals transversal to a manifold. Amer. J. Math. 60, 423—435 (1938).

CARATHÉODORY, C.: [1] Variationsrechnung und partielle Differentialgleichungen erster Ordnung. Berlin und Leipzig 1935. — [2] Über die diskontinuierlichen Lösungen in der Variationsrechnung. Dissertation, Göttingen 1904, 71 pp. — [3] Über die starken Maxima und Minima bei einfachen Integralen. Math. Ann. 62, 449—503 (1906). — [4] Bemerkung über die Eulerschen Differentialgleichungen der Variationsrechnung. Nachr. Ges. Wiss. Göttg., Math.-phys. Kl. H. 1, 40—42 (1931). — [5] Geometrische Optik. Ergebnisse d. Mathematik, Vol. 4, No. 5, Berlin (1937).

CARTAN, E.: [1] Les espaces de Finsler. Actualités 79. Paris 1934. — [2] Sur les espaces de Finsler. C. R. Acad. Sci. (Paris) 196, 582—586 (1933). — [3] Les espaces de Finsler. Abh. aus dem Seminar für Vektor- und Tensoranalysis 4, 70—81, 82—94 (1937). (French and Russian.) — [4] Observations sur la communication précédente (Referring to GOLAB [4]). C. R. Acad. Sci. (Paris) 196, 27—28 (1933). — [5] Sur une problème d'équivalence et la théorie des espaces métriques généralisés. Math. Cluj 4, 111—136 (1930). — [6] Les problèmes d'équivalence. Selecta de M. CARTAN, 113—136. Paris 1939. — [7] Leçons sur la géométrie des espaces de Riemann. Second Edition. Paris 1951. — [8] Les systémes différentiels extérieurs et leurs applications géométriques. Actualités 994. Paris 1945.

CHERN, S. S.: [1] On the euclidean connections in a Finsler space. Proc. nat. Acad. Sci. USA 29, 33—37 (1943). — [2] Local equivalence and euclidean connections in Finsler spaces. Sci. Rep. Nat. Tsing Hua Univ. Ser. A. 5, 95—121 (1948).

CHOQUET, G.: [1] Étude métrique des espaces de Finsler. Nouvelles méthodes pour les théorèmes d'existence en calcul des variations. C. R. Acad. Sci. (Paris) 219, 476—478 (1944). — [2] Étude différentielle des minimisantes dans les problèmes réguliers du calcul des variations. C. R. Acad. Sci. (Paris) 218, 540—542 (1944).

DAMKÖHLER, W.: [1] Zur Frage der Äquivalenz indefiniter Variationsprobleme mit definiten. S.-B. Math.-Nat. Abt. bayer. Akad. Wiss. 1940, 1—14.

—, u. E. HOPF: [1] Über einige Eigenschaften von Kurvenintegralen und über die Äquivalenz von indefiniten mit definiten Variationsproblemen. Math. Ann. 120, 12—20 (1947).

DAVIES, E. T.: [1] Lie derivation in generalised metric spaces. Ann. Mat. Pura Appl. 18, 261—274 (1939). — [2] Subspaces of a Finsler space. Proc. London math. Soc. (2) 49, 19—39 (1945). — [3] On the second variation of a simple integral with movable end-points. J. London math. Soc. 24, 241—247 (1949). — [4] On the second and third fundamental forms of a subspace. J. London math. Soc. 12, 289—295 (1937). — [5] On the invariant theory of contact transformations. Math. Z. 57, 415—427 (1953). — [6] Sur la théorie invariante des transformations de contact. Géométrie Différentielle, Colloques int. Centre nat. Rech. Scient. 11—15, C. N. R. Paris 1953. See also YANO and DAVIES.

DAVIS, D. R.: [1] The inverse problem in the calculus of variations in higher space. Trans. Amer. math. Soc. 30, 710—736 (1928). — [2] Integrals whose extremals are a given 2-parameter family of curves. Trans. Amer. math. Soc. 33,311—314 (1931).

DONDER, TH. DE: Théorie invariantive du calcul des variations. 2nd ed. Paris 1935.

DEICKE, A.: [1] Über die Finsler-Räume mit $A_i = 0$. Arch. Math. 4, 45—51 (1953).— [2] Über die Darstellung von Finsler-Räumen durch nicht-holonome Mannigfaltigkeiten in Riemannschen Räumen. Arch. Math. 4, 234—238 (1953). — [3] Finsler spaces as non-holonomic subspaces of Riemannian spaces. J. London math. Soc. 30, 53—58 (1955).

DELENS, P.: [1] Sur certains problèmes relatifs aux espaces de Finsler. C. R. Acad. Sci. (Paris) 196, 1356—1358 (1933). — [2] La métrique angulaire des espaces de Finsler et la géométrie différentielle projective. Actualités 80, Paris 1934.

—, et J. DEVISME: [1] Sur certaines formes différentielles et les métriques associées. C. R. Acad. Sci. (Paris) 196, 518—521 (1933).

DEVISME, J.: [1] Sur une espace dont l'élément linéaire est défini par $ds^3 = dx^3 + dy^3 + dz^3 - 3\,dx\,dy\,dz$. J. Math. pures appl. (9) 19, 359—393 (1940). — [2] Sur quelques propriétés des trièdres d'Appell. C. R. Acad. Sci. (Paris) 212 43—45 (1941).

DIRAC, P. A. M.: [1] Homogeneous variables in classical mechanics. Proc. Camb. Phil. Soc. 29, 389—400 (1933).

DOUGLAS, J.: [1] The general geometry of paths. Ann. Math. (2) 29, 143—168 (1928). — [2] The transversality relative to a surface of $\int F(x, y, z; x', y'\, z')\, dt$ = Minimum. Bull. Amer. math. Soc. 32, 669—674 (1926).

DOYLE, T. C.: [1] Tensor decompositions with applications to the contact and complex groups. Ann. Math. (2) 42, 698—722 (1941).

DUSCHEK, A.: Über geometrische Variationsrechnung. Abh. Semin. Vektor- und Tensoranalysis 4, 95—99 (1937).

—, u. W. MAYER: [1] Lehrbuch der Differentialgeometrie. Vol. I and II. Leipzig u. Berlin 1930. — [2] Zur geometrischen Variationsrechnung. II. Über die zweite Variation des eindimensionalen Problems. Mh. Math. Phys. 40, 294—308 (1933).

DUTKA, J.: [1] Transversality in higher space. J. Math. Phys. Mass. Inst. Tech. 23, 126—133 (1944).

EISENHART, L. P.: [1] Riemannian geometry. Princeton 1926. — [2] Non-Riemannian geometry. Amer. math. Soc. Coll. Publ., New York 1927. — [3] Finsler spaces derived from Riemann spaces by contact transformations. Ann. of Math. (2) 49, 227—254 (1948).

—, and M. S. KNEBELMAN: [1] Invariant theory of homogeneous contact transformations. Ann. of Math. (2) 37, 747—765 (1936).

ELIOPOULOS, H. A.: [1] Methods of generalised metric geometry with applications to mathematical physics. Thesis, Toronto 1956, 112 pp. — [2] The theory of subspaces of a generalised metric space. To appear in Canad. J. Math.

FINSLER, P.: [1] Über Kurven und Flächen in allgemeinen Räumen. Dissertation, Göttingen 1918. Unveränderter Nachdruck: Basel: Birkhäuser 1951. — [2] Über die Krümmungen der Kurven und Flächen. Reale Accad. Ital., Fondazione Alessandro Volta, IX Convegno Volta, Roma 1940. — [3] Über eine Verallgemeinerung des Satzes von MEUSNIER. Vierteljschr. naturforsch. Ges. Zürich 85, Beibl. (Festschr. RUDOLF FUETER), 155—164 (1940).

FRANK, P., u. R. VON MISES: [1] Die Differential- und Integralgleichungen der Mechanik und Physik. 2nd Edition, Braunschweig 1930.

FREEMAN, J. G.: [1] First and second variations of the length integral in a generalised metric space. Quart. J. Math. (Oxford Series) 15, 70—83 (1944).

FRIESECKE, H.: [1] Vektorübertragung, Richtungsübertragung, Metrik. Math. Ann. 94, 101—118 (1925).

FUJINAKA, M.: [1] On Finsler spaces and dynamics with special reference to the equations of hunting. Proc. 3rd Jap. Nat. Congress appl. Mechanics 433—436, Tokyo 1954.

FUNK, P.: [1] Über den Begriff „extremale Krümmung" und eine kennzeichnende Eigenschaft der Ellipse. Math. Z. 3, 87—92 (1919). —[2] Über zweidimensionale Räume, insbesondere über solche mit geradlinigen Extremalen und positiver konstanter Krümmung. Math. Z. 40, 86—93 (1935). — [3] Über die Geometrien bei denen die Geraden die Kürzesten sind. Math. Ann. 101, 226—237 (1929). — [4] Über Geometrien, bei denen die Geraden die kürzesten Linien sind und die Äquidistanten zu einer Geraden wieder Gerade sind. Mh. Math. Phys. 37, 153—158 (1930). — [5] Beiträge zur zweidimensionalen Finslerschen Geometrie. Mh. Math. Phys. 52, 194—216 (1948).

—, u. L. BERWALD: [1] Flächeninhalt und Winkel in der Variationsrechnung. Lotos (Prag) 67/68, 45—56 (1919/1920).

GALVANI, O.: [1] Sur la réalisation des espaces de Finsler. C. R. Acad. Sci. (Paris) 222, 1067—1069 (1946). — [2] Les connexions finslériennes de congruences de droites. C. R. Acad. Sci. (Paris) 222, 1200—1202 (1946). — [3] Sur l'immersion du plan de Finsler dans certains espaces de Riemann à trois dimensions. C. R. Acad. Sci. (Paris) 223, 1088—1090 (1946). — [4] La réalisation des connexions euclidiennes d'éléments linéaires et des espaces de Finsler. Ann. Inst. Fourier Grenoble 2, 123—146 (1951). — [5] La réalisation des espaces de Finsler. C. R. Congrès Soc. Savantes Paris et des Départements tenu à Grenoble en 1952, 57—60 (1952). — [6] Réalisations euclidiennes des plans de Finsler. Ann. Inst. Fourier Grenoble 5, 421—454 (1955).

GOLAB, S.: [1] Einige Bemerkungen über Winkelmetrik in Finslerschen Räumen. Verh. int. Math. Kongr. Zürich 2, 178—179 (1932). — [2] Quelques problèmes métriques de la géométrie de Minkowski. Trav. Acad. Mines Cracovie Fasc. 6, 1—79 (1932) (Polish). — [3] Sur un invariant intégral relatif aux espaces métriques généralisés. Atti Accad. naz. Lincei, Rend. (6), 17, 515—518 (1933). — [4] Sur la représentation conforme de l'espace de Finsler sur l'espace euclidien. C. R. Acad. Sci. (Paris) 196, 25—27 (1933). — [5] Sur la représentation conforme de deux espaces de Finsler. C. R. Acad. Sci. (Paris) 196, 986—988 (1933). — [6] Contribution à un théorème de M. M. S. KNEBELMANN. Prace mat. fiz. 41, 97—100 (1934). — [7] Sur la mesure des aires dans les espaces de Finsler. C. R. Acad. Sci. (Paris) 200, 197—199 (1935). — [8] Sur une condition nécessaire et suffisante afin qu'un espace de Finsler soit un espace Riemannien. Atti Accad. naz. Lincei Rend. (6), 21, 133—137 (1935). — [9] Les transformations par polaires réciproques dans la géométrie de Finsler. C. R. Acad. Sci. (Paris) 200, 1462—1464 (1935). — [10] Sur le rapport entre les notions des mesures des angles et des aires dans les espaces de Finsler. C. R. Acad. Sci. (Paris) 201,

250—251 (1935). — [11] On Finsler's measurement of angle. Ann. Soc. pol. Math. **24**, 78—84 (1954). — [12] Zur Theorie der Übertragungen. Bericht von der Riemann-Tagung des Forschungsinstituts für Mathematik, 162—177. Berlin 1957.

—, u. H. HÄRLEN: [1] Minkowskische Geometrie. I u. II. Mh. Math. Physik **38**, 387—398 (1938).

GOLDSTINE, H. H.: [1] The calculus of variations in abstract spaces. Duke Math. J. **9**, 811—822 (1942).

GRÜSS, G.: [1] Über Gewebe auf Flächen in dreidimensionalen allgemeinen metrischen Räumen. Math. Ann. **100**, 1—31 (1928). — [2] Eine Bemerkung über geodätische Kegelschnitte auf Flächen allgemeiner Metrik. Jber. dtsch. Math.-Ver. **38**, 83—91 (1929). — [3] Beiträge zur Differentialgeometrie zweidimensionaler allgemein-metrischer Räume. Math. Ann. **103**, 162—184 (1930).

HAANTJES, J.: [1] Distance geometry. Curvature in abstract metric spaces. Ned. Akad. Wet. Proc. Ser. A **50**, 496—508 (1947).

HAAR, A.: [1] Der Maßbegriff in der Theorie der kontinuierlichen Gruppen. Ann. Math. (2) **34**, 147—169 (1933).

HADAMARD, J.: [1] Leçons sur le calcul des variations. Paris 1910.

HAIMOVICI, M.: [1] Formules fondamentales dans la théorie des hypersurfaces d'un espace de Finsler. C. R. Acad. Sci. (Paris) **198**, 426—427 (1934). — [2] Sur les espaces généraux qui se correspondent point par point avec conservation du parallélisme de M. CARTAN. C. R. Acad. Sci. (Paris) **198**, 1105—1107 (1934). — [3] Sur quelques types de métriques de Finsler. C. R. Acad. Sci. (Paris) **199**, 1091—1093 (1934). — [4] Les formules fondamentales dans la théorie des hypersurfaces d'un espace général. Ann. Sci. Univ. Jassy **20**, 39—58 (1935). — [5] Sur les espaces de Finsler à connexion affine. C. R. Acad. Sci. (Paris) **204**, 837—839 (1937). — [6] Le parallélisme dans les espaces de Finsler et la différentiation invariante de M. LEVI-CIVITA. Ann. Sci. Univ. Jassy, I. Math. **24**, 214—218 (1938). — [7] Sulle superficie totalmente geodetiche negli spazi di Finsler. Atti Accad. naz. Lincei Rend. (6), **27**, 633—641 (1938).

HAMEL, G.: [1] Über die Geometrien, in denen die Geraden die kürzesten Linien sind. Math. Ann. **57**, 231—264 (1903).

HARDY, G. H., J. E. LITTLEWOOD and G. PÓLYA: [1] Inequalities. Cambridge 1934.

HÄRLEN, H.: See GOLAB u. HÄRLEN.

HIRAMATU, H.: [1] Groups of homothetic transformations in a Finsler space. Tensor (N. S.) **3**, 131—143 (1954). — [2] On some properties of groups of homothetic transformations in Riemannian and Finslerian spaces. Tensor (N. S.) **4**, 28—39 (1954).

HLAVATÝ, V.: See SCHOUTEN u. HLAVATÝ.

HOKARI, S.: [1] Winkeltreue Transformationen und Bewegungen im Finslerschen Raum. J. Fac. Sci. Hokkaido Univ., Ser. I. Math. **5**, 1—8 (1936).

HÖLDER, E.: [1] Auf Extremalintegrale gegründete metrische Räume. Bericht von der Riemann-Tagung des Forschungsinstituts für Mathematik, 178—193. Berlin 1957.

HOMBU, H.: [1] Konforme Invarianten im Finslerschen Raume. J. Fac. Sci. Hokkaido Univ., Ser. I. Math. **2**, 157—168 (1934). — [2] Konforme Invarianten im Finslerschen Raum. J. Fac. Sci. Hokkaido Univ., Ser. I. Math. **4**, 51—66 (1935). — [3] Die Krümmungstheorie im Finslerschen Raum. J. Fac. Sci. Hokkaido Univ., Ser. I. Math. **5**, 67—94 (1936).

HOPF, E.: See DAMKÖHLER u. HOPF.

HORVÁTH, J. I., and A. MOÓR: [1] Entwicklung einer einheitlichen Feldtheorie begründet auf die Finslersche Geometrie. Z. Physik **131**, 544—570 (1952). —

[2] Entwicklung einer Feldtheorie begründet auf einen allgemeinen metrischen Linienelementraum. I. II. Ned. Akad. Wet. Proc. Ser. A **58**, 421—429; 581—587 (1955).

HOSOKAWA, T.: [1] Conformal property of a manifold B_n. Jap. J. Math. **9**, 59—62 (1932). — [2] Über nicht-holonome Übertragung in allgemeiner Mannigfaltigkeit T_n. J. Fac. Sci. Hokkaido Univ., Ser. I. Math. **2**, 1—11 (1934). — [3] On the various linear displacements in the Berwald-Finsler's manifold. Sci. Rep. Tokyo **19**, 37—51 (1930). — [4] Finslerian wave geometry and Milne's worldstructure. J. Sci. Hiroshima Univ. (A) **8**, 249—270 (1938). — [5] Connections in the manifold admitting contact transformations. Journ. Fac. Sci. Hokkaido Imp. Univ. **2**, 169—176 (1934).

HOUSEHOLDER, A. S.: [1] The dependence of a focal point upon curvature in the calculus of variations. Contributions to the calculus of variations 1933—1937, 485—526. Chicago 1937.

HUMBERT, P.: [1] Sur certaines figures planes de l'espace attaché à l'opérateur Δ_3. C. R. Acad. Sci. (Paris) **209**, 590—591 (1939); **211**, 530—531 (1940). — [2] Sur une extension de la notion d'angle: angles d'un faisceau de trois droites. C. R. Acad. Sci. (Paris) **213**, 970—971 (1941). — [3] Sur certaines figures planes de l'espace attaché à l'opérateur Δ_3. Bull. Sci. Math. (2) **66**, 145—154 (1942); J. Math. pures appl. (9) **21**, 141—153 (1942); Ann. Univ. Lyon. Sect. A (3) **4**, 93 (1941). — [4] Bitétraèdres de l'espace attaché à l'opérateur Δ_3. Bull. Sci. Math. (2) **68**, 50—59 (1944). — [5] Formules trigonométriques dans le plan et l'espace attachés à l'opérateur Δ_3. Ann. Soc. Sci. Bruxelles. Sér. I **60**, 196—199 (1946).

INGARDEN, R. S.: [1] Über die Einbettung eines Finslerschen Raumes in einem Minkowskischen Raum. Bull. Acad. Polon. Sci. Cl. III. **2**, 305—308 (1954).

ISPAS, C. I.: [1] Les identités de Veblen dans les espaces généralisés. Acad. Repub. Pop. Române. Bul. Şti. Sect. Şti. Mat. Fiz. **4**, 533—539 (1952). — [2] Identités de type Ricci dans l'espace de Finsler. Com. Acad. R. P. Române **2**, 13—18 (1952).

JOHNSON, M. M.: [1] Tensors of the calculus of variations. Amer. J. Math. **53**, 103—116 (1931).

KÄHLER, E.: [1] Einführung in die Theorie der Systeme von Differentialgleichungen. Hamburger math. Einzelschr., H. 16 (1934).

KAMKE, E.: [1] Differentialgleichungen, Lösungsmethoden und Lösungen. Leipzig 1942.

KAMPEN, E. R. VAN: See SCHOUTEN u. VAN KAMPEN.

KATSURADA, Y.: [1] On the connection parameters in a non-holonomic space of line-elements. J. Fac. Sci. Hokkaido Univ. Ser. I, **11**, 129—149 (1950). — [2] On the theory of non-holonomic systems in the Finsler space. Tôhoku math. J. (2) **3**, 140—148 (1951).

KAWAGUCHI, A.: [1] Beziehung zwischen einer metrischen linearen Übertragung und einer nicht-metrischen in einem allgemeinen metrischen Raume. Ned. Akad. Wet. Proc. Ser. A **40**, 596—601 (1937). — [2] On the theory of nonlinear connections. I. Introduction to the theory of general non-linear connections. Tensor (N. S.) **2**, 123—142 (1952). — [3] On the theory of non-linear connections. Convegno Int. Geometria differenziale, Italia, 27—32, Roma 1954. — [4] Generalised direction transformation and generalised homogeneous function. Tensor (N. S.) **5**, 68—70 (1955). — [5] On the theory of non-linear connections. II. Theory of Minkowski space and of non-linear connections in a Finsler space. Tensor (N. S.) **6**, 165—199 (1956).

KIKUCHI, S.: [1] On the theory of subspace in a Finsler space. Tensor (N. S.) **2**, 67—79 (1952).

KNEBELMAN, M. S.: [1] Collineations and motions in generalised space. Amer. J. Math. **51**, 527—564 (1929). — [2] Conformal geometry of generalised metric spaces. Proc. nat. Acad. Sci. USA **15**, 376—379 (1929). — [3] Groups of collineations in the space of paths. Proc. nat. Acad. Sci. USA **13**, 396—400 (1927). — [4] Motions and collineations in general path space. Proc. nat. Acad. Sci. USA **13**, 607—611 (1927).

KONDO, K.: [1] On the theoretical investigation based on abstract geometry of dynamical systems appearing in engineering. Proc. 3rd Jap. Nat. Congress appl. Mechanics, 425—432, Tokyo 1954.

KOSAMBI, D. D.: [1] Géométrie différentielle et calcul des variations. Accad. naz. Lincei Rend. (6) **16**, 410—415 (1932). — [2] Parallelism and path-spaces. Math. Z. **37**, 608—618 (1933). — [3] Systems of differential equations of second order. Quart. J. Math. (Oxford Ser.) **6**, 1—12 (1935). — [4] Les espaces des paths généralisés qu'on peut associer avec un espace de Finsler. C. R. Acad. Sci. (Paris) **206**, 1538—1541 (1938).

KOSCHMIEDER, L.: [1] Die neuere formale Variationsrechnung. Jber. dtsch. Math.-Ver. **40**, 109—132 (1931).

KULK, W. VAN DER: See SCHOUTEN and VAN DER KULK.

LANDSBERG, G.: [1] Über die Totalkrümmung. Jber. dtsch. Math.-Ver. **16**, 36—46 (1907). — [2] Über die Krümmung in der Variationsrechnung. Math. Ann. **65**, 313—349 (1908). — [3] Krümmungstheorie und Variationsrechnung. Jber. dtsch. Math.-Ver. **16**, 547—551 (1907).

LA PAZ, L.: [1] Problems of the calculus of variations with prescribed transversality conditions. Bull. Amer. math. Soc. **36**, 674—680 (1930). — [2] The Euler equations of problems of the calculus of variations with prescribed transversality conditions. Proc. nat. Acad. Sci. USA **17**, 459—463 (1931).

LAPTEW, B.: [1] Une forme invariante de la variation et la dérivée de S. LIE. Bull. Soc. Phys.-Math. Kazan (3) **12**, 3—8 (1940) (Russian). — [2] L'integration covariante dans les espaces de Finsler à trois dimensions. Bull. Soc. Phys.-Math. Kazan (3) **9**, 62—76 (1938).

LAUGWITZ, D.: [1] Konvexe Mittelpunktsbereiche und normierte Räume. Math. Z. **61**, 235—244 (1955). — [2]. Zur geometrischen Begründung der Parallelverschiebung in Finslerschen Räumen. Arch. Math. **6**, 448—453 (1955). — [3] Die Vektorübertragung in der Finslerschen Geometrie und der Wegegeometrie. Indag. Math. **18**, 21—28 (1956). — [4] Zur projektiven und konformen Geometrie der Finsler-Räume. Arch. Math. **7**, 74—77 (1956). — [5] Grundlagen für die Geometrie der unendlichdimensionalen Finsler-Räume. Ann. Mat. pura appl. (4) **41**, 21—41 (1956). — [6] Eine Beziehung zwischen affiner und Minkowskischer Differentialgeometrie. Publ. Math. Debrecen **5**, 72—76 (1957).

LEBESGUE, H.: [1] Leçons sur l'integration et la recherche des fonctions primitives. Second edition, Paris 1928.

LEVASOV, A.: [1] La géométrie finslérienne généralisée et la mécanique classique. Bull. Univ. Asie centr. Taskent **22**, 109—118 (1938).

LEVI-CIVITA, T.: [1] Sur l'écart géodésique. Math. Ann. **97**, 291—320 (1926). — [2] The absolute differential calculus. Trans. M. Long, London 1929.

LEVINE, J.: [1] Collineations in generalised spaces. Proc. Amer. math. Soc. **2**, 447—455 (1951).

LEWIS, D. C.: [1] Metric properties of differential equations. Amer. J. Math. **71**, 294—312 (1949).

LIBER, A. E.: [1] On two-dimensional spaces with an algebraic metric. Trudy Sem. Vektor. Tenzor. Analizu **9**, 319—350 (1952) (Russian).

LICHNEROWICZ, A.: [1] Les espaces à connexion semi-symétrique et la mécanique. C. R. Acad. Sci. (Paris) **212**, 328—331 (1941). — [2] Sur une généralisation des espaces de Finsler. C. R. Acad. Sci. (Paris) **214**, 599—601 (1942). — [3] Sur une extension du calcul des variations. C. R. Acad. Sci. Paris **216**, 25—28 (1943). — [4] Les espaces variationnels généralisés. C. R. Acad. Sci. (Paris) **217**, 415—418 (1943). — [5] Les espaces variationnels généralisés. Ann. Sci. École norm. sup. (3) **62**, 339—384 (1945). — [6] Sur une extension de la formule d'Allendoerfer-Weil à certaines variétés finslériennes. C. R. Acad. Sci. (Paris) **223**, 12—14 (1946). [7] Quelques théorèmes de géométrie différentielle globale. Comment. Math. Helv. **22**, 271—301 (1949). — [8] Théorie globale des connexions et des groupes d'holonomie. Rome 1955.

—, et Y. THIRY: [1] Problèmes de calcul des variations liés à la dynamique classique et à la théorie unitaire du champ. C. R. Acad. Sci. (Paris) **224**, 529—531 (1947).

LIPKA, J.: [1] On the relative curvature of two curves in V_n. Bull. Amer. math. Soc. **29**, 345—348 (1923).

LIPSCHITZ, R.: [1] Untersuchungen in betreff der ganzen homogenen Funktionen von n Differentialen. J. Math. (Crelle) **70**, 71—102 (1869). — [2] Fortgesetzte Untersuchungen in betreff der ganzen homogenen Funktionen von n Differentialen. J. Math. (Crelle) **72**, 1—56 (1870).

LITTAUER, S. B.: See MORSE and LITTAUER.

MANCILL, J. D.: [1] Problems in the calculus of variations with prescribed transversality conditions. Amer. J. Math. **61**, 330—334 (1939).

MAURIN, K.: [1] Eingliedrige Gruppen der homogenen kanonischen Transformationen und Finslersche Räume. Ann. Polon. Math. **2**, 97—102 (1955).

MAYER, W.: [1] Beitrag zur geometrischen Variationsrechnung. Jber. dtsch. Math.-Ver. **38**, 260—281 (1929).

McFARLAN, L. H.: [1] Problems of the calculus of variations in several dependent variables and their derivatives of higher order. Tôhoku math. J. **33**, 204—218 (1931).

MENGER, K.: [1] Untersuchungen über allgemeine Metrik. Vierte Untersuchung. Math. Ann. **103**, 466—501 (1930).

MIKAMI, M.: [1] On the theory of non-linear direction displacements. Jap. J. Math. **17**, 541—568 (1948).

MINKOWSKI, H.: [1] Geometrie der Zahlen. Leipzig u. Berlin 1910. Reproduction by Chelsea Publications. — [2] Theorie der konvexen Körper, insbesondere Begründung ihres Oberflächenbegriffs. Ges. Abh. Bd. 2, S. 131—229. Leipzig u. Berlin 1911.

MOÓR, A.: [1] Généralisation du scalaire de courbure et du scalaire principal d'un espace finslérien à n dimensions. Canad. J. Math. **2**, 307—313 (1950). — [2] Espaces métriques dont le scalaire de courbure est constant. Bull. Sci. Math. (2) **74**, 13—32 (1950). — [3] Finslersche Räume mit der Grundfunktion $L = f/g$. Comment. Math. Helv. **24**, 188—195 (1950). — [4] Einführung des invarianten Differentials und Integrals in allgemeinen metrischen Räumen. Acta math. **86**, 71—83 (1951). — [5] Quelques remarques sur la généralisation du scalaire de courbure et du scalaire principal. Canad. J. Math. **4**, 189—197 (1952). — [6] Über die Dualität von Finslerschen und Cartanschen Räumen. Acta math. **88**, 347—370 (1952); „Ergänzung" Acta math. **91**, 187—188 (1954). — [7] Finslersche Räume mit algebraischer Grundfunktion. Publ. Math. Debrecen **2**, 178—190 (1952).

MORSE, M.: [1] The calculus of variations in the large. Amer. math. Soc. Coll. Publ., Vol. XVIII. New York 1934.

—, and S. B. LITTAUER: [1] A characterisation of fields in the calculus of variations. Proc. nat. Acad. Sci. USA **18**, 724—730 (1932).

Muto, Y.: [1] On the connections in the manifold admitting homogeneous contact transformations. Proc. Phys.-Math. Soc. Japan 20, 451—457 (1938).

—, et K. Yano: [1] Sur les transformations de contact et les espaces de Finsler. Tôhoku math. J. 45, 295—307 (1939).

Myers, S. B.: [1] Arc length in metric and Finsler manifolds. Ann. Math. (2) 39, 463—471 (1938). — [2] Arcs and geodesics in metric spaces. Trans. Amer. math. Soc. 57, 217—227 (1945). — [3] Riemannian manifolds with positive mean curvature. Duke math. J. 8, 401—404 (1941).

Nagai, T.: See Nobuhara and Nagai.

Nagata, Y.: [1] Normal curvature of a vector field in a hypersurface in a Finsler space. Tensor (N. S.) 5, 17—22 (1955).

Nalli, P.: [1] Trasporti rigidi di vettori negli spazi di Riemann. Atti Accad. naz. Lincei Rend. (6) 13, 669—675 (1931).

Nakajima, S.: [1] Geometrische Untersuchungen zur Variationsrechnung. Jap. J. Math. 3, 61—64 (1926).

Nasu, Y.: [1] On the torse-forming directions in a Finsler space. Tôhoku math. J. (2) 4, 99—102 (1952). — [2] Non-euclidean geometry in Finsler spaces. Kumamoto J. Sci. Ser. A. 1, 8—12 (1952).

Nazim, A.: [1] Über Finslersche Räume. Dissertation München 1936, 56 pp. — [2] Über den Satz von Gauss-Bonnet im Finslerschen Raum. Univ. d'Istanbul, 26—32 (1948).

Nobuhara, T., and T. Nagai: [1] On the special Finsler space of three dimensions. Tensor (N. S.) 2, 175—180 (1952).

Noether, E.: [1] Invarianten beliebiger Differentialausdrücke. Gött. Nachr. 25, 37—44 (1918).

Norden, A. P.: [1] Affinely connected spaces. Moscow-Leningrad 1950 (Russian).

Ohkubo, T.: [1] On a metric displacement in a Finsler space (Japanese). Tensor 4, 53—55 (1941). — [2] On relations among various connections in Finslerian space. Kumamoto J. Sci. Ser. A. 1, No. 3, 1—6 (1954). — [3] Geometry in a space with a generalised metric. Tensor 3, 48—55 (1940) (Japanese). — [4] Geometry in a space with generalised metrics. II. J. Fac. Sci. Hokkaido Imp. Univ. Ser. I. 10, 157—178 (1941). — [5] On a symmetric displacement in a Finsler space. Tensor 4, 53—55 (1941) (Japanese).

Otsuki, T.: [1] On geodesic coordinates in Finsler spaces. Math. J. Okayama Univ. 6, 135—145 (1957).

Owens, O. G.: [1] The integral geometry definition of arc-length for two-dimensional Finsler spaces. Trans. Amer. math. Soc. 73, 198—210 (1952).

Pauc, C. Y.: [1] L'inégalité triangulaire dans les espaces de Minkowski généralisés. Atti Accad. naz. Lincei Rend. (6), 26, 369—374 (1937). — [2] Les méthodes directes en calcul des variations et en géométrie différentielle. Paris 1941. — [3] Les théorèmes fort et faible de Vitali et les conditions d'évanescence de halos. C. R. Acad. Sci. (Paris) 232, 1727—1729 (1951).

Perron, O.: [1] Algebra I: Die Grundlagen. Berlin u. Leipzig 1927.

Pihl, M.: [1] Classical mechanics in a geometrical description (Danish). Danske Vid. Selsk. Mat.-Fys. Medd. 30, No. 12, 26 pp. (1955).

Rachevsky, P. K.: [1] Dualité métrique dans la géométrie à deux dimensions de Finsler, en particulier, sur une surface arbitraire. C. R. Acad. Sci. URSS, N. s. 3, 147—150 (1935). — [2] Systèmes trimétriques et la métrique de Finsler généralisée. C. R. Acad. Sci. (Paris) 202, 1237—1239 (1936).

Radon, H.: [1] Über eine besondere Art ebener konvexer Kurven. Leipz. Ber. 68, 123—128 (1916).

RAPCSÁK, A.: [1] Kurven auf Hyperflächen im Finslerschen Raume. Hungarica
   Acta math. 1, 21—27 (1949). — [2] Invariante Taylorsche Reihe in einem
   Finslerschen Raum. Publ. Math. Debrecen 4, 49—60 (1955). — [3] Eine neue
   Definition der Normalkoordinaten im Finslerschen Raum. Acta Univ. Debrecen
   1, 109—116 (1954) (Hungarian). — [4] Eine neue Charakterisierung Finslerscher
   Räume skalarer und konstanter Krümmung, und projektiv-ebene Räume. Acta
   math. Acad. Sci. Hungar. 8, 1—18 (1957).

REEB, G.: [1] Quelques propriétés globales de géodésiques d'un espace de Finsler
   et des variétés minima d'un espace de Cartan. Coll. Topologie Strasbourg 1951,
   No. II; La Bibliothéque Nationale et Universitaire de Strasbourg 1952. —
   [2] Sur certaines propriétés globales des trajectoires de la dynamique, dues à
   l'existence de l'invariant intégral de M. ELIE CARTAN. Coll. Topologie Strasbourg
   1951, No. III; La Bibliothéque Nationale et Universitaire de Strasbourg 1952. —
   [3] Sur les espaces de Finsler et les espaces de Cartan. Géométrie différentielle,
   Coll. int. Centre nat. Recherche scient. Strasbourg, p. 35—40, C. N. R. Paris
   1953.

RIDER, P. R.: [1] The figuratrix in the calculus of variations. Trans. Amer. math.
   Soc. 28, 640—653 (1926).

RIEMANN, B.: [1] Über die Hypothesen, welche der Geometrie zugrunde liegen.
   Habilitationsvortrag 1854. Ges. math. Werke, 272—287, Leipzig 1892. Re-
   produced by Dover Publications 1953.

RINOW, W.: [1] Über vollständige differentialgeometrische Räume. Dtsch. Math. 1,
   46—63 (1936).

RUND, H.: [1] Zur Begründung der Differentialgeometrie der Minkowskischen
   Räume. Arch. Math. 3, 60—69 (1952). — [2] Die Hamiltonsche Funktion bei
   allgemeinen dynamischen Systemen. Arch. Math. 3, 207—215 (1952). — [3]
   Finsler spaces considered as locally Minkowskian spaces. Thesis, Cape Town
   1950, 107 pp. — [4] Über die Parallelverschiebung in Finslerschen Räumen.
   Math. Z. 54, 115—128 (1951). — [5] On the analytical properties of curvature
   tensors in Finsler spaces. Math. Ann. 127, 82—104 (1954). — [6] On the geometry
   of generalised metric spaces. Convegno Int. Geometria differenziale, Italia,
   p. 114—121. Roma 1954. — [7] Eine Krümmungstheorie der Finslerschen
   Räume. Math. Ann. 125, 1—18 (1952). — [8] Theory of curvature in Finsler
   spaces. Coll. Topologie Strasbourg 1951, No. IV. La Bibliothéque Nationale et
   Universitaire de Strasbourg 1952. — [9] Hypersurfaces of a Finsler space.
   Canad. J. Math. 8, 487—503 (1956). — [10] The theory of subspaces of a Finsler
   space. I. Math. Z. 56, 363—375 (1952); II. Math. Z. 57, 193—210 (1953). —
   [11] The scalar form of Jacobi's equations in the calculus of variations. Ann.
   Math. pura appl. (4), 35, 183—202 (1953). — [12] Über nicht-holonome allgemeine metrische Geometrie. Math. Nachr. 11, 61—80 (1954). — [13] Application des méthodes de la géométrie métrique généralisée à la dynamique théorique.
   Géométrie différentielle, Coll. int. Centre nat. Recherche scient., Strasbourg,
   p. 41—51, C. N. R. Paris 1953. — [14] Über allgemeine nicht-holonome und
   dissipative Systeme. Bericht von der Riemann-Tagung des Forschungsinstituts
   für Mathematik, S. 269—279, Berlin 1957. — [15] Some remarks concerning the
   theory of non-linear connections. Nederl. Akad. Wetensch. Proc. Ser. A 61,
   341—347 (1958). — [16] On Carathéodory's method of equivalent integrals in
   the calculus of variations. Nederl. Akad. Wetensch. Proc. Ser. A 62.

SASAKI, S.: [1] Non-euclidean geometry in generalised space. Sci. Rep. Tôhoku
   Univ. I. s. 26, 313—322 (1937).

SAVAGE, L. J.: [1] On the crossing of extremals at focal points. Bull. Amer. math.
   Soc. 49, 467—469 (1943).

SCHOENBERG, I. J.: [1] Some applications of the calculus of variations to Riemannian geometry. Ann. Math. 33, 485—495 (1932).

SCHOUTEN, J. A.: [1] Ricci calculus. 2nd edition. Berlin 1954. — [2] Erlanger Programm und Übertragungslehre. Neue Gesichtspunkte zur Grundlegung der Geometrie. Rend. Circ. Mat. Palermo 50, 142—169 (1926).

—, u. J. HAANTJES: [1] Über die Festlegung von allgemeinen Maßbestimmungen und Übertragungen in bezug auf ko- und kontravariante Vektordichten. Mh. Math. Phys. 43, 161—176 (1936).

—, u. V. HLAVATÝ: [1] Zur Theorie der allgemeinen linearen Übertragung. Math. Z. 30, 414—432 (1929).

—, and W. VAN DER KULK: [1] PFAFF's problem and its generalisations. Oxford 1949.

—, u. E. R. VAN KAMPEN: [1] Beiträge zur Theorie der Deformation. Prace Mat. Fiz. Warszawa 41, 1—19 (1933).

SEGRE, B.: [1] Geometria non euclidea ed ottica geometrica. I. Atti Accad. naz. Lincei, Rend. Cl. Sci. Fis. Mat. Nat. (8) 7, 16—19 (1949); II. Atti Accad. naz. Lincei, Rend. Cl. Sci. Fis. Mat. Nat. (8) 7, 20—26 (1949).

SHANKS, S.: [1] Homothetic correspondences between Riemannian spaces. Duke math. J. 17, 299—311 (1950).

ŠIROKOV, A. P.: [1] A gonometric system in the geometry of Finsler. Trudy Sem. Vektor. Tenzor. Analizu 8, 414—424 (1950) (Russian).

'SLEBODZINSKI, W.: [1] Sur deux connexions affines généralisées. Prace mat. fiz. 43, 1—39 (1935). — [2] Note sur les variétés métriques. Prace mat. fiz. 36 (2), 61—63 (1929).

SOMERVILLE, D. M. Y.: [1] An introduction to the geometry of N dimensions. London 1929.

SOÓS, G.: [1] Über Gruppen von Affinitäten und Bewegungen in Finslerschen Räumen. Acta Math. Acad. Sci. Hungar. 5, 73—84 (1954). — [2] Über Gruppen von Automorphismen in affinzusammenhängenden Räumen von Linienelementen. Publ. Math. Debrecen 4, 294—302 (1956).

SPERNER, E.: [1] Einführung in die analytische Geometrie und Algebra. 1. Teil. Oberwolfach 1948.

STOKES, E. C.: [1] Applications of the covariant derivative of CARTAN in the calculus of variations. Contributions to the calculus of variations 1938—1941, 139—174. Chicago 1942.

SU, B.: [1] Geodesic deviation in generalised metric spaces. Acad. Sinica Science Rec. 2, 220—226 (1949).

SUBRAMANIAN, S.: [1] On SYNGE's paper. Bull. Acad. Sci. Allahabad 3, 61—64 (1933).

SUGURI, T., and S. NAKAYAMA: [1] Note on Riemann spaces and contact transformations. Tensor (N. S.) 5, 1—16 (1955).

SULANKE, R.: [1] Die eindeutige Bestimmtheit des von HANNO RUND eingeführten Zusammenhanges in Finsler-Räumen. Wiss. Z. Humboldt-Univ., Math.-naturwiss. R. 4, 229—233 (1955). — [2] „Anmerkung" to [1]: Wiss. Z. Humboldt-Univ., Math.-naturwiss. R. 5, 269 (1956).

SYNGE, J. L.: [1] A generalisation of the Riemannian line-element. Trans. Amer. math. Soc. 27, 61—67 (1925). — [2] On the first and second variations of the line integral in Riemannian geometry. Proc. London math. Soc. (2) 25, 247—264 (1926). — [3] On the connectivity of spaces of constant curvature. Quart. Math. (Oxford Series) 7, 316—320 (1936). — [4] Geometrical mechanics and de Broglie waves. Cambridge 1954.

—, and A. SCHILD: [1] Tensor calculus. Toronto 1949.

SZMYDT, Z.: [1] Un théorème de M. KNEBELMAN. Prace mat. fiz. **48**, 101—103 (1952).

TACHIBANA, S.: [1] On Finsler spaces which admit a concurrent vector field. Tensor (N. S.) **1**, 1—5 (1950).

TAKANO, K.: [1] Homothetic transformations in Finsler spaces. Rep. Univ. Electro-Commun. **1952**, 61—70. — [2] Contact transformations and generalised metric spaces. Tensor (N. S.) **4**, 51—66 (1954).

TAYLOR, J. H.: [1] A generalisation of Levi-Civita's parallelism and the Frenet formulas. Trans. Amer. math. Soc. **27**, 246—264 (1925). — [2] Reduction of Euler's equations to a canonical form. Bull. Amer. math. Soc. **31**, 257—262 (1925). — [3] Parallelism and transversality in a subspace of a general (Finsler) space. Ann. Math. (2) **28**, 620—628 (1927).

THÉODORESCO, N.: [1] Sur les géodésiques de longueur nulle de certains éléments linéaires finslériens. Bull. École Polytech. Bucarest **12**, 9—16 (1941). — [2] Géométrie finslérienne et propagation des ondes. Acad. Roumaine Bull. Sect. Sci. **23**, 138—144 (1942). — [3] Introduction physico-mathématique à la théorie invariante de la propagation des ondes. Rev. Univ. ,,C. I. Parhon'' Politehn. Bucureşti. Ser. Şti. Nat. **1**, 25—51 (1952) (Roumanian).

THIRY, Y.: See LICHNEROWICZ et THIRY.

THOMAS, T. Y.: [1] Differential invariants of generalised spaces. Cambridge 1934.

TUCKER, A. W.: [1] On tensor invariance in the calculus of variations. Ann. Math. (2) **35**, 341—350 (1934). — [2] On generalised covariant differentiation. Ann. Math. (2) **32**, 451—460 (1931).

UNDERHILL, A. L.: [1] Invariants of the function $F(x, y, x', y')$ under point and parameter transformations, connected with the calculus of variations. Thesis, Chicago 1907. — [2] Invariants of the function $f(x, y, x', y')$ in the calculus of variations. Trans. Amer. math. Soc. **9**, 316—338 (1908).

VAGNER, V. V.: [1] Über Berwaldsche Räume. Rec. Math. Moscou, N. S. **3**, 655—662 (1938). — [2] On the imbedding of a field of local surfaces in X_n in a constant field of surfaces in affine space. Doklady Akad. Nauk SSSR (N. S.) **66**, 785—788 (1949) (Russian). — [3] A generalisation of non-holonomic manifolds in Finslerian space. Abh. Tschernyschwesky Univ. Saratow. Serie Phys.-Math. Forschungsinst. **1** (14), No. 2, 67—97 (1938). — [4] On generalised Berwald spaces. C. R. (Doklady) Acad. Sci. URSS (N. S.) **39**, 3—5 (1943). — [5] Les espaces de Finsler à deux dimensions à groupes d'holonomie finis et continus. C. R. (Doklady) Acad. Sci. URSS (N. S.) **39**, 210—212 (1943). — [6] The inner geometry of non-linear non-holonomic manifolds. Rec. Math. [Mat. Sbornik] N. S. **13** (55), 135—167 (1943). — [7] Homological transformations of Finslerian metric. C. R. (Doklady) Acad. Sci. URSS (N. S.) **46**, 263—265 (1945). — [8] Geometry of fields of local curves in X_3 and the simplest case of Lagrange's problem in the calculus of variations. C. R. (Doklady) Acad. Sci. URSS (N. S.) **48**, 229—232 (1945). — [9] Geometry of fields of local central plane curves in X_3. C. R. (Doklady) Acad. Sci. URSS (N. S.) **48**, 382—384 (1945). — [10] The geometrical theory of the simplest n-dimensional singular problem in the calculus of variations. Mat. Sbornik N. S. **21** (63), 321—364 (1947) (Russian). — [11] The theory of a field of local hyperstrips in X_n and its application to the mechanics of a system with non-linear anholonomic connection. Doklady Akad. Nauk SSSR (N. S.) **66**, 1033—1036 (1949) (Russian). — [12] Theory of a field of local $(n-2)$-dimensional surfaces in X_n and its application to the problem of LAGRANGE in the calculus of variations. Ann. Math. (2) **49**, 141—188 (1948). — [13] The geometry of Finsler as a theory of the field of local hypersurfaces in X_n. Trudy Sem. Vektor. Tenzor Analizu **7**, 65—166 (1949) (Russian). — [14] The theory of a field of local hyperstrips. Trudy Sem. Vektor. Tenzor. Analizu **8**, 197—272 (1950) (Russian). — [15] The theory of composite manifolds. Trudy Sem. Vektor. Tenzor. Analizu **8**, 11—72 (1950) (Russian).

VARGA, O.: [1] Zur Begründung der Minkowskischen Geometrie. Acta Szeged. Math. **10**, 149—163 (1943). — [2] Zur Herleitung des invarianten Differentials in Finslerschen Räumen. Mh. Math. Phys. **50**, 165—175 (1941). — [3] Beiträge zur Theorie der Finslerschen Räume und der affinzusammenhängenden Räume von Linienelementen. Lotos **84**, 1—4 (1936). — [4] Aufbau der Finslerschen Geometrie mit Hilfe einer oskulierenden Minkowskischen Maßbestimmung. S.-B. ung. Akad. Wiss. **61**, 14—22 (1942). — [5] Zur Differentialgeometrie der Hyperflächen in Finslerschen Räumen. Dtsch. Math. **6**, 192—212 (1941). — [6] Bestimmung des invarianten Differentials in Finslerschen Räumen. Mat. Fiz. Lapok **48**, 423—435 (1941) (Hungarian). — [7] Über eine Klasse von Finslerschen Räumen, die die nichteuklidischen verallgemeinern. Comment. Math. Helv. **19**, 367—380 (1947). — [8] Die Krümmung der Eichfläche des Minkowskischen Raumes und die geometrische Deutung des einen Krümmungstensors des Finslerschen Raumes. Abh. Math. Sem. Univ. Hamburg **20**, 41—51 (1955). — [9] Eine geometrische Charakterisierung der Finslerschen Räume konstanter Krümmung. Acta Math. Acad. Sci. Hung. **2**, 143—156 (1951). — [10] Über das Krümmungsmaß in Finslerschen Räumen. Publ. Math. (Debrecen) **1**, 116—122 (1949). — [11] Über affinzusammenhängende Mannigfaltigkeiten von Linienelementen, insbesondere deren Äquivalenz. Publ. Math. Debrecen **1**, 7—17 (1949). — [12] Über den Zusammenhang der Krümmungsaffinoren in zwei eineindeutig aufeinander abgebildeten Finslerschen Räumen. Acta Sci. Szeged **12**, 132—135 (1950). — [13] Normalkoordinaten in allgemeinen differentialgeometrischen Räumen und ihre Verwendung zur Bestimmung sämtlicher Differentialinvarianten. C. R. Premier Congrès Math. Hongrois, p. 131 to 162, Budapest 1952. — [14] Eine Charakterisierung der Finslerschen Räume mit absolutem Parallelismus der Linienelemente. Arch. Math. **5**, 128—131 (1954). — [15] Eine Charakterisierung der Finslerschen Räume mit absolutem Parallelismus der Linienelemente. Acta Univ. Debrecen **1**, 105—108 (1954) (Hungarian). — [16] Bedingungen für die Metrisierbarkeit von affinzusammenhängenden Linienelementmannigfaltigkeiten. Acta Math. Sci. Hungar. **5**, 7—16 (1954).

VEBLEN, O.: [1] Projektive Relativitätstheorie. Ergebn. Math. **2**, H. 1 (1933). — [2] A generalisation of the quadratic differential form. Quart. J. Math. (Oxford Series) **1**, 60—76 (1930).

—, and T. Y. THOMAS: [1] The geometry of paths. Trans. Amer. math. Soc. **25**, 551—608 (1923).

—, and J. H. C. WHITEHEAD: [1] The foundations of differential geometry. Cambridge Tracts No. 19 (1932).

VELTE, W.: [1] Bemerkung zu einer Arbeit von H. RUND. Arch. Math. **4**, 343—345 (1953).

WAERDEN, B. L. VAN DER: [1] Differentialkovarianten von n-dimensionalen Mannigfaltigkeiten in Riemannschen m-dimensionalen Räumen. Abh. Math. Sem. Hamburg **5**, 153—160 (1927).

WALKER, A. G.: [1] Completely symmetric spaces. J. London math. Soc. **19**, 219—226 (1944).

WANG, H.-C.: [1] On Finsler spaces with completely integrable Killing equations. J. London math. Soc. **22**, 5—9 (1947).

WEGENER, J. M.: [1] Untersuchung der zwei- und dreidimensionalen Räume mit der Grundfunktion $L = (a_{ikj}\, \dot{x}^i\, \dot{x}^k\, \dot{x}^j)^{1/3}$. Nederl. Akad. Wetensch. Proc. Ser. A

38, 949—955 (1935). — [2] Hyperflächen in Finslerschen Räumen als Transversalflächen einer Schar von Extremalen. Mh. Math. Phys. **44**, 115—130 (1936). — [3] Untersuchungen über Finslersche Räume. Lotos **84**, 4—7 (1936).

WEIL, A.: [1] L'intégration dans les groupes topologiques et ses applications. Actualités Sci. Ind., Paris 1940.

WEYL, H.: [1] Zur Infinitesimalgeometrie: Einordnung der projektiven und der konformen Auffassung. Göttinger Nachr. **1921**, 99—112.

WHITEHEAD, J. H. C.: [1] Convex regions in the geometry of paths. Quart. J. Math. (Oxford Series) **3**, 33—42 (1932). — [2] Convex regions in the geometry of paths — addendum. Quart. J. Math. (Oxford Series) **4**, 226—227 (1933). — [3] The Weierstrass E-function in differential metric geometry. Quart. J. Math. (Oxford Series) **4**, 291—296 (1933). — [4] On the covering of a complete space by the geodesics through a point. Ann. Math. **36**, 679—704 (1935).

WINTERNITZ, A.: [1] Über die affine Grundlage der Metrik eines Variationsproblems. S.-B. preuß. Akad. Wiss. **26**, 457—469 (1930).

WIRTINGER, W.: [1] Über allgemeine Maßbestimmungen, in welchen die geodätischen Linien durch lineare Gleichungen bestimmt sind. Mh. Math. Phys. **33**, 1—14 (1923).

WRONA, W.: [1] Neues Beispiel einer Finslerschen Geometrie. Prace mat. fiz. **46**, 281—290 (1939).

WUNDHEILER, A.: [1] Über Variationsgleichungen für affine geodätische Linien und nicht-holonome nicht-konservative dynamische Systeme. Prace mat. fiz. **38**, 129—147 (1931). — [2] Kovariante Ableitung und die Cesàroschen Unbeweglichkeitsbedingungen. Math. Z. **36**, 104—109 (1932).

YANO, K.: [1] On the linear displacements in the generalised manifold. Proc. Math.-Phys. Soc. Japan (3) **16**, 318—326 (1934). — [2] On the theory of linear connections in the manifold admitting homogeneous contact transformations. Proc. Math.-Phys. Soc. Japan (3) **17**, 39—47 (1935). — [3] Groups of transformations in generalized spaces. Tokyo 1949.

—, and E. T. DAVIES: [1] Contact tensor calculus. Ann. Mat. pura appl. (4) **37**, 1—36 (1954). — [2] On the connection in Finsler space as an induced connection. Rend. Circ. Mat. Palermo (2) **3**, 409—417 (1955).

ZHANG, M. Y.: [1] Die mittlere Krümmung einer Fläche im dreidimensionalen Finslerschen Raum. Acad. Sinica Science Rec. **3**, 35—39 (1950). — [2] Mean curvatures of a subspace in a Finsler space. Ann. Mat. pura appl. (4) **31**, 297—302 (1950).

Index

(Numbers refer to pages)

Symbols

FSC
www.fsc.org
MIX
Papier aus verantwortungsvollen Quellen
Paper from responsible sources
FSC® C105338